NORTH AMERICAN

OWLS

SECOND EDITION

BIOLOGY
AND
NATURAL
HISTORY

PAUL A. JOHNSGARD

SMITHSONIAN INSTITUTION PRESS
Washington and London

To those who know owls to be something more
than ordinary birds
if something less than gods,
deserving our respect and love

Copy editor: Anne R. Gibbons
Production editor: Duke Johns

Library of Congress Cataloging-in-Publication Data

Johnsgard, Paul A.
    North American owls : biology and natural history / Paul A. Johnsgard.—2nd ed.
        p.   cm.
    Includes bibliographical references (p.).
    ISBN 1-56098-939-4 (alk. paper)
        1. Owls—North America.   I. Title.
    QL696.S8 J64 2002
    598.9′7′097—dc21   2002021015

Manufactured in the United States of America

Color plates printed in Hong Kong

09 08 07 06 05 04 03 02   5 4 3 2 1

♾ The paper in this publication meets the minimum requirements of the American
National Standard for Information Sciences—Permanence of Paper for Printed Library
Materials ANSI Z39.48-1984.

Contents

Contents

Foreword

Owls have both delighted and horrified people ever since the dawn of our species. This interest was likely initiated and sustained by the depiction of owl images on cave walls and by the telling of colorful legends involving owls with mysterious and supernatural powers. You can imagine youngsters listening to such tales at night, told by animated storytellers around fires, perhaps even while the eerie calls of distant owls floated in on evening breezes. Thus were people's first encounters with owls often biased by their social preconceptions. Recent research, as well a my own experience, suggests that oral owl-related legends and stories still thrive, even in this modern age of scientific enlightenment and wireless electronic communication. This seemingly disjunct clash of myth and knowledge warrants discussion, as it relates directly to the ongoing critical need for Paul A. Johnsgard's book to help us obtain the wisdom to conserve owls as part of the rich diversity of life on Earth.

I know that this odd dichotomy of science and lore still exists because I am a self-confessed owlaholic who expresses his obsession by studying great gray owls in Manitoba, Canada, in the fine tradition of science and natural history started by my major professor Robert W. Nero. When I started studying owls in 1985, Bob Nero showed my wife—fellow owl biologist Patsy Duncan—and me that there was great value in sharing our information and enthusiasm for owls with those we met in the field and with the general public. We not only learned about the presence of local owls for banding and study but also helped people understand the relationship between owls and the environment. Often these people had a direct influence on the habitats that support owls.

We also learned that regardless of whether they consider owls good or bad, everyone has an owl story! When people find out you study owls, most feel compelled to tell you their favorite story. Listening to these personal experiences is always a pleasure, but people's interpretation of what they see or experience is often the result of cultural myths that are now known to be false. Some people still feel that seeing an owl means bad things are going to happen, yet others talk to owls in a friendly manner as if they were kindred spirits. Some still consider *all* owls as predators too willing to help themselves to chickens, pet cats, or small dogs. I have "lost" a chicken or two to a great horned owl living on my farm, but locking the chickens up in their coop each night solved that problem. Another myth suggests that owls "clean out" local populations of grouse or pheasants and must be eliminated if humans are to enjoy hunting these game birds. Conversely, it is false that owls benefit farmers by eating *all* the mice and rats in agricultural areas. Even in scientific publications some myths are still perpetuated—many owls are incorrectly described as nomadic only because detailed studies have not been conducted on their dispersal patterns. While romantic and entertaining, persistent myths and misinformation about owls may only accelerate the decline of some species into extinction.

As Earth's resources upon which we depend for survival continue to diminish under an ever-increasing human population, owls and other species are increasingly threatened. The need for this book has never been greater. The complex challenge facing us to ensure the continued survival of North American owls requires an interaction between research, education, and conservation. Research is needed to better understand owls. Education is needed to enhance people's appreciation of owls, which may be the most effective path to owl conservation. Toward this end, Paul's book provides a great service to researchers, educators, and conservationists by providing in one place a benchmark of the state of our collective wisdom on owl biology.

Bob Nero wrote the foreword to the first edition of Paul Johnsgard's book in 1988. It is no coincidence that I was invited by Paul to write the foreword for this wonderfully enhanced second edition. Paul had first asked Nero to do a new foreword, but he kindly referred Paul to me, his former student, claiming that I was more current with owl conservation and research. For the record, although I am honored to craft the foreword for this book, Bob Nero remains productive and up-to-date. He continues to work with long-time owl field biologist Herb Copland and with Patsy Duncan and me on studies of great gray and northern hawk owls in Manitoba and northern Minnesota. My writing this foreword is simply another example of the means by which Bob Nero has encouraged and supported many careers in owl conservation and biology.

In the foreword to the first edition of this book, Bob Nero refers to Paul's attending the 1987 Northern Forest Owls Symposium, a major international gathering of owl biologists. Patsy and I were there, too, young and somewhat nervous owl researchers in awe of the prominent owl biologists and authors in attendance. Imagine our surprise when Paul Johnsgard, author of numerous significant books on birds, introduced himself and asked if he could consult with us and take notes on our great gray owl research to update his soon-to-be-published manuscript on North American owls! I was astonished that he considered our work important. In such a manner Paul has helped reinforce the confidence and self-esteem among owl biologists, both young and old. His interest, skill, and commitment to produce this book help confirm that what we are doing is important.

Science is but another set of cultural values that those engaged in it endeavor to perpetuate through oral tradition, the written word, and through its very practice. Science is an attempt to learn the truth about our experience. The subjects of science vary widely from the smallest, invisible subatomic particles that make up matter to the largest scales of organization that we are aware of—the origin and processes of the entire universe. The study of owls lies somewhere in between, and imagination plays no lesser role as a tool in our efforts to know more about their lives and their role in our shared ecology. But human imagination is a fickle partner in our quest to understand. Imagination plays indiscriminately on both sides of the true-false dichotomy. That myths harmful to owls and their habitats still flourish in the world means that we have work to do. Paul's book on the owls of North America provides us all with a means to imagine the lives of owls more accurately. This, in turn, can help us do the right things to ensure the continued existence of owls on our planet. So here's a "cheers" to Paul for his effort and to owls for stimulating our imaginations.

James R. Duncan
Balmoral, Manitoba

It is perhaps safe to suggest that the only persons willing to assume the writing of a book on raptors, especially owls, are likely to be those who are either masochists or mentally deficient. Raptoriphiles, and especially strigiphiles, are so attached to their subjects that few books escape the wrath of reviewers who feel that their particular species or subject area has been slighted or badly dealt with. Owls have the additional emotional baggage of being birds that are usually either loved or hated, with little middle ground apparent. I admit to being in the owl-lover's camp; I think that few if any other birds are so beautiful, graceful, or mysterious as these, and my first book on the owls of North America attempted to capture that emotional and intellectual attraction. Yet in the past my vision has often proved to be greater than my actual eyesight, and keen eyesight is the first requisite of raptors or their students. I thus approached this revision with all the fascination of a moth that is fatally attracted to a candle, nevertheless hoping against all odds to survive the experience.

During the winter of 1999–2000 I was approached by the Smithsonian Institution Press about writing a second edition of my *North American Owls,* then more than ten years past its 1988 publication date. Much has been published relative to owls and owl biology since then. *The Birds of North America,* a series of single-species monographs sponsored by the American Ornithologists' Union and Philadelphia's Academy of Natural Sciences, was well under way by the late 1990s. In addition, a second International Symposium on the Biology of Northern Hemisphere Owls had been held and its proceedings published (Duncan, Johnson, and Nicholls, 1997). Furthermore, a comprehensive book on the taxonomy, distribution, and identification of the owls of the world had just appeared (König, Weick, and Becking, 1999). And volume 5 of the *Handbook of Birds of the World* (Hoyo, Elliott, Sargatal, 2000), with its much-anticipated Strigiformes section, was soon to be published. Relevant popular to semitechnical books that had appeared since my first edition included an excellent and beautifully illustrated survey of the northern hemisphere owls by Karalus Voous (1988), a similarly illustrated world survey of owls by Rob Hume (1991), and a photographic study and narrative survey of North American owls by Julio de la Torre (1990).

Partly because I felt that the species accounts of the U.S. and Canadian owls in *The Birds of North America* would probably be completed before I could finish a revision and partly because these other useful references had already been or were about to be published, I initially declined revising my book. Yet the idea kept niggling at me, and I soon thought that if I added the dozen-odd Mexican owl species occurring north of the Isthmus of Tehuantepec, the book would encompass all the owls occurring in continental North America, thus separating it somewhat from the more restricted species coverage of *Birds of North America.* This prospect had become much more practical than was the case a decade earlier, because of the landmark publication of *A Guide to the Birds of Mexico and Central North America* by Howell and Webb (1995), with its excellent range maps. The perplexing taxonomies of the Mexican species of screech-owls and pygmy-owls were by then also finally starting to make sense, largely through comparative behavioral studies and some recent biochemical approaches.

As a result, I called the Press's science editor about a month after receiving his initial request, saying I had reconsidered my position and was now willing to undertake the revision. I suggested that the size of the proposed new edition could be kept relatively manageable in length, in spite of the increased species total and more than a decade of new owl literature, by eliminating the detailed plumage descriptions for all species. This proved to be a rather more painful process than I had anticipated, so I have tried to compensate for this loss by including more information on ageing and sexing criteria than appeared in the first edition. The text was thereby ultimately increased by about 15 percent, and there are 24 entirely new figures, plus modifications of others, including many of the range maps. There are also 12 new color plates, mainly of Mexican owls. Because of increased public interest in owls and other raptors as well as an ever-expanding litera-

ture base, I have nearly doubled the number of literature citations (from 470 to about 900), including some recent titles I did not cite in the text.

In my preoccupation with other book projects, by 1999 I had not only stopped gathering owl references but also in preparation for retirement given away to friends and colleagues all my major journals and my collection of separates dealing with raptors. I thus had to try to recover at least some of them and to reactivate communication with those who had helped me with the first edition. Early in my quest I was provided a copy of the important proceedings of the Second International Symposium on Northern Hemisphere Owls through the kindness of one of its editors, Tom Nicholls. This was the first of many reprints and other publications I was able to beg or borrow from old friends in the owl research community, and other owl biologists provided me with unpublished information. These people include Elaine Bachel, Fred Gehlbach, Betsy Hancock, Janet Hinshaw, Lloyd Kiff, Josef Kren, Scott Rashid, Bill Scharf, and many others. Two outside reviewers, Stuart Houston and an anonymous reader, evaluated the shortcomings of the first edition and offered helpful suggestions for the second. Frederick Gehlbach provided a similar review of the second edition manuscript. A computer search of relevant recent owl literature was very kindly done for me by Barbara Voeltz of the Nebraska Game and Parks Commission. Museum specimen records of Mexican owls were provided by David Willard of the Field Museum of Natural History, and egg records were assembled by the Western Foundation of Vertebrate Zoology. I was also assisted again by several wildlife photographer friends, such as Ken Fink, Alan Nelson, and Heidi Velaga. I especially thank Jim Culbertson, Richard Gerhardt, Mark Kasprzyk, and Jose Luis Rangel Salazar for offering the use of their color photos of Neotropical owls. James R. Duncan kindly provided a new preface for this second edition. The splendid one that Bob Nero wrote for the first edition tended to date both him and me, as well as the book itself.

In North America the Raptor Research Foundation (117th Street South, Hastings, Minn. 55033; http://biology.boisestate.edu/raptor/) offers the best informational platform for persons interested in learning more about current research on the biology of owls and other raptors. Much of the information on which this revision is based comes directly or indirectly from the continuing efforts of the foundation's members, as published in the *Journal of Raptor Research.* I thank them collectively for their efforts on behalf of raptors everywhere. A related source of informa-

tion, and especially valuable for literature searches on owls and hawks is the Raptor Information System of the USGS (http://nighthawk.boisestate.edu/). The Owl Roost (http://www.geocities.com/raptor_res/) is a useful source for more general information on owls and their biology. A similar website with general information, photos, and calls of nearly 30 species of North American owls is Owling.com (http://www.owling.com/).

The owls of the world, being mainly forest-adapted birds, are increasingly feeling the pressures of worldwide deforestation trends. An esti-mated 83 percent of the world's nearly 200 surviving species of owls are associated with old, dense, or undisturbed forests (Marcot, 1995), the very ecosystems that are now being demolished at ever increasing rates. It was once widely believed that the voices of owls emanating from the dark forests were an omen of impending ill fortune if not death. Perhaps now we must realize that the increasing absence of owl voices should be taken as an omen of impending ill fortunes for the human species. The future will be a stressful one for owls as well as for all of us!

Preface to the First Edition

This book had its genesis over lunch at the Smithsonian Institution in November 1985, when Ted Rivinus proposed to me that I write a book for the Smithsonian Institution Press to follow up my earlier one on the hummingbirds of North America. Specifically he suggested that a book on owls might make an attractive offering for the Press, because of the nearly universal appeal of owls and the success of an earlier Smithsonian Institution Press book on the great gray owl written by Robert Nero. I began to think back on the various books that had been written on North American owls, mentally tallying their strengths and weaknesses. I suggested that a modern, not-too-technical treatment of the owls of North America was indeed a promising topic for a book and that I would immediately begin to look into its feasibility.

In particular, I wanted to inquire of Cornell's Laboratory of Ornithology whether a set of L. A. Fuertes's owl paintings I remembered having seen there many years previously might be available for reproduction. I felt that this set of paintings might make a very nice illustrative keystone for the book, which could be completed with other paintings or photos as might be needed. Within a few months I had completed an agreement with the Laboratory of Ornithology allowing me to reproduce the entire set of ten Fuertes owl paintings, and I began arranging for necessary additional paintings to be made and collecting literature references.

I was able to begin serious work on the book during the winter of 1985–1986. Writing was nearly completed by the end of 1986, and only a few references to later research could be added, primarily to research reported upon at the Symposium on Northern Forest Owls held at Winnipeg, Manitoba, early in 1987.

From the outset I felt that the book should represent a compromise between the many overly simplified and often highly erroneous books on owls that have appeared in the past few decades, and a highly technical publication likely to repel the average reader. A recent (1983) book on the owls of Europe, by Heimo Mikkola, provided not only a rich source of information on those 17 species (7 of which are shared with North America) but also a general organizational approach to emulate, including several chapters on comparative biology and a series of accounts of individual species, emphasizing breeding biology. That book offers a somewhat more ecological orientation than the one that I have written, which tends to be more heavily oriented toward general and reproductive behavior and also provides various taxonomic keys, weights, measurements, and plumage descriptions that I felt were important inclusions for any basic ornithological reference. I have also included a chapter on owls in myth and legend, a topic that seemed to me so rich and fascinating as to make it impossible to overlook. However, I have attempted to make my book as complementary as feasible to Mikkola's, not only in a geographic but also in a contextual sense.

Another very useful if not indispensable book for all owl researchers is the *Working Bibliography of Owls of the World,* by Clark, Smith, and Kelso (1978), which contains more than 6500 literature citations, and which greatly assisted in my literature review. Clark and Smith (1987) have since accumulated about 3500 additional references on the owls of the world. Of the several thousand literature references I accumulated on North American owls, I have restricted my choice of listed citations to only about 500, but I hope that they have been well chosen, and nearly half of them postdate those in the *Working Bibliography.* For providing me with literature help, interlibrary loans, or both I thank the Van Tyne Memorial Library of the University of Michigan, the University of Nebraska libraries, the Museum of Natural History of the University of Kansas, Scott Johnsgard, and David Rimlinger. Unpublished biological or distributional information, manuscripts, or other useful materials and data were provided by numerous people, including Chris Adam, Harriet Allen, Evelyn Bull, Wayne Campbell, Richard Cannings, Vincent Conners, Jim Duncan, Betsy Hancock, Greg Hayward, Denver Holt, Rick Howie, Jeff Marks, Tim Osborne, Richard Reynolds, Ronald Ryder, Wolfgang Scherzinger, Dwight Smith, Ann and Scott Swengel, and John Winter. Photos were offered or provided by Hans Aschenbrenner, A. J. Borodayko, Rick Bowers, Richard Cannings, Ken Fink, Greg Hayward, Edgar Jones, Tom Mangelsen, Alan Nelson, David Palmer, David Rintoul, B. J. Rose, Wolfgang Scherzinger, and Bill Shuster.

One or two additional acknowledgments to earlier literature need to be made. First, I have generally used the American Ornithologists' Union *Check-list of North American Birds,* 5th and 6th editions, as a basis for range descriptions, although in most cases these have been abbreviated and modified on the basis of alternative or more recent information available to me. Second, my plumage descriptions are based on those by Robert Ridgway, as published in the *Birds of North and Middle America,* Part 6 (1914), again somewhat abbreviated and occasionally supplemented with additional information. Unless otherwise indicated, all anatomical measurements are in millimeters, and weights are in grams. Various linear

or areal measurements, originally cited in the literature as yards, acres, square miles, and the like, have been converted in the text to metric equivalents.

Besides the cooperation and assistance I obtained from the Laboratory of Ornithology in my use of their Fuertes paintings, I was also provided certain useful data from their Nest Record Card scheme. Particularly important and extensive nest and clutch information on several little-studied owl species was provided me by Lloyd Kiff of the Western Foundation of Vertebrate Zoology. Additional egg data were provided by the National Museum of Natural History, through the courtesy of James Dean. Tom Labedz of the Nebraska State Museum helped me in numerous ways, as did Kim Larson and Edwin Minnick, who made various anatomical measurements for me. Betsy Hancock, of the Raptor Rehabilitation Center of Lincoln, Nebraska, was invariably helpful and generous with her time. When I visited the Owl Rehabilitation Research Foundation, Vineland Station, Ontario, Mrs. Katherine McKeever graciously allowed me to photograph the several rare owl species in her care and provided me with an unlimited supply of useful information, advice, coffee, and cookies. She also reinforced my belief that a book on the biology of North American owls is badly needed, particularly one that might help increase the general level of public understanding of and sympathy for owls. The sight of such wonderful creatures suffering in these and many other rehabilitation centers from gunshot wounds unlawfully inflicted by ignorant "sportsmen" or other unenlightened persons is a sickening one, and represents a situation that should not be allowed to persist. If this book serves to educate only a few people as to the ecological value of owls as the most efficient (and cost-free!) natural controllers of rodent populations available, as well as to their enormous aesthetic appeal and value as scientific research subjects in such important areas as the physiology of vision and hearing, and thereby gains for owls a slightly increased level of protection, it will easily have been worth my time and effort.

As a final postscript, for any who might doubt my statement about owls as effective rodent predators, consider that a common barn owl, with an appetite averaging about 90 grams of animal food per day, is likely to consume about 33 kilograms of animals, almost entirely small rodents, per year. Assuming a potential 10-year lifespan, this works out to 330 kilograms (about 725 pounds) of mice. At 30 grams per average-sized house mouse, this is equal to about 11,000 mice consumed in a single owl's lifetime. Each of these mice eats about 10 percent of its weight in food per day, to say nothing of its other potential undesirable effects in spreading disease, fouling human foods, and the like. In the course of a year, these 11,000 mice might thus have consumed about 12,000 kilograms of growing crops, seeds, and grain, or about 13 tons of potential crops, hay, or grain. Clearly, every barn owl living on a farmer's property is worth hundreds of dollars in reduced crop damage and other benefits, and they should be guarded as zealously as a prized watchdog. Yet I have driven past farms where the carcasses of buckshot-riddled barn owls have been impaled, wings outstretched, on barbed-wire fences and left for all who pass to see, perhaps to give dire warning to any other owls in the vicinity not to trespass! At such times one can only ponder the shortsightedness of farmers who spend thousands of dollars a year in fertilizers, pesticides, and rodenticides, thereby endangering both themselves and their neighbors, and yet destroy the very birds that would be most able to save them from their own foolishness.

PART ONE

Comparative Biology of Owls

> *And thorns shall come up in [Babylon's]*
> *palaces, nettles and brambles in the fortresses*
> *thereof; and it shall be a habitation of dragons,*
> *and a court for owls.*
>
> —Isaiah 34:13

There can be little doubt that owls have silently been flying through the earth's skies for a very long time indeed; their fossil record is one of the longest of all groups of living birds. There is no absolute certainty as to how owls originated, but DNA hybridization data provided by Sibley and Ahlquist (1985) suggest that perhaps at least 70–80 million years ago their progenitors separated from the phyletic line that produced the other large group of nocturnal predators, the nightjars, or Caprimulgiformes. This relationship is supported by the similar syringeal structures of both and similarities in their feather pattern arrangements (pteryloses), especially between the aberrant cave-dwelling oilbird (*Steatornis*) of South America and the principal modern owl family, the Strigidae. Several points of similarity in feather structure also occur between the caprimulgiform and strigiform groups, and in a few other anatomical points. However, the chances of morphological convergence in two nocturnally adapted groups of birds are so great as to place severe restraints on many such sorts of evidence.

If we accept the idea of a common strigiform-caprimulgiform ancestor, it is reasonable to assume that adaptations for nocturnality in owls occurred prior to the separation of the two groups, and that following separation the Caprimulgiformes radiated largely in the direction of nocturnal insect catching during aerial foraging, while in owls the primary trends were toward the capturing of relatively large prey by pouncing on it from above, with consequent tendencies toward evolving improved prey-killing abilities. These involved the development of raptorial talons and beaks, predatory behavior associated with immobilizing and rapidly killing large and potentially dangerous prey, and improvements in nocturnal sight, hearing, or both that might provide additional advantages in capturing and killing prey under dim light conditions.

The evolution of a raptorial existence brought the owls into direct competition with the hawklike birds (Falconiformes). The latter have undergone parallel or convergent evolution with the owls in many of their behavioral and morphological adaptations but are largely diurnal in their hunting behavior. Table 1 illustrates some of the major anatomical and behavioral similarities and dissimilarities that exist between these two major groups of avian raptors, with most or all of the similarities attributable to the processes of convergent evolution during the long period in which these two groups have been phyletically separated. For example, the killing behavior of owls and some hawks (falcons) is

Table 1

Comparative Traits of Owls and Hawklike (Falconiform) Birds

Trait	Strigiformes	Falconiformes
Dissimilarities		
Activity pattern	Mostly nocturnal	Mostly diurnal
Eyes	Frontal, larger	Lateral, smaller
Facial disk	In all species	Only in harriers
Ear flaps (opercula)	In some species	Lacking in all
Reversible fourth toe	In all species	Only in osprey
Mandibular edge	Never irregular	Often irregular
Nares location	In front of cere	Within cere
Feather aftershafts	Lacking in all	Usually present
Crop	Lacking in all	Present in all
Intestinal ceca	Large in all	Reduced or absent
Similarities		
Reversed sex dimorphism	In nearly all species	In most species
Head bobbing used for distance judging	Yes	In falcons
Mantling of prey	Yes	Yes
Prey killed with beak	Yes	In falcons
Prey carried by 1 foot	Often (if large)	In falcons
Lack of nest building	Yes	In falcons
Incubation mostly or entirely by female	Yes	Yes
Asynchronous hatching	Yes	Yes
Young downy, nidicolous	Yes	Yes
Hissing by young	Yes	In falcons
Biparental care	Yes	Yes

very similar, with the birds typically severing the neck vertebrae while covering (mantling) the prey with extended wings, after having immobilized it with the talons. In both groups the birds are typically monogamous, with the male remaining to provide food for both the female and young during the relatively long incubation and fledging periods. In both groups incubation typically begins with the laying of the first egg, resulting in a staggered period of hatching and fledging of young, which tends to prolong the nesting period substantially and tends to reduce food-gathering demands on the adults during this critical period. Additionally in both groups the female is as large as or even substantially larger than the male, providing a relatively rare example of reversed sexual dimorphism among birds. The possible significance of this condition is still being debated and is discussed in Chapter 2. The anatomical and behavioral dissimilarities between owls and falconiform birds are substantial and probably reflect differing modes of predation as well as long-term, fundamental differences in the phyletic history of the two groups. For example, owls lack crops for short-term storage of ingested food but do have well-developed intestinal ceca, whereas hawks show the reverse situation. However, members of both groups regurgitate pellets of hair, feathers, bone, and other undigested materials. Owl pellets are especially likely to contain intact, largely undigested bones.

The fossil record of owls is indeed a long one, although highly fragmentary, extending back to at least the Paleogene of Europe and North America. One even older fossil group has been described by Harrison and Walker (1975) as a new owl family (Bradycnemidae) from the Upper Cretaceous of Romania. The reliability of these two workers' paleontological conclusions has been severely criticized (Steadman, 1981), and it now appears that these early fossils actually represent small dinosaurs!

Nevertheless, at least two and perhaps three families of fossil owls are known from the Paleogene. The earliest of these is represented by *Ogygoptynx*, from the early Paleocene of North America, which has been assigned to a special family, the Ogygoptyngidae (Rich and Bohaska, 1976, 1981). The slender foot of this owl is not greatly different in size or shape from that of a modern burrowing owl, and it is possible that small arthropods may have been a significant part of its diet. A second Paleogene owl family from North America, the Protostrigidae, includes *Eostrix* and *Protostrix* (Rich, 1982), as well as the more recently redescribed *Minerva,* a Middle Eocene fossil from Wyoming that had been originally misidentified as an edentate mammal (Mourer-Chauvire, 1983). Compared with *Ogygoptynx, Protostrix* (now considered a synonym of *Minerva*) had a more robust foot and toe structure, and probably was a more effective predator of small vertebrates. Some additional genera have been found in European Paleogene sediments, and all of these early owls appear to be quite distinct from the North American *Ogygoptynx* fossil type.

It seems likely that fairly early in the evolution of the modern owls adaptive radiation began to establish two distinct lineages, one of which led to the modern barn owls and bay owls (Tytonidae), and another apparently later branch leading to the remaining large assemblage of owls (Strigidae). Fossil evidence suggests that the earliest members of the Tytonidae can be traced back to as early as the Eocene, when as many as five fossil genera (*Necrobyas, Nocturnavis, Paleobyas, Paleotyto,* and *Seleornis*) occurred in what is now France. A related Eocene fossil form, *Paleoglaux,* shows intermediate traits between the Tytonidae and Strigidae and has been separated as a distinct family Paleoglaucidae (Mourer-Chauvire, 1987; D. Peters, 1992). A similar rather intermediate fossil (*Sophiornis*) from these same Paleogene French strata became, together with a similar form *Berruornis,* the basis for later recognizing another distinct owl lineage, the family Sophiornithidae (Mourer-Chauvire, 1987; Feduccia, 1996).

Excluding some early misidentified fossil types, the first good evidence of the origin of the family Strigidae occurs much more recently, in early Miocene times. This occurred in the form of *Bubo poirrieri* (France) and *Strix brevis* (North America), which were both roughly contemporaneous with the first known forms representing the modern genus *Tyto.* A substantial number of structural differences exist between the surviving members of these two extant owl families (see the section on the Tytonidae in Part Two), but it is likely that convergent or parallel evolution in the two has also occurred, especially in the development of hearing specializations (Feduccia and Ferree, 1978).

A substantial and interesting but much more recent radiation of strigid owls occurred on the Hawaiian Islands during Holocene times. At least four species of long-legged, short-winged, and *Strix*-like owls grouped in the genus *Grallistrix* then occurred, each confined to a different island

(Olson and James, 1991). These mostly ground-dwelling birds probably survived right up to the time of initial human colonization of these islands, or about 1600 years ago. Another fairly recent (Pleistocene) owl of special interest is *Ornimegalonyx oteroi,* a *Strix*-like and very large and flightless or nearly flightless owl of Cuba (Arrendondo, 1976, 1982). These great owls, with feet twice as long as a great horned owl's, and probably much more powerful, evidently preyed on ground sloths and other large mammals. This was done in the company of three species of giant barn owls, a massive eagle owl (*Bubo*), and several other extremely large but now extinct birds of prey. Giant barn owl fossils are also known from widely scattered deposits elsewhere in the Caribbean region as well as in the Mediterranean region.

Nearly all modern classifications of the extant owls are at least partly based on that proposed by J. Peters (1940). He followed tradition in recognizing the barn owls and bay owls as comprising a family (Tytonidae) separate from the remaining owls (Strigidae) and utilized variations in the external ear structure for further subdivision of the latter group. In particular, Peters used the relative size of the ear opening, the development of the facial disk, the presence or absence of a ligamentous bridge crossing the ear conch, and the presence or absence of an ear flap or operculum as characters for distinguishing two subfamilies of the Strigidae. Species of his subfamily Buboninae, which includes the majority of species, may be characterized as having relatively small external ears, no ear flaps, and a facial disk that is more extensive below the eye than above. The second subfamily, the Striginae, include a few genera of owls with more specialized hearing adaptations. These consist of a relatively large external ear, with not only a large ear opening (external auditory meatus) but also an area of bare skin around the meatus (the ear conch), bounded by specialized feathers and sometimes crossed by a ligamentous bridge. Dermal ear flaps are also present in front of and behind the meatus, which provide at least some capabilities for adjusting the facial disk of feathers and apparently thus influence efficiency of sound reception. These owls not only have large ears and well-developed facial disks that are equally developed above and below the eyes but also often have relatively large eyes, a coevolved trait associated with efficient hunting under near-dark conditions.

A species-level taxonomic summary for the owls of North America is presented below, based on the classification adopted by the American Ornithologists' Union (AOU) in their most recent (1998) *Check-list of North American Birds.* The latter classification has been used at the species level in this book and is the one most widely utilized in North America. The subspecies recognized here are largely those employed in this and earlier editions of the AOU *Check-list* A more complete synoptic species-level classification of the owls of the world was provided by König, Weick, and Becking (1999).

ORDER STRIGIFORMES

Family: Tytonidae (Barn and Grass Owls)
    Genus: *Tyto* Billberg 1828 (Barn Owls)
        Tyto alba Barn Owl
Family: Strigidae (Typical Owls)
    Genus: *Otus* Pennant 1769 (Scops Owls and Screech-Owls)
        Otus flammeolus Flammulated Owl
        Otus asio Eastern Screech-Owl
        Otus kennicottii Western Screech-Owl
        Otus seductus Balsas Screech-Owl (previously considered part of *kennicottii*)
        Otus cooperi Pacific Screech-Owl (previously considered part of *kennicottii*)
        Otus trichopsis Whiskered Screech-Owl
        Otus guatemalae Vermiculated (Central American) Screech-Owl (sometimes separated into two species, including typical *guatemalae* [occurring south to Costa Rica] and *vermiculatus* [occurring from Costa Rica southward])
    Genus: *Lophostrix* Lesson 1836 (Crested Owl)
        Lophostrix cristata Crested Owl
    Genus: *Pulsatrix* Kaup 1848 (Spectacled Owls)
        Pulsatrix perspicillata Spectacled Owl
    Genus: *Bubo* Dumeril 1806 (Eagle Owls)
        Bubo virginianus Great Horned Owl
    Genus: *Nyctea* Stephens 1826 (Snowy Owl)
        Nyctea scandiaca Snowy Owl
    Genus: *Surnia* Dumeril 1806 (Northern Hawk Owl)
        Surnia ulula Northern Hawk Owl
    Genus: *Glaucidium* Boie 1826 (Pygmy-Owls)
        Glaucidium gnoma Northern Pygmy-Owl (including *californicum* and *hoskinsii*)
        Glaucidium griseiceps Central American Pygmy-Owl (previously considered part of *minutissimum*)
        Glaucidium sanchezi Tamaulipas Pygmy-Owl (previously considered part of *minutissimum*)
        Glaucidium palmarum Colima Pygmy-Owl (previously considered part of *minutissimum*)
        Glaucidium brasilianum Ferruginous Pygmy-Owl
    Genus: *Micrathene* Coues 1866 (Elf Owl)
        Micrathene whitneyi Elf Owl
    Genus: *Athene* Boie 1822 (Little Owls)
        Athene cunicularia Burrowing Owl
    Genus: *Ciccaba* Wagler 1832 (Tropical Wood Owls)
        Ciccaba virgata Mottled Owl
        Ciccaba nigrolineata Black-and-White Owl
    Genus: *Strix* Linnaeus 1758 (Wood Owls)
        Strix occidentalis Spotted Owl
        Strix varia Barred Owl
        Strix nebulosa Great Gray Owl
    Genus: *Asio* Brisson 1760 (Eared Owls)
        Asio otus Long-Eared Owl
        Asio stygius Stygian Owl
        Asio flammeus Short-Eared Owl
    Genus: *Pseudoscops* Kaup 1848 (False Scops Owls)
        Pseudoscops clamator Striped Owl
    Genus: *Aegolius* Kaup 1829 (Forest Owls)
        Aegolius funereus Boreal Owl
        Aegolius acadicus Northern Saw-whet Owl

After the appearance of J. Peters's (1940) landmark classification of owls, a growing amount of evidence has gradually developed to suggest that his widely adopted upper-level division of the Strigidae (into two subfamilies Buboninae and Striginae) on the basis of external ear structure might be an artificial one (Kelso, 1940; Voous, 1964). The most complete morphological analysis of this question was that of Ford (1967), who examined the osteology of 75 owl species, representing 23 of Peters's 29 genera. He confirmed that the barn owls and bay owls are more closely related to one another than to any other owls and concluded that they might best be considered subfamilies of a distinct family. However, he also found that the family Strigidae consists of three osteologically distinct evolutionary assemblages, specifically an *Otus-Bubo-Strix* group, a *Surnia-Aegolius-Ninox* group, and an *Asio* group. Ford believed that these apparent phyletic relationships might best be expressed by recognizing three subfamilies of Strigidae, with the first two subfamilies having three tribes each, corresponding to the just-mentioned generic assemblages that occur within each.

Although Ford considered it reasonable to assume that those owls with modified external ears probably represent derived types from those with simpler ear anatomy, he declined to advocate this position formally. His conclusions include some surprising taxonomic changes from the Peters sequence, such as the association of the "large-eared" *Aegolius* with a group of otherwise "small-eared" owls in the subfamily Surniinae. And he not only shifted *Strix* and the more tropically distributed wood owls *Ciccaba* into the same subfamily but also regarded them as adjacent, closely related genera, a conclusion that had been reached earlier by Voous (1964) and was later supported by Cannell (1985). Finally, Ford found evidence that the large ears and associated hearing specializations of such genera as *Strix, Asio,* and *Aegolius* represent quite different anatomical modifications that probably reflect independent evolutionary pathways in these three groups. S. Olson (1995), also using morphological traits, came to rather similar taxonomic conclusions to those of Ford. Ford's suggested sequence of North American owl genera is shown below:

Family TYTONIDAE
   Subfamily TYTONINAE
      Genus *Tyto*
Family STRIGIDAE
   Subfamily STRIGINAE
      Tribe OTINI
         Genus *Otus*

      Tribe BUBONINI
         Genus *Bubo*
         Genus *Nyctea*
      Tribe STRIGINI
         Genus *Strix* (including *Ciccaba*)
         Genus *Lophostrix*
         Genus *Pulsatrix*
   Subfamily SURNIINAE
      Tribe SURNIINI
         Genus *Surnia*
         Genus *Glaucidium*
         Genus *Micrathene*
         Genus *Athene* (including *Speotyto*)
      Tribe AEGOLIINI
         Genus *Aegolius*
   Subfamily ASIONINAE
      Genus *Asio* (including *Pseudoscops*)

Since the 1970s, most taxonomic work on the owls has been addressed to the molecular level of variability, an approach initially spearheaded by Charles Sibley through his use of egg-white protein electrophoresis and later by his DNA–DNA hybridization studies. This work, and similar approaches of others, was first fully summarized in book form by Sibley and Ahlquist (1985) and Sibley and Monroe (1990). These authors retained the basic upper-level separation of the barn and bay owls from the typical owls (designating them respectively as the parvorders Tytonida and Strigida). However, they abandoned Peters's subfamilial subdivision of the Strigidae, while at the same time retaining much of his generic-level sequencing within the Strigidae. They also expanded the order Strigiformes to include the caprimulgiform birds as well. However, other recent biochemical and morphological analyses have retained the traditional ordinal separation of these two groups.

Among the several molecular approaches to avian taxonomy on recent years are those of Mindell et al. (1997), who used mitochondrial DNA to address several major problems of higher-level groupings. With regard to the owls, these authors found *Tyto* to consistently fall outside the limits of the typical owls, recognizing these two groups as discrete subfamilies. They also concluded that the scops owls they investigated (*Otus* and *Mimizuku*) are a monophyletic group, that these owls are in turn closely related to a *Nyctea/Bubo* clade, and that *Ninox* and *Aegolius* are sister taxa.

The most recent and comprehensive approach to molecular taxonomy of the owls is that of Wink and Heidrich (1999), using the mitochondrial cytochrome b gene sequencing data. Their data support a familial separation of the Tytonidae and Strigidae, and a surprising amount of genetic variation within the populations traditionally included within *Tyto alba*. If confirmed by others, a splitting of this single, nearly cosmopolitan species into several geographically distinct allospecies may be warranted.

These authors also found surprises in other seemingly monophyletic groups, especially the genera *Glaucidium* and *Otus*. Thus the Old World and New World forms of *Glaucidium* cluster to two discrete clades, separated by an estimated 7–8 million years of evolutionary history. As such, the New World forms may need to be removed from the genus *Glaucidium*, while some of the Old World species may also have to be shifted elsewhere. Additionally, the New World species *G. gnoma/californicum* is clearly not conspecific with the Old World *G. passerinum*, as has sometimes been suggested (Sibley and Monroe, 1990).

Likewise, the New World *Otus* species, with the single curious exception of *flammeolus*, differ from the Old World forms of this genus by a genetic distance estimated to represent 6–8 million years of isolation. In this case *Megascops* appears to be the most appropriate name for application to the New World clade, while the Old World *Otus* group may require some further subdivision. Although these authors estimated that the New World burrowing owl may be separated from the Old World *Athene* group by as much as 6 million years, they suggest that *Speotyto* may nonetheless be merged within *Athene* because of their many morphological and behavioral similarities. With regard to *Aegolius*, genetic differences suggest that the North American taxon *acadicus* diverged from the Old World *funereus* more than 6 million years ago, with the South American representative *harrisii* falling closer genetically to the North American than the Old World form.

Other major findings of Wink and Heidrich with regard to North American owls are that the Neotropical genus *Ciccaba* can be merged within *Strix*, that the Neotropical genus *Pulsatrix* is a valid genus with still unresolved affinities, and that *Bubo* and *Nyctea* are closely related genera, with a separation time estimated as something over 4 million years ago. Finally, these authors concluded from their data that the owls are obviously related neither to the diurnal raptors nor to the caprimulgiform birds, but also judged that mitochondrial DNA differences may be unable to resolve these higher-level groupings accurately.

Additional evidence as to multiple evolutionary pathways in the owls comes from the work of Feduccia and Ferree (1978), who examined variations in the stapes (columella) bone of the middle ear. They found that the condition of the stapes in *Tyto*, which exhibits an extended footplate specialization, is similar to that of some strigids having highly specialized ears, such as certain species of *Strix*. Thus during the evolution of the owls convergence has occurred between the Tytonidae and some of the more highly specialized forms of the Strigidae in the specialization of the stapes's shape (see Figure 12). Correspondingly, evolutionary adaptations associated with improved nocturnal vision and with reduced self-generated wing noise during flight must gradually have improved the nocturnal prey-catching abilities of owls.

Embodied silence, velvet soft, the owl slips through the night.
With Wisdom's eyes, Athena's bird turns darkness into light.

—Joel Peters, "The Birds of Wisdom"

*He [the scritch-owl] keepeth ever in the deserts
and loveth not only such unpeopled places, but
also those that are horribly hard of access.*

—Pliny, *Natural History*

Owls are large predatory animals having high metabolic rates and
a consequent high demand for food in the form of prey species
that often can be captured only with a high degree of skill. Hence
their ecology is of special interest. Each species of owl has evolved
a well-defined ecological niche, which is a composite of many aspects of its ecological profession or evolutionary strategy. These include such things as its many anatomical adaptations associated
with survival and reproduction, its innate or learned behavior required for individual survival as well as for coping with intra- and
interspecific interactions, and its physiological abilities to deal
with both its abiotic and biotic environments. In a short review
such as this one it is impossible to deal fully with all these interesting aspects of owl ecology; instead this chapter concentrates on
only a few selected topics of general comparative interest, with
discussions of individual species ecologies reserved for the separate species accounts.

Habitat and Food Selection

Owls, unlike many other North American birds, tend to be relatively sedentary, with only the smaller and highly insectivorous
species showing predictable migratory tendencies. Thus habitats
used during the breeding season are often essentially the same as
those used during the rest of the year, except that in the case of
several northern species the birds may be variably displaced southward during the coldest parts of the year. In any case, a comparison of breeding habitats utilized by owls provides a useful device
for judging ecological affinities of owls.

    Based on general descriptions in the literature, the North
American owls can for the most part be associated with one or two
primary breeding habitats each (Table 2). Collectively these represent most of the major habitat types available on the continent. Major habitat types that are underrepresented in or absent from this
table, and perhaps thus are underexploited by breeding owls, are
the alpine tundra and timberline zones, whereas forests and woodlands would seem to be highly utilized by owls of all sizes. A similar
predominance of arboreally adapted forms occurs among 13
species of European owls, according to a habitat selection analysis
by Mikkola (1983). Insectivorous species of North American owls
are largely those that occur in fairly open, arid environments of the
Southwest, whereas larger owls concentrating primarily on birds

Table 2

Preferred Habitats and Predominant Prey of North American Owls

	Large[1]	Medium[2]	Small[3]
Preferred Habitats			
Arctic tundra	Snowy		
Forest and woodland			
Northern coniferous			
Dense forest edge	Great gray		Boreal
Muskeg, open woods		N. hawk-owl	
Western forests			
Climax coniferous	Spotted		
Open lower montane			Flammulated
Open coniferous/mixed			N. pygmy-owl
Eastern and western			
Mature forests	Barred		N. saw-whet
Forest edges	Barred	Long-eared	N. saw-whet
Southwestern woodlands			
Dense pine-oaks			Whiskered
Desert and thorn forest			
Saguaro desert, oaks			Elf
Thorn forest and mesquite			F. pygmy-owl
Grasslands			
Prairies and marshes		Short-eared	
Steppes and semideserts		Short-eared	Burrowing
No single habitat type			
Pandemic	Great horned		
Mainly southern		Barn owl	
Mainly eastern			E. screech-owl
Mainly western			W. screech-owl
Predominant Prey and Daily Hunting Periods[4]			
Arthropods (esp. insects)			Elf (N, C)
			Flammulated (N)
			Whiskered (N, C)
Mixed prey[5]			Burrowing (N, D)
			E. screech-owl (N, C)
			W. screech-owl (N, C)
Small vertebrates (esp. birds)			N. pygmy-owl (C)
			F. pygmy-owl (C)
Small vertebrates (esp. rodents)	Great gray (N, D)	Long-eared (N)	N. saw-whet (N)
	Barred (N)		Boreal (N)
		N. hawk-owl (D)	
	Spotted (N)	Short-eared (N, D)	
		C. Barn owl (N)	
Larger vertebrates (mammals and birds)	Snowy (N, D)		
	Great horned (N, C)		

[1]Averaging over 500 g.

[2]Averaging 250–500 g.

[3]Averaging under 250 g.

[4]D = diurnal; N = nocturnal; C = crepuscular (dusk and dawn).

[5]Arthropods and vertebrates, varying seasonally or locally.

and mammals are generally associated with boreal to arctic habitats.

In the hot southwestern deserts, owls would seem to have an ecologic advantage over hawks and other diurnal predators; they can use the cool of the night to reduce evaporative water loss and avoid overheating. Most desert life is also mainly active at night, increasing the owls' prey base. It is thus not surprising that perhaps the common raptor in most North American deserts is the elf owl, which greatly outnumbers such diurnal avian predators as kestrels, shrikes, and roadrunners. Other typical North American desert owls include the burrowing owl and western screech-owl. All these are relatively small owls that often consume insects and other arthropods.

Seemingly the great horned owl is the owl species in North America having the greatest ecological flexibility in its breeding habitats. It also has perhaps one of the widest ranges of acceptable prey types, albeit with a distinct tendency for selecting the largest available prey. Its European counterpart species, the eagle owl (*Bubo bubo*), is also notable for its opportunistic tendencies in habitat and prey selection (Mikkola, 1983).

In an interesting early analysis, J. Craighead and Craighead (1956) compared the habitats and foraging ecologies of several owl and hawk species of North America, including the eastern screech-owl, the great horned owl, the short-eared owl, and the barred owl. They observed not only that each of these owl species tended to use a different habitat for its nocturnal foraging but also that each had one or more fairly close ecological counterparts among the diurnally hunting hawks. Individually these species either hunted in different habitats or used the same general habitat at a different time or in a different manner than others; but collectively they effectively exploited a wide range of available prey resources. In a similar study of 12 breeding species of raptors occurring in the Great Basin area of Utah, those raptors most likely to compete with one another differed in their choice of nest sites, daily activity (hunting) periods, prey species, or nesting timetables (D. Smith, 1971; D. Smith and Murphy, 1973). In both of these studies it was found that the tendency of established pairs to reoccupy old nesting sites and territories tends to stabilize the overall raptor population; some Utah nest sites known to have been present in the early 1940s were found to be still present and in some cases being used by the same species in the late 1960s.

Species-specific habitat and nest site selection differences among European owls have been discussed by Mikkola (1983), who calculated ecological overlap values (index of community similarity) for three species of *Strix*. He found that the overlap values were lower for the species' nest site characteristics than for their nesting habitats in general, which in turn were somewhat lower than for their dietary similarities. Competition among these three closely related species was perhaps reduced in part by differing degrees of nocturnality, with the great gray owl the most diurnal of the three, the tawny owl (*Strix aluco*) the most nocturnal, and the Ural owl (*Strix uralensis*) intermediate. Considering all four measured characteristics, the calculated ecological overlaps were greatest between the tawny and Ural owls, which also overlapped most in average body sizes, geographic distributions, and hunting methods; the least overlap occurred between the tawny and great gray owls. In general, the estimated degree of ecological overlap and consequent potentially severe interspecific competition among these three species was unexpectedly high, which Mikkola tried to explain in part by suggesting that these owls' food resources may normally not be in short supply, and indeed at times may be superabundant.

In another study of the food niches of European owls, Herrera and Hiraldo (1976) similarly found that calculated dietary overlaps were very high among middle and northern European owls, and tended to be related to food supplies, which were mostly microtine rodents. In middle latitudes the largest foraging guild consisted of medium-sized owls (200–500 grams) that all concentrated on this usually abundant prey source. In southern Europe an absence of abundant food in the form of such microtines apparently causes (1) some species elimination, (2) dietary shifts in species reaching that region, and (3) adjustments in niche breadth of the species occurring there.

For North American owls the most complete analysis of possible ecological competition for food owing to similarities in body size and associated overlapping food habits was that of Earhart and Johnson (1970). These authors found that in most owls the smaller species tend to exhibit a lower degree of sexual dimorphism in body weight than do larger ones, and this same trend is evident in various races of screech-owls and of the great horned owl. Those owls feeding predominantly on vertebrates (generally the larger ones) show the greatest degree of dimorphism, and those concentrating on arthropods (the smaller owls) are either essentially monomorphic or exhibit low degrees of sexual dimorphism. Similarly, Snyder and Wiley (1976) found a moderately strong correlation between the percentage of vertebrates in the diet of North American raptors (hawks and owls) and the degree of reversed sexual dimorphism, as well as a highly significant correlation between the percentage of bird prey

Table 3

Estimated Percentage Prey Consumption of North American Owls

Species (D.I.)[1]	Invertebrates	Lower Vertebrates	Mammals	Birds	Reference
Elf (0.3)	98.1	0	1.9	0	Snyder and Wiley, 1976
Flammulated (0.8)	100	0	0	0	McCallum, 1994b
F. pygmy-owl (4.2)	58	22.5	8.6	10.5	Proudfoot and Beasom, 1997
N. pygmy-owl (3.9)	61.4	2.5	23.3	12.8	Snyder and Wiley, 1976
N. saw-whet (6.4)	1.4	tr.[2]	96.8	1.6	Snyder and Wiley, 1976
	<1.0	0	96–98	2–3	Cannings, 1993
Whiskered (1.6)	97.4	tr.	1.8	tr.	Snyder and Wiley, 1976
	86–100	0–7	0–7	0	Gehlbach and Gehlbach, 2000
Boreal (8.2)					
North America	0	0	93.6	6.4	Snyder and Wiley, 1976
Europe	13	0	82	5	Hayward, Hayward, and Garton, 1993
Burrowing (−1.5)	90.9	2.0	6.9	tr.	Snyder and Wiley, 1976
Screech-owls (2.3)					
Eastern and western	30.7	0.6	65.5	3.3	Snyder and Wiley, 1976
Eastern	0–49	1.7–27	2.5–60.3	22.2–74.1	Gehlbach, 1995b
Long-eared (2.8)					
North America	tr.	0	97.1	2.3	Snyder and Wiley, 1976
Europe	tr.	0	85–98	2–5	Mikkola, 1983
N. hawk-owl (4.2)					
North America	0	2.5	93.7	3.8	Snyder and Wiley, 1976
Europe	2.3	0	68.2	29.6	Mikkola, 1983
Short-eared (2.5)					
North America	0–4.7	0–4.8	83–99.8	0.2–15.1	Holt and Leasure, 1993
Europe (nest)	<2.6	tr.	95–99	<1.9	Mikkola, 1983
Barn owl (3.1)					
North America	1.6	tr.	91.4	6.9	Snyder and Wiley, 1976
Europe	<4.3	<4.4	87–99	<4.4	Mikkola, 1983
Barred (4.6)	15.8	2.5	76.0	5.8	Snyder and Wiley, 1976
Spotted (2.3)					
North America	57.4	tr.	37.5	4.5	Snyder and Wiley, 1976
Oregon	tr.	tr.	92.6	5.3	Forsman, Meslow, and Wight, 1984
Great gray (6.3)					
North America	0	0	93.3	6.7	Snyder andWiley, 1976
Europe (nest)	tr.	tr.	98.4	1.0	Mikkola, 1983
Great horned (7.0)	14.7	1.6	77.6	6.1	Snyder and Wiley, 1976
Snowy (6.6)					
North America	tr.	tr.	78.0	21.3	Snyder and Wiley, 1976
Finland	tr.	0	96.9	2.8	Mikkola, 1983

Note: Prey consumption based on frequency analysis of regurgitated pellets; species arranged by increasing average adult weight.

[1]D.I. = Dimorphism index (larger positive numbers indicate proportionately larger females; negative numbers indicate larger males).

[2]tr. = trace (under 1%).

in the diet and the degree of dimorphism. (See Table 3, which is a summary of the authors' food intake data for owls and includes dimorphism index values based on body weights.) These data tend to support the idea that sexual size dimorphism may be related to differential niche utilization, with those species that feed mainly on vertebrates tending to evolve divergences in bodily size relative to the intensity of intersexual food competition. Each sex might thereby specialize on a size range or type of prey that tends to minimize intersexual competition as well as interspecific competition with other raptors. Mikkola (1983) reviewed the available information on this possibility with respect to European owls and stated that the currently available data are too limited to support such a conclusion with respect to inter-sexual competition. However, he presented data indicating that the average prey weight of 9 male boreal (Tengmalm's) owls was 12 grams, compared with 21 grams for 20 females, while for 25 male and 29 female great gray owls the average prey weights were 20 and 24 grams, respectively. In a thorough survey of reversed sexual dimorphism among the western Eurasian species of hawks and eagles, Mueller and Meyer (1985) could not find adequate evidence supporting the idea of differential prey utilization by the two sexes as a cause of this dimorphism. However, they did find high reversed sexual dimorphism to be correlated with large clutch size and high clutch weight (but not egg weight), prevalence of the female in direct feeding of the young, predation on alert and elusive prey, female dominance over her mate, and increased involvement of the female in territorial defense. At least some of these correlations are probably artifacts rather than possible causal explanations, and these two authors regarded diet as a limiting but not selective force in the evolution of reversed sexual dimorphism. Later, Mueller (1986) extended his analysis to the North American and Eurasian owls, and generally found additional support for his view that prey specialization may influence the degree of, but not cause, reversed sexual dimorphism.

Other ideas as to the potential ecological advantages of reversed sexual dimorphism include such possibilities as relative food stress toward the end of the breeding season (Snyder and Wiley, 1976), the advantages to females in laying larger eggs and bringing larger prey to their young, thus reducing the number of foraging trips (Nilsson and von Schantz, 1982), the possible advantage of large body size in generating incubation heat and surviving incubation stresses during cold temperatures often encountered by high-latitude nesters (Mikkola, 1983), and others. Thus the effects of relative female size on feeding or breeding ecology define a large group of hypotheses set forward to explain reversed sexual dimorphism in raptors (see reviews in D. Smith, 1982, and Mikkola, 1983). Gerhardt and Gerhardt (1997) have questioned whether any of these explain reversed sexual dimorphism in either temperate or tropical owls. A second group of hypotheses, involving sex-related behavioral traits in territorial and pair-forming behavior, is mentioned in Chapter 4.

In a Colorado study, Marti (1974) examined the feeding ecology of four species of sympatric owls (great horned, long-eared, burrowing, and common barn owl). He found that the four species selected different prey species, and that they also selected prey of significantly different mean weights. The great horned owl exploited not only the heaviest average prey but also the largest range in prey size, while the other three species utilized prey of considerably smaller average weights (in the same relative sequence as the species' average adult weights). Additionally the four varied in their relative nocturnality (the barn owl and long-eared entirely nocturnal, the great horned largely crepuscular, and the burrowing active during both day and night). The long-eared was best at finding prey visually under low light conditions, the burrowing the poorest, and the barn owl and great horned intermediate. The great horned owl hunted primarily by making flights from observation posts and had the highest wing loading (body weight per surface area of wing) of the four species; the long-eared and barn owls had lower wing loadings and apparently were adapted for hunting on the wing; and the burrowing owl used varied hunting techniques. Barn owls and great horned owls exhibited significant variations in prey composition in different habitats and at different times, but no such habitat or temporal differences were found for the other two species. Of all the parameters measured, prey size was judged by Marti to be probably the single most important one in effecting foraging niche segregation. Marti's studies confirmed the general ecological prediction that smaller predators will tend to spend less time in searching for relatively abundant prey and also will tend to have a more restricted diet, whereas larger predators feed more on prey representing the rarer (larger) end of the available food resource spectrum, and so spend more time in hunting and also exercise less selectivity in their prey choice.

In a comparative study of the great horned owl and common barn owl in northern California, Rudolph (1978) also found that these two species exhibited differences in hunting methods and habitat preferences that reduced interspecific spatial overlap, such as the barn owl hunting more while remaining on the wing, and the great horned initiating more foraging flights from perches. Additionally, although substantial dietary overlap existed, the relative proportions of prey species taken varied somewhat, and the barn owl was apparently somewhat more nocturnal in its foraging behavior. In a similar study undertaken in central Washington, Knight and Jackman (1984) found that great horned owls not only took prey of larger average weight but also exhibited a broader food niche than did barn owls. However, both studies indicated a very high estimated level of food-niche overlap, which may in part have resulted, at least in the area of the latter study, from a relatively recently acquired sympatry of the two species. Marks and Marti (1984) compared the feeding ecology of the common barn owl and long-eared owl in Idaho, finding that barn owls took prey that were significantly heavier than those of long-eared owls and concentrated on *Microtus* voles. Long-eared owls fed mainly on *Peromyscus* and heteromyid rodents. Dietary overlap between the species ranged from 48 to 61 percent during two consecutive years.

In line with these findings, a comparative analysis of various North American and European hawks and owls indicated that those species of owls that primarily forage by actively searching while remaining airborne (Striginae and Tytoninae) tend to have lighter average wing-loading characteristics than do those species (mainly Buboninae, including the great horned owl) employing mainly sit-and-wait hunting tactics from convenient perches, making periodic forays upon detected prey (Jaksic and Carothers, 1985).

Clearly the potential parameters for measuring the levels of interspecific competition are complex. A somewhat simplified approach on a broad scale is to consider the degrees to which owl species interact by (1) having overlapping (sympatric) breeding ranges, (2) having similar daily (diel) patterns of hunting activity (diurnal, nocturnal, crepuscular, or some combination of these), (3) having overlapping average adult weights, (4) having overlapping breeding habitat preferences, and (5) having similar prey requirements. Using ecological information presented in

Tables 2 and 3 and weight data from summaries in the species accounts, and extracting information on range overlap from the individual distribution maps, such a collective comparison can be produced (Figure 1). Some potential competition-reducing factors, such as having different nest site preferences or having highly specific prey selection adaptations that are more restricted than the categories used in Table 2, might serve to provide a more fine-grained analysis than the one offered here. Species of owls that are geographically isolated (allopatric) during the breeding season are considered noncompetitors for purposes of this analysis, although it is of course apparent that seasonal competition might sometimes occur outside the breeding period. Also included in Figure 1 is an indication of predatory interactions among owl species (the preying on smaller species of owls by larger ones), based on records encountered in the literature. Like the eagle owl among Eurasian owls (Mikkola, 1983), it would seem that only the great horned owl is an ecologically significant predator on other North American owls.

With such limitations in mind, the diagram suggests that four pairs of sympatric species overlap in all four niche-overlap traits that tend to promote competition, namely the barred and spotted owls (which have only recently developed sympatric contact), the eastern and western screech-owls (which are also only very slightly sympatric), the snowy and great horned owls, and the burrowing and western screech-owl. If one

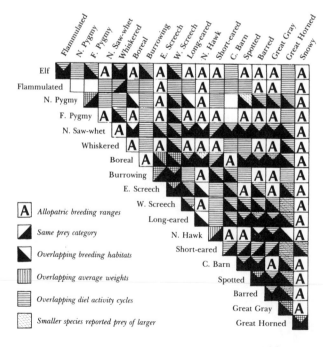

Figure 1. Interspecific ecological interactions among North American (excluding Mexican) owls.

	Allopatric breeding ranges
	Same prey category
	Overlapping breeding habitats
	Overlapping average weights
	Overlapping diel activity cycles
	Smaller species reported prey of larger

considers potential collective interspecific competition, the snowy owl is the species most effectively ecologically isolated from others, competing only with the short-eared and great horned owls and in a total of 6 out of 72 potential niche overlap categories, whereas the western screech-owl has the greatest potential overall competitive interactions, involving 16 other owl species and 34 of 72 potential niche overlap categories.

Comparative Biogeography and Species Densities

In a broad continental scope, the ecologies of North American owls can be viewed in terms of the actual numbers of owl species occupying specific geographic regions, irrespective of their ecological habitat preferences. Such an analysis offers an insight into "species packing" and possible areas of potentially high interspecific competition. Mikkola (1983) determined that in Europe the 13 species of breeding owls tend to reach a maximum of species density at about 57° north latitude (10 species), and from there species density decreases both toward the arctic and toward the Mediterranean region before finally increasing once again south of the Sahara Desert. In North America the number of breeding species is somewhat higher (19) than in Europe, although both areas share a considerable number of the same species (7) as well as having several closely related ecological replacement forms (e.g., tawny and barred owls, common and northern pygmy-owls, scops owl *Otus scops,* and flammulated owl). In Figure 2 a species-density map for North and Central America is provided, showing numbers of breeding species present in various regions (based on the distribution maps presented elsewhere in the book and a literature survey for Central and South American areas).

The general pattern that emerges from Figure 2 is that species density tends to increase in North America from east to west and from north to south, and is generally higher in mountainous areas than elsewhere. Maximum species diversity in North America occurs along the Pacific Coast, but perhaps reaches a hemispheric peak in Panama, where 14 species breed in an area of about 77,000 square kilometers. Species-density averages decline from there northward toward Mexico, where 26 species breed in an area of some 2 million square kilometers. This is still a very high species-density level compared with Canada, where 14 species breed in an area of 10.5 million square kilometers, or with the continental United States (excluding Alaska), where 17 species breed in an area of 8 million square kilometers. What one might expect from such a situation is that species occurring in areas of low species density should tend to be ecological generalists, exploiting a broad resource base, whereas areas of high species density should support greater numbers of ecological specialists. To a considerable degree all owls must be considered specialists, and data from northern Europe suggest that several species of owls having high levels of ecological overlap can coexist indefinitely in areas where food supplies are usually relatively high (Mikkola, 1983; Herrera and Hiraldo, 1976).

Ecological Aspects of Body Size

Size differences among owls are important in several respects. In general, owl species of very similar body weight tend to consume very similar foods (see Table 3), and thereby often offer potential ecological competition with similar-sized species. Second, large owls often tend to prey on smaller owls, sometimes to a considerable degree (see review by Mikkola, 1983, and his Table 56 for European species). Finally, there is an inverse relationship between body size and relative daily food intake requirements, with the smallest species such as pygmy-owls requiring food averaging up to about 50 percent of their body weight daily and the largest species such as eagle owls requiring only about 15 percent. This of course means that smaller owls in general must spend a greater percentage of their waking hours in hunting or must concentrate on taking more common or easily obtained but smaller prey. Larger species can more easily cope with day-to-day variations in prey availability and can perhaps afford to spend more energy in capturing relatively rare and more difficult or dangerous larger prey species.

Body weight of birds is also of significance in terms of its effect on ease of flight, as affected by wing surface area relative to total body weight. This trait is usually measured by judging a species' wing loading (see Poole, 1938). Representative wing-loading data have been provided by Mikkola (1983) for 9 species of European owls, and similar data were provided by D. R. Johnson (1978) for 8 North American owls. Highest wing loadings of more than 0.5 grams per square centimeter have thus been obtained for the largest owls (snowy, great horned, and eagle owls), and relatively low wing loadings of less than 0.3 grams per square centimeter have been calculated for the long-eared, boreal (Tengmalm's), northern saw-whet, Eurasian pygmy-owl (*Glaucidium passerinum*), and common barn owl. Although useful, the wing-loading statistic is only usable when two species of essentially the same body weight are compared, owing to the fact that estimates of body mass are based on three-dimensional or vol-

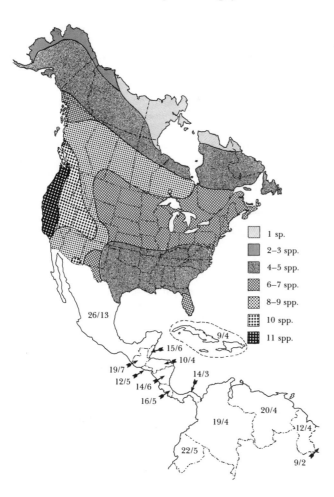

Figure 2. Species-density map of North American owls. Numerals shown for areas south of U.S. border indicate total number of native owl species, followed by the number of these species that occur north of the U.S.-Mexican border.

▢	1 sp.
▨	2–3 spp.
▨	4–5 spp.
▨	6–7 spp.
▨	8–9 spp.
⊞	10 spp.
▦	11 spp.

26/13

9/4

15/6

19/7   10/4

12/5   14/3

14/6

16/5

20/4

19/4   12/4

22/5

9/2

umetric measurements, whereas wing area data reflect two-dimensional measurements, and thus the two do not exhibit a linear relationship with one another. Jaksic and Carothers (1985) have avoided this problem by utilizing a linearized wing-loading statistic, by which the cube root of body mass (in grams) is divided by the square root of the wing area (in square centimeters). Of the 8 North American owl species included in their analysis, the highest linearized wing loadings were determined for the great horned owl, and the lowest for the short-eared and long-eared owls. The former species tends to utilize sit-and-wait hunting tactics, whereas the latter two tend to hunt during prolonged coursing flights. Norberg (1987) reported that forest-dwelling owls tend to have short and broad wings with low wing loading, facilitating prey transport and reduced wing noise during flight. Forest owls are also primarily search-and-pounce foragers, spending most of their time searching for prey and little time chasing and catching it. Many utilize hearing to a greater degree than vision during such

searches, and asymmetrical external ear anatomy is best developed in forest owls.

In Table 4 some data are presented on owl specimens brought into the Nebraska State Museum; all measurements are from fully grown birds of seemingly normal weight and body condition. The collective weight of the breast muscles relative to total body weight (the "power-weight ratio" of Clark, 1975) should provide an indication of relative work that can be expended during flight, representing a statistic independent of wing-loading measurements. These data suggest that both eastern screech-owls and common barn owls have small breast muscles relative to adult body weights and that long-eared and short-eared owls are similar to the great horned owl in this regard, which does not agree very closely with wing-loading data. The data on eye weight relative to body weight are included as a possible index to the relative significance of vision during hunting; these data suggest that the eastern screech-owl (which has relatively small and unspecialized external ears) has a substantially larger relative eye

Table 4

Some North American Owl Morphometric Traits of Ecological Interest

Species	Sample Size	Average Total Weight (g)	Average Breast Weight (g)	Average Weight of Eyes (g)	Average Wing Area (sq cm)	Linear Wing Loading (g/sq cm)	Wing Loading
Great horned	5	1670	170.9 (10.2%)	25.7 (1.5%)	2087	0.80	0.259
Barred owl	1	874	111.5 (12.8%)	16.0 (1.8%)	—	—	—
Barn owl	1	368	27.2 (7.4%)	4.1 (1.1%)	1148	0.320	212
Short-eared	1	353	37.5 (10.6%)	3.9 (1.1%)	936	0.38	0.231
Long-eared	1	222	26.9 (12.1%)	3.8 (1.7%)	846	0.26	0.208
E. screech-owl	3	138	11.5 (8.4%)	7.4) (5.4%	373	0.37	0.268

Note: Species arranged by decreasing body weight; "breast weight" includes total of both breast muscles (*pectoralis major* and *supracoracoideus*) in fresh specimens; "eye weight" includes total weight of both eyes in fresh specimens. Wing areas were determined by using outline drawings of extended wings.

size than any of the other included species, and that the common barn owl and short-eared owl (which both have highly specialized external ears) have the smallest relative eye sizes. In general these results agree with the relative importance of vision and hearing during hunting that is believed to be typical of these species. A more detailed discussion of the role of vision in owl survival and ecology is provided in Chapter 3.

The owl is not accounted the wiser for living retiredly.

—T. Fuller, *Gnomologia*

The screech-owl, screeching loud,
Puts the wretch that lies in woe
In remembrance of a shroud.

—Shakespeare, *A Midsummer Night's Dream*

General Morphological Characteristics

Owls differ from other North American birds in a wide variety of
ways, some of which can be detected in their external features
(Figure 3). They are generally large-headed birds, which relates to
their relatively large brains, their very large, frontally oriented
eyes, and the presence of a surrounding facial disk of feathers that
sometimes includes a pair of "horns" or "ears" at the top of the
disk. The actual ears are in fact well hidden within the facial disk,
the purpose of which is probably to increase the effectiveness of
binaural (stereophonic) sound reception and the bird's related un-
paralleled abilities to localize point sources of sound. In all owls the
rather short, decurved, and raptorial beak is at least partly hidden
by long rictal bristles that extend down and forward on either side
from the eyes, often forming a moustachelike structure. Similar
loral feathers extend forward and upward from the eyes to form
"eyebrows" that, like the rictal bristles, are of uncertain function,
but that help impart a humanlike expression to the face. Indeed, at
least some owls can modify the position of these feathers in a way
that greatly influences the appearance of the face. Hidden along
the lateral edge of the facial rim is an opening in the feathers (the
ear conch) at the base of which is the actual opening of the external ear. Dermal ear flaps, or opercula, are frequently associated
with these feathers, the movement of which can markedly alter the
shape of the facial disk. In some owls the ear flaps are connected by
a tensor membrane or ligament that is probably related to the
bird's ability to alter the positions of these feather areas.

The facial disk of owls such as the great gray owl and other
owls of the genus *Strix* is especially notable (Figure 4), as the
feathers that comprise the disk and nearby areas are highly distinctive, and vary from tightly packed and short bristlelike structures that extend out from the edges of the ear flaps (seemingly
increasing their effective length) to amazingly soft and open
feathers that seem to be designed for letting the maximum
amount of sound through. In the great gray owl the shape of the
skull itself is somewhat modified by the specializations for directional hearing and sound localization, so that the right side of the
skull appears somewhat inflated as compared with the left side
(see Figure 4). This skull asymmetry reaches an extreme in the
boreal owl, in which the skull appears to be grossly misshapen
when viewed from almost any angle (see Figure 67). Its relationship to hearing is discussed in Chapter 4.

Figure 3. External features of owls.
Adapted in part from Mikkola (1983).

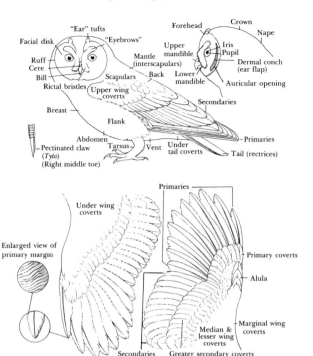

Lower Wing Surface   **Upper Wing Surface**

Nearly all the feathers of owls tend to be fairly soft, and this is especially true of the body feathers. Even the relatively rigid and strong primary feathers are notable for their specialized leading edges of softened feathers, which are generally believed to be responsible for the silent flight of most owls (although the elf, burrowing, and pygmy-owls are all fairly noisy during flight, perhaps because they are either relatively diurnal or feed largely on insect prey). The middle toe of the common barn owl is notable for its comblike claw, which serves for grooming and maintaining the plumage, although the other North American owls seemingly manage to maintain their feathers equally well without this adaptation. Among owls, true down feathers are quite restricted in their distribution, and aftershafts are lacking, but most of the normal contour feathers covering the body have downy bases.

In nearly all owls the feathers are cryptically colored with tones of brown, gray, and sometimes black, often in a manner that allows the birds to blend remarkably well with their surroundings. Usually little sex-related dichromatism of plumage occurs in owls, but in some (especially the snowy owl) the sexes are readily recognizable by their plumage differences, the female being more pigmented. Additionally, there are usually few if any postjuvenal age differences evident in

the plumages of owls. In some species (such as the eastern screech-owl) there are, however, two distinct color morphs or "phases" that are genetically determined (the red-determining genetic factors of the screech-owl dominant to those producing gray) but independent of sex. The ecological significance of such color phases in owls is uncertain, but in the case of the eastern screech-owl the more rusty phase is associated with primarily deciduous forests and fairly humid, generally warmer climates of the eastern United States. The gray morph (which is the typical morph of the western screech-owl) is associated with coniferous forest habitats having generally cooler and drier climates, but it also occurs in southeastern evergreen forests rich in the grayish epiphyte Spanish moss (*Tillandsia*) (Hasbrouck, 1893a, b). Possibly the gray color provides better concealment in a coniferous forest environment. Similarly distributed gray and rusty brown plumage morphs also occur in the ruffed grouse (*Bonasa umbellus*), which occupies comparable forest habitats and probably is exposed to similar predatory risks. Several other small woodland owls such as the flammulated owl and northern pygmy-owl have similar types of plumage dimorphism. It is unlikely that the owls themselves are even aware of these differences, judging from what we believe to be their relatively poorly developed sense of color discrimination. It has been suggested that the plumage dimorphism of the eastern screech-owl is not directly related to regional variations in humidity (Owen, 1963a), as had been suggested earlier (Hasbrouck, 1893b), but rather represents a case of balanced genetic polymorphism, with selection for bimodal variation occurring in most areas and intermediate types thus relatively rare. Moser and Henry (1976) found that red-morph birds have higher average metabolic rates and seem to suffer higher winter mortality than do gray-morph individuals, which they believed makes the gray phase better adapted to northern climates. However, the low incidence of red-morph birds in the Gulf states is not explained by this hypothesis, nor is the virtual absence of red-morph birds in the western screech-owl population. Gehlbach (1994a) reported that the rufous morph may be preadapted to city and suburban life, with their warmer, wetter, and more climatically stable microclimates, and the gray morph to cool-dry years of climatic cycles.

All owls have highly developed talons on all four of their toes, and although three of the toes may sometimes be directed forward in the usual (anisodactyl) manner of most birds, owls more commonly swing their outer (fourth) toes backward when perched so as to produce a two-in-front, two-behind (zygodactyl) arrangement. This toe arrangement is also always assumed when the bird is reaching for or carrying prey, thereby producing a maximum spread of the talons and probably also improving the owl's ability to clutch and carry heavy prey, the weight of which is equally distributed in front of and behind the owl's feet. Smaller prey are often carried in the bill.

Although the basic leg and foot musculature of owls is nearly identical to that of other birds, a few points are worth noting (Figure 5). The anterior toes are collectively flexed by a large tendon, the Achilles tendon, that extends from the very large gastrocnemius muscle originating on the upper midleg (drumstick). The tendon of this muscle extends down, passes over the rear joint of the "heel," and sends flexor tendons to each of the anterior toes. This produces a powerful flexing action, which is automatically aided by the increased tension on the associated tendon that is generated as the heel is bent forward during contact with a prey animal. Prior to such contact the outer (fourth) toe is moved posteriorly by an abductor muscle, bringing it into opposition with the two other front toes and nearly in line with the hallux. The second toe is likewise spread away from the third toe by another abductor muscle, further spreading the effective grasp. The hallux is extended and flexed by a pair of short muscles (extensor and flexor hallucis) originating on the proximal end of the tarsometatarsus. The tendons of these two opposing muscles respectively insert on the top (extensor) and bottom (flexor) of the hallux. The flexor tendon is unusually large, and its muscle contraction is responsible for driving the large hallux talon strongly into the prey's body. It gains mechanical advantage by passing over a pulleylike groove at the base of the hallux.

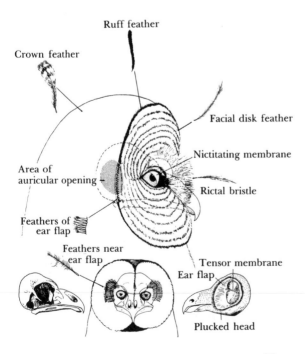

Ruff feather

Crown feather

Facial disk feather

Nictitating membrane

Area of auricular opening

Rictal bristle

Feathers of ear flap

Feathers near ear flap

Tensor membrane

Ear flap

Plucked head

Figure 4. Variations in facial feather structure and skull specializations of the great gray owl. Note skull asymmetry in anterior view. In part after skull drawings in Shuffeldt (1900).

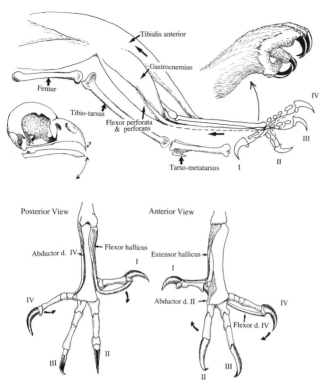

Figure 5. Talon and toe-closing mechanism of owls, as illustrated by barred owl (*above*, after Heinrich, 1987). Bones and tendons are diagrammatic; proportions are not all to scale. Shown underneath are corresponding foot and leg bones drawn to actual scale, plus a barred owl skull, showing its kinetic actions (from museum specimens). At bottom are posterior and anterior views of great horned owl foot bones and a few selected tendons, illustrating extension and flexion mechanisms for the hallux (digit *I*), plus abduction and protraction mechanisms for two of the anterior digits, with arrows indicating associated movements. Adapted from Hudson (1937).

Additional general adaptations of owls are evident when their skeletons are examined (Figure 6). In common with other birds, their skeletons are relatively light in weight; the weight of 10 eastern screech-owls' averaged 7.5 percent of their whole body weight and 16 adult great horned owl skeletons in the Nebraska State Museum collection averaged 8.6 percent of body weight (proportional skeletal weights tending to increase with increasing body mass). Owl skeletons are notable for their broad skulls and massive imbedded bony ring of sclerotic ossicles surrounding the eye. The neck is fairly short and composed of 14 cervical vertebrae, which allows for enough rotation of the neck that an owl can turn its head and peer directly over its back. The sternum is fairly small in its relative size, in correlation with the relatively small mass of associated breast muscles (see Table 2), and in all North American owls but one the posterior edge of the sternum is indented with four deep sternal notches. (Only two are present in the common barn owl.) Additionally, in the common barn owl the furcula (wishbone) is fused to the anterior edge of the sternum, but in the other owls the bones are separate. The foot bones, or tarsometatarsi, are relatively short and stout in owls, probably in conjunction with their important raptorial role in the efficient killing and carrying of prey. However, the wing bones are relatively long

and the associated wing surface area relatively broad, producing a low wing loading and associated ease of taking and maintaining flight, even when carrying prey. Open-country owls, such as short-eared owls, have noticeably longer and more slender wings than do owls of woodland habitats, such as screech-owls. These differences in wing shape are apparent among a group of similar size owls (Figure 7), in which the screech-owls, flammulated owl, and northern saw-whet owl are all forest-dwelling species with similar wing shapes. The northern pygmy-owl has notably short primaries and longer secondaries, perhaps related to its needs for aerial maneuverability, whereas the open-country (and strongly migratory) burrowing owl's wings are relatively long and narrow.

The pattern of wing molt in owls is notably complex, with replacement of the flight feathers being done over a very prolonged period that may often require at least three years, and perhaps as many as six, for the replacement of a complete set of these feathers. The first generation of flight feathers is associated with the acquisition of the bird's juvenal plumage (which follows two increasingly downy plumages on the nest). This molt results in a wing having uniformly colored and patterned primaries and secondaries, except that the outermost primaries tend gradually to become somewhat worn and more pointed

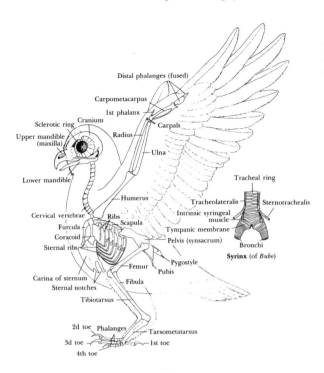

Figure 6. Skeletal and syringeal anatomy of owls. Adapted in part from Mikkola (1983); syrinx shown (*inset*) in ventral aspect.

through their first year. During the second summer of life the flight feathers and rectrices begin to be slowly replaced, the primary molt curiously often starting about in the middle (the sixth primary), and the molt proceeding in both directions. However, some juvenal remiges (the innermost primaries and secondaries) may persist until the fourth year of life, so that flight feathers representing up to three feather generations may be simultaneously present (Figure 8). Juvenal alular feathers may be carried to at least the third year,

at least in barn owls. Detailed accounts of flight-feather molts in North American owls can be found in Pyle (1997a, b).

Tail molt in owls is also quite irregular, and in some small owls this molt may be almost simultaneous, whereas in other owls it may be extended over a two-year period. At least some of these complexities of wing and tail molt in owls can be related to the importance of maintaining minimal life and propulsion needs throughout the year. Additionally wing and tail molt occurs ear-

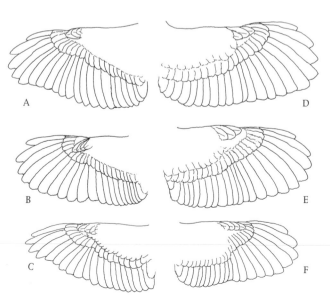

Figure 7. Wing shapes among small owls, including (*A*) northern pygmy-owl, (*B*) northern saw-whet owl, (*C*) burrowing owl, (*D*) eastern screech-owl, (*E*) western screech-owl, and (*F*) flammulated owl. Drawn to scale, except for *A* (× 1.55) and *C* (× 0.73). After specimens.

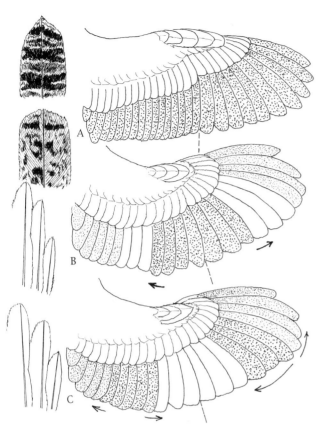

Figure 8. Molting patterns of owl wings, showing the right wing of a hatch-year/second-year (*A*) great horned owl, (*B*) an adult (possibly third-year/fourth-year) northern saw-whet owl, (*C*) and an adult (possibly third-year/fourth-year) boreal owl. Possible molting sequences of flight feathers are indicated by arrows in the two lower drawings. Heavy stippling indicates most recently grown remiges, light stippling represents presumed second-year remiges, unstippled ones presumed third-year remiges. Also shown (*left*) are the more pointed central rectrices typical of hatch-year/second-year (*above*) vs. older (*below*) great horned owls, and the respective shapes of three outermost primaries of various owl species representing these same age-classes. After Pyle (1997b).

lier in breeding females than in males, as it may be advantageous for females to replace as many of these feathers as possible early in the nesting phase, when they are on the nest nearly all the time. Wing molt in breeding males tends to occur after the young have fledged, by which time hunting demands on them have been reduced (I. Taylor, 1994).

Other internal features of interest include a well-developed nictitating membrane that is quickly pulled across the eye to protect the cornea as well as to help keep it clean and moist, an unusually fleshy tongue, and the absence of an esophageal crop for temporarily storing swallowed food but a well-developed intestinal cecum. All owls consume various undigestable materials such as feathers, hair, and similar materials. These materials accumulate in the gizzard, and periodically are regurgitated in the form of pellets. Substantial amounts of bony materials are also usually present in these pellets, probably because of a less acidic gastric environment than occurs in, for example, hawks. In hawks the acid gastric fluids are roughly six times more concentrated than in owls. As a result a smaller percentage of the bones is broken down while in the digestive tract of owls. Many of the larger appendicular bones and skull bones are often regurgitated virtually intact, and thus may be readily identified for prey analysis.

The tracheal anatomy of owls is fairly simple, with an associated relatively simple non-passerine type of sound-producing syrinx (see Figures 71 and 72). It may be characterized as having an associated pair of extrinsic muscles (sternotrachealis) originating on the sternum and inserting near the anterior end of the syrinx, as well as a second pair of muscles (tracheolateralis) that are confined to the lateral surface of the trachea itself and also terminate slightly anterior to the syrinx. A third pair of muscles pass down from a point anterior to the syrinx on the sides of the trachea to terminate on the syrinx itself (as in *Bubo*) or extend some distance beyond and insert on the rings of each of the bronchi (as in *Otus*). Presumably these three pairs of muscles operate interdependently to regulate tension on the sound-producing (tympaniform) membranes of the syrinx. These are evidently set into motion when the bronchial tubes are constricted at the posterior ends of the syrinx, throwing their transmitted air against the surfaces of the dorsomedial syringeal membranes (A. Miller, 1934).

Eyes and Vision in Owls

Of all birds, probably no group exceeds owls in their ability to see under dim light conditions or to localize sound sources with high levels of accuracy. Indeed, the heads of owls are basically little more than brains with raptorial beaks and the largest possible eyes and ears attached. The eyes of owls are so large that they are essentially immobile in their sockets (Figure 9); thus the birds must move their heads to bring objects into focus at their points of sharpest retinal reception (foveae). Additionally, owls have shifted the axis of their eyes to allow for maximum binocular (stereoscopic) perception. D. Martin (1986) estimated a 48-degree area of binocularity for the tawny owl. This species has its eyes diverging at an angle of 55 degrees, and an estimated overall visual field of about 200 degrees, of which 24 percent would represent binocular vision. By comparison, the maximum visual field of humans approaches 180 degrees, of which nearly 80 percent represents binocular vision, as both eyes are oriented almost directly forward. The eyes of owls are so large (see Table 4) that the combined weight of the eyes of an adult screech-owl relative to its body weight, for example, is greater than the total relative weight of the brain in adult humans.

Besides its enormous relative size, the eye of an owl has a number of remarkable if not unique characteristics (Figure 9). It is both relatively long (which provides for a long focal length and associated large projected image size on the retina) and has very large corneal surface and pupillary areas (the equivalent of the lens aperture in a camera), providing a potentially high light-gathering capability. Hocking and Mitchell (1961) compared these aspects of vision between humans and owls, reporting that in human eyes there is an average focal length of 2.5 centimeters, an effective maximum diaphragm opening of $f/2.8$, and a relative retinal illumination of 0.13 (calculated as the reciprocal of the maximum diaphragm f value squared). By comparison, the little owl (*Athene noctua*) has a focal length of 1.2 centimeters, a maximum diaphragm opening of $f/1.7$, and thus a relative retinal illumination of 0.35, or an eye that is potentially almost 3 times as efficient in light-gathering ability as the human eye. In the highly nocturnal tawny owl the focal length is 2.26 centimeters, the maximum diaphragm is $f/1.16$, and the relative retinal illumination is 0.74, or potentially nearly 6 times better in light-gathering power than that of humans. Other studies have suggested that this ratio of light-gathering ability may be closer to 2.7:1

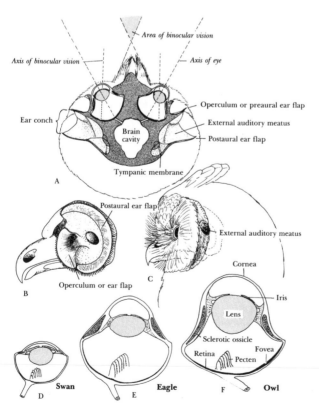

Figure 9. Eye and external ear anatomy of owls, showing (*A*) position and sizes of these areas (partly after Schwartzkopff, 1963), (*B*) external ear flaps of *Asio*, (*C*) facial disk feathering of *Asio* (after J. Sparks and Soper, 1970), and (*D–F*) relative eye morphology in swans, eagles, and owls (after Walls, 1942).

(Hoyo, Elliott, and Sargatal, 1999). Hocking and Mitchell also calculated an average retinal rod:cone ratio of 12.6:1 for the little owl, as compared to 2.4:1 for domestic fowl (*Gallus domesticus*), but a very similar number of visual elements per unit area in the two. This suggests that at least the little owl has evolved a rod-rich retina that is highly adapted for monochromatic vision under low-light intensities (high visual sensitivity) at the expense of reduced capacity for color vision and its associated characteristically high visual acuity. Owls generally have foveae containing a mixture of rods and cones, suggesting that their color-perception abilities and visual acuity are indeed rather limited, at least as compared with more diurnal birds, and experimental studies have shown that this would seem to be the case. Their daytime visual acuity is about 1.5–5 minutes of arc, about 5 times poorer than that of humans and diurnal raptors but roughly comparable to that of domestic cats and rats (Hoyo, Elliott, and Sargatal, 1999).

Research by G. Martin (1982) has suggested that image-processing (neural integration mechanism) characteristics that are dependent upon a large-sized retinal image, rather than increased light-gathering power or unusually high rod sensitivity, may be the basis for the remarkable low-level visual capabilities of the tawny owl. Martin judged the pupillary area of the tawny owl to be approximately 3 times as large as that of humans and about 13 times that of the domestic pigeon or rock dove (*Columba livia*). The estimated lens maximum aperture (f value) was calculated as $f/0.85$ for the domestic cat $f/0.92$ for the tawny owl, $f/1.98$ for the pigeon, and $f/2.13$ for humans. The resulting improved retinal image illumination relative to man (1.0) for point sources of light was calculated to be 3.1 for the domestic cat, 2.8 for the tawny owl, and 0.25 for the pigeon. Thus, tawny owls and domestic cats evidently have comparable nocturnal visual sensitivity, both of which probably approach the absolute limit possible in the vertebrate eye.

The similarity in light-sensitivity values estimated for these two species is noteworthy inasmuch as the cat's eye has a well-developed retinal tapetum lucidum, causing it to shine in the dark with reflected light. This light-reflecting layer at the rear of the retina effectively serves as an image-amplifier under low-light conditions, and is apparently quite variably developed in most owls. Van Rossem (1927) reported a bright orange red eye shine in the barred owl, and similar intense reddish to golden reflections appear in most flash-exposed photos that I have made of spotted, barred, boreal, and great gray owls, all of which

are variably nocturnal. Walker (1974) listed eight relatively nocturnal North American owls in which he had observed weak to strong eye shine (strongest in spotted, barred, and long-eared), compared to the more diurnal burrowing owl that lacks it.

Perhaps because of the cat's well-developed tapetum, the absolute visual threshold illumination value determined for it by Martin was significantly lower (approximately half the light intensity) than that estimated to be required for threshold vision by the tawny owl. The tawny owl was estimated to have generally lower visual acuity than do humans at low levels of illumination, and the rate of change in visual acuity of humans and tawny owls was generally parallel over the measured luminance range above the absolute threshold of vision. In the tawny owl this limit was estimated to be at about the illumination level typical of that reaching the substrate below a broad-leaved woodland canopy at minimum starlight, whereas the human visual limit under the same starlight conditions is reached in open-country habitats. Light levels occurring below woodland canopies under maximum cloud conditions fell beneath the absolute thresholds of both the tawny owl and the domestic cat. Probably the tawny owl approaches the absolute visual sensitivity and the maximum spacial resolution at low levels that are physically and physiologically possible for vertebrate eyes (D. Martin, 1986).

In an equally interesting study, Murphy and Howland (1983) examined 15 species of owls in an effort to estimate their relative capabilities for visual accommodation (focusing abilities), optical performance associated with corneal curvature, and related aspects of vision. All the species were found to have eyes of high optical quality, relatively free of any form of astigmatism, but they differed widely in their accommodation abilities. Nearly all the species studied could focus on distant targets (near optical infinity), but their capacity to focus on objects closer than 1 meter was correlated with small body size. This may be related to the fact that smaller owls are often insectivorous and must be able to focus on and capture small prey items at close range. However, the common barn owl was notable for its extremely high (over 10 diopters) accommodation range as well as its unusually close (0.1 meter) near-point of focus. This in part may be related to the relatively small eyes typical of barn owls, the focusing of which by their optical characteristics alone can be more easily attained.

By comparison, the relatively large-eyed great horned owl has an estimated accommodation range of only 2.2 diopters and an estimated near-point of focus at 0.85 meters. However, this species is known to have a well-developed temporal fovea containing both rods and cones (Fite, 1973), which presumably is efficient at focusing on more distant targets. The amazing speed of accommodation of the northern hawk-owl was found to be an order of magnitude faster than that previously reported for human accommodation, and presumably results from the striated rather than smooth ciliary musculature associated with the avian focusing mechanism. The great horned owl has also been tested for its rates of iris response to light; minimum pupillary diameter (also controlled by striated muscle) is attained within 176 milliseconds of a light flash. Pupil dilation is much slower, requiring about a second (Oliphant et al., 1983).

Experiments in prey-finding abilities of owls under minimal illumination conditions were conducted by Dice (1945) and more extensively by Curtis (1952). Dice used four species of owls and tested their abilities at finding dead mice at night, using different-colored substrates and pelage colors. In one experiment, a common barn owl was able to find a mouse in one of 10 trials when the incident illumination on the floor was judged to be 3.1×10^{-7} foot-candles. In a more fully controlled set of experiments, Curtis found that a flying common barn owl could consistently (in 19 of 20 trials) avoid obstacles (white barriers 2 inches wide) when the illumination level of these barriers was only 1×10^{-8} millilamberts, or appreciably lower than the light levels judged by Dice to be minimally effective for locating prey. However, the owl could not see objects at a brightness of 2×10^{-9} millilamberts, and the apparent lowest level of illumination that could sometimes be perceived effectively enough to avoid obstacles was 3×10^{-9} millilamberts. The estimated maximum visual threshold value $(0.43 \times 10^{-7}$ millilamberts) was nearly 35 times less than the lowest reported human visual threshold value Curtis could locate and about two-thirds the reported threshold value for domestic cats.

In a separate estimate of visual threshold for various owls, Lindblad (1967; data summarized by Mikkola, 1983) reported that the lowest levels of illumination by which various owls could find dead prey at night ranged from 1.45×10^{-4} foot-candles (Eurasian pygmy-owl) to 2.5×10^{-7} foot-candles (long-eared owl). Lindblad himself had a threshold level of 7.5×10^{-5} foot-candles, or somewhat better than the nocturnal sensitivity of the Eurasian pygmy-owl but appreciably poorer than those of any of the other four species of owls he tested.

Owl Ears and Hearing

No group of birds other than owls has external ears with such remarkable structural adaptations as movable ear flaps that may be present both in front of and behind the opening of the external ear canal (auditory meatus) and an asymmetrical development of the external ears on the two sides that may involve both their size and vertical positioning on either side of the head. Although a thorough review of the anatomy and physiology of hearing in owls is impractical in such a short survey as the present one, it would be unthinkable to ignore these topics completely.

Of the several aspects of hearing that might be discussed, especial importance lies in such measurements as the range of frequency response that owls can detect, their thresholds of sound reception across this frequency range, and the capability for point-source localization of sounds by nonvisual means. D. Martin (1986) has reviewed the physiology of owl hearing and suggested that their auditory sensitivity has reached the absolute useful limits as dictated by ambient (background) sound levels.

Most of the work on the hearing of owls has been done with the common barn owl, a species notable for its well-developed facial disk and an asymmetrically placed pair of external ear flaps or opercula (Figure 10). During the 1950s a series of clever experiments by Payne (1962) established that at a high level of efficiency this species could locate and capture live mice as they ran across the floor of totally darkened rooms and could also similarly locate loudspeakers broadcasting mouse-generated rustling sounds. When such broadcast sounds were filtered in such a way that frequencies above 8,500 cycles per second (8.5 kHz) were filtered out, the owl refused to strike, suggesting that the bird was depending on high-frequency sounds for localizing its prey. Payne hypothesized that such sound-source information analysis must be dependent upon one or more of three possible methods. The Doppler effect, or changes in component sound frequencies of the target, was dismissed as a possible source of information, leaving (1) differences in relative arrival time or phase differences between the sounds received at each ear, and (2) differences in the relative intensities of sound in each ear, which vary with the angle at which the sound is received. Payne considered the last of these possibilities as the most likely explanation for the owl's localizational abilities. He believed that an owl need only orient its head in such a way as to hear all frequencies at maximum intensity in both ears and thus automatically face the sound source within a

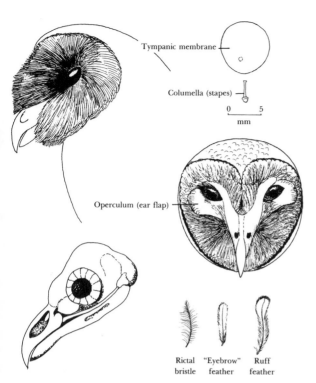

Tympanic membrane

Columella (stapes)

0   5
mm

Operculum (ear flap)

Rictal   "Eyebrow"   Ruff
bristle   feather   feather

Figure 10. Skull, facial feather variation, and middle ear anatomy of the barn owl in part after Schwartzkopff (1955).

potential error of less than 1 degree. Work by Rice (1982) has confirmed that common barn owls can consistently acoustically locate prey with a minimum horizontal resolution ability of as little as 1–2 degrees, as well as with a coarse (40 degree) vertical resolution ability.

Payne's pioneering work was later taken up by Konishi (1973, 1983, 1993), who contributed substantially to our understanding of owl hearing. Konishi replicated some of Payne's experiments and confirmed that the prey's rustling noises are all that is required for the owl to locate it accurately in space. He compared the minimum audible fields of the barn owl with humans (Figure 11), and found that barn owls can detect sound levels inaudible to humans, especially in the frequency range of about 0.5–9 kHz, whereas at about 12 kHz humans are more sensitive than barn owls. By removing most of the feathers of the owl's facial disk, Konishi established that such owls tended to make large errors in prey catching, landing far short of their targets. Thus the facial disk probably serves as an effective sound amplifier, by focusing the sound on the opening of the external ear. He estimated that, given the diameter of the facial disk, only sounds having wavelengths of less than about 7 centimeters (or above about 5 kHz) could thus be focused. Konishi established that, contrary to Payne's views, the owl does not require sound frequencies

in excess of 8.5 kHz for locating prey, but instead can localize sound source frequencies in the range of 5–8.5 kHz very effectively. From these and other observations he thus judged as untenable Payne's conclusion that the bird used the method of maximizing signal intensity in both ears to localize the horizontal location of prey. Although the bird does indeed turn its head rapidly toward a sound source, Konishi found that fairly effective localization is still possible for sounds originating within 30 degrees of either side of the owl's axis even before turning its head, the bird apparently rapidly converting binaural acoustic clues into an approximate azimuth location. Even when the head is turned toward the source the difference in external ear location means that the shape of the facial disk is such that it is oriented slightly downward on the right side and upward on the list, thus producing extremely slight differences in sound arrival times in the two ears that might help azimuth localization.

Konishi pointed out that, if sound-intensity differences are to be used to localize sound sources, the shorter wavelengths are of little use, since these bend around the head of a large-headed animal without generating any sound-intensity differences. The shorter wavelengths of high frequencies do, however, produce sound shadows in animals with heads larger than the wavelengths of the sound, which means that hu-

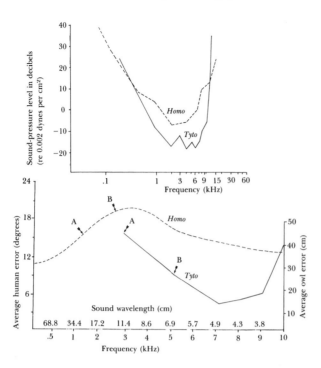

Figure 11. Hearing thresholds (*above*) and relative average errors in binaural localization capabilities (*below*) of humans and barn owls, based on data in Konishi (1973). *A* indicates approximate upper frequency limits for binaural phase-discrimination potential, and *B* indicates approximate lower limits for binaural intensity discrimination, based on adult head widths of each species.

mans can judge the locations of high frequency sounds (above about 5 kHz) much more accurately than sounds in the frequency range of 2–4 kHz, even though these frequencies are perceived at lower energy thresholds than are the higher pitched sounds (Figure 11). Low frequency sound sources can be located by an alternative method, namely by detecting phase differences between the ears, which result from differences in the paths traveled by the sound in order to reach the two ears.

In order for this phase difference method to be utilized, the wavelength of the tone must be longer than twice the distance between the ears; thus the upper limit of human sound localization using this method is below 2 kHz (Figure 11). As a result, in humans there is an informational "gap" between about 2–4 kHz, within which the perceived sounds are too high pitched to be useful for phase discrimination but too low pitched to be analyzed by intensity discrimination. In the barn owl, with its smaller head width, these corresponding sound frequency limits are somewhat higher. Based on minimum error rates, the most effective sound localization in barn owls evidently occurs at frequencies of about 5 kHz, suggesting that this species in fact uses some method other than recognizing phase differences for locating its prey accurately.

For barn owls to respond with speed and accuracy in localizing sound sources, both ears are needed; with one ear plugged they are unable to localize sounds at all. Additionally, research by

Konishi (1983) indicates that precise sound localization in the barn owl is attained by measuring binaural differences in the time required for the sound to reach each of the two ears. For barn owls, this difference amounts only to about 150 microseconds (a microsecond being a millionth of a second), as compared with 570 microseconds in humans. Although this seems an astonishing ability, Konishi's research suggests that owls can detect binaural time lags as brief as only 10 microseconds. Binaural sound intensity differences are believed by Konishi to be used by the owl to estimate vertical displacement of a sound source from the horizontal, based on the asymmetrical positioning of the ears. By lowering its head until the perceived sound level is equally loud in both ears, the owl "knows" that the sound source is in line with its eyes. In conjunction with this remarkable three-dimensional sound perception, the portion of the barn owl's brain that is devoted to binaural hearing is much larger than in any other bird so far studied. Additionally, the cells of this area have been found to conform spatially with the area in the bird's external environment that it is interpreting; thus the brain of the owl represents a kind of three-dimensional map of its environmental acoustical space (Konishi, 1983).

When considering ear asymmetry the case of the boreal owl should be mentioned, as its ear asymmetry is the most extreme of all owls (see Figure 67) and markedly affects the shape of the skull. Studies by Norberg (1968, 1978) on this species indicate that its vertical ear asymmetry allows the owl to localize sounds in the horizontal and vertical planes simultaneously, especially if the owl can orient its head so that both high and low frequency sounds are being received in the ears simultaneously and in phase (with no elapsed time or wavelength phase differences between the ears). The owl's directional sensitivity in the vertical plane is especially pronounced with sounds of very high (12–15 kHz) frequency. The external ear structures of several other owl genera having asymmetrical structures have evidently been achieved in various ways, probably by convergent evolution, to attain similar functional capabilities for acoustic analysis. In the great gray owl, which has a highly asymmetrical ear structure, the birds are able to locate and capture rodents accurately under 45 centimeters of snow or more than 2 centimeters of soil (Hoyo, Elliott, and Sargatal, 1999).

In considering the possible value of specialized external ear structure in sound reception, the observations of Van Dijk (1973) are of interest. He studied hearing in two medium-sized species of owls (tawny and long-eared) having highly modified external ear characteristics, and eight additional species of owls with less highly specialized external ears. He found the tawny and long-eared owls to have among the best high frequency reception reported for such large birds, extending to about 13 kHz, and with extremely high auditory sensitivity to weak sounds at frequency ranges of 0.4–7 kHz and 0.5–8 kHz, respectively. The other species of owls with less specialized external ears had similar general ranges of auditory sensitivity, but the frequency ranges associated with very high sensitivity in these species never exceeded 6 kHz. Generally, strigid owls have high sensitivity to sound frequencies extending down to about 250 Hz. Their high frequency sensitivity thresholds extend only to about 7 kHz in very large species such as the great horned owl, which has a zone of maximum sound sensitivity at about 500 Hz, or substantially lower than that of smaller owls (Hoyo, Elliott, and Sargatal, 1999).

The Evolution of Hearing in Owls

It is impossible to judge just how long selection favoring increased hearing capabilities has been occurring in owls, but probably even the earliest owls were crepuscular in their activity patterns, if as generally believed they evolved from common ancestors with the nightjars and their relatives. Sibley and Ahlquist (1985) estimated the timing of this separation as 70–80 million years ago, during the late Mesozoic era, and thus it is likely that adaptations relating to low-light vision and associ-

ated hearing abilities have been going on for most or all of the Cenozoic era. Feduccia and Ferree (1978) proposed a hypothetical phylogeny of the strigiform birds, based in part on the morphology of the columella (stapes), the single middle ear ossicle that connects the tympanic membrane (eardrum) with the inner ear. They judged that shortly after the separation of the stock that gave rise to the present-day Tytonidae from that producing the Strigidae, a derived type of stapes, with a modified extended footplate, evolved in the Tytonidae, while the ancestral Strigidae retained the primitive condition. Following separation of the early Buboninae stock, which has retained the primitive condition of the stapes, various lines of the Striginae adaptively altered the shape of the stapes, with some of the forms of *Strix* showing convergent similarities to the condition typical of the Tytonidae. This evolutionary scenario is illustrated in Figure 12, which has been redrawn and modified from a similar illustration by Feduccia and Ferree but has an added overlay of geologic time units and associated known fossil owls. The postulated relationships among these early owl groups are extremely hypothetical and are likely to be greatly modified as new paleontological information becomes available. For example, the Cretaceous family Bradycnemidae that is shown at the base of this diagram is now considered to be reptilian rather than avian.

Owl Vocalizations

In contrast to many songbirds, there is little correlation between the frequency range of owl hearing and the range of sounds produced by these same birds. Whereas in songbirds the dominant vocal frequency range is usually above the zone of maximum hearing sensitivity, in owls so far tested the dominant frequencies of vocalizations are lower than those of their best hearing range (Van Dijk, 1973). This would of course suggest that hearing in owls has evolved largely in association with adaptive prey-localization functions, whereas vocalizations have probably been adaptively adjusted to provide for efficient intraspecific information transfer under varying ecological conditions.

In a series of papers, A. Miller (1934, 1935, 1947) described the syringeal anatomy of various North American owl species. These studies allowed him to make several generalizations about owl vocalizations. First, the larger owl species tend to have larger cross-sectional areas of bronchial and syringeal air passages and longer vibratile syringeal (tympanic) membranes. These longer tympanic membranes of the larger owl

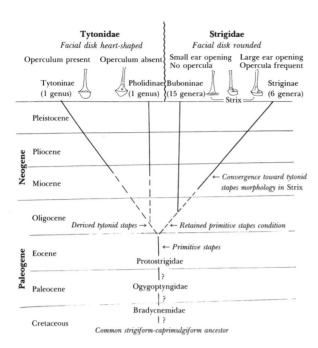

Figure 12. Evolutionary trends in external and middle ear anatomy of owls. Adapted from Feduccia and Ferree (1978), but with fossil forms and geological periods added. Unbroken lines indicate length of known fossil records for taxa. Suggested fossil family relationships are highly speculative and should be considered as temporal sequences rather than as known phyletic lineages. A seemingly ancestral Cretaceous taxon, the Bradycnemidae, is now considered to be reptilian rather than avian.

species tend to vibrate more slowly and produce lower pitched sounds than in smaller species. However, these sounds can be modified by the size of the syrinx, which ranges from 203 to 238 percent of the diameter of the bronchus in males of the 10 species studied by Miller, compared with from 190 to 209 percent in females. The tension on the tympanic membranes is also partially under muscular control; thus pitch can presumably be altered voluntarily. Finally, although the cross-sectional area of the bronchus increases roughly in proportion to the size of the species, the diameter of the bronchus, to which the vibratory tympanic membranes correspond, increases at a slower rate. Thus the pitch of the call in larger owl species is relatively higher, considering the mass of the bird, than one would expect.

One exception to this general trend is in the flammulated owl, which, although very small, has a remarkably low-pitched voice (see Figure 75). A. Miller reported (1947) that in this species the vibratory membranes of the syrinx are unusually thickened, and there is a peculiar swelling of the throat skin, which noticeably bulges out with each note uttered. Such neck swelling, although not so extreme, is evident in many owls during their territorial hooting, and in some, such as the great horned owl, it is made especially conspicuous by the contrasting white feathers that are exposed during calling. Evidently the inflated throat acts as a resonating device for the flammulated owl's unusually low-pitched call, tending to amplify it and making it "as impressive as possible" (A. Miller, 1947). An example of the syringeal

structure of an adult great horned owl is shown in Figure 5. Relatively few differences in gross syringeal structure occur in the other North American owls (see Figures 71 and 72).

A. Miller (1934) believed that the syringeal tympaniform membranes act as an acoustical "driver" and set up a given vibration rate (which is dependent upon such variables as the membranes' thickness, area, and tension), regardless of whether or not the acoustic characteristics of the tracheal tube help resonate the fundamental sound frequencies that are source-generated by the syrinx. In other words, the activity of the syringeal membranes is not influenced by the length or other resonating characteristics of the sound "carrier," or tracheal tube. Miller was led to this conclusion because he found that the pitch produced by experimentally activating the syrinx of a freshly killed great horned owl was not altered when the trachea was shortened, and because female owls, although having tracheae as long as or longer than males, produce sounds of higher average pitch. It would be of interest to learn whether the throat inflation apparent when a male great horned owl (or other male owl) calls might be responsible for facilitating the resonance of the low-pitched sounds typical of advertising males, thus making the total length of the trachea an irrelevant consideration.

Since Miller's early studies many additional ones have been devoted to the acoustical mechanisms of avian vocalizations, but several basic questions still remain unanswered. Gaunt, Gaunt, and Casey (1982) reviewed and brought into question the major accepted theories of syringeal mechanics, utilizing the call of the domesticated ring dove (*Streptopelia risoria*) as a data source. This species has a syrinx quite similar to that of an owl, as well as a similar cooing vocalization, even to the point of having a noticeably inflated neck during cooing. This species' cooing call is lacking in harmonics, and varies little if at all in frequency. However, its variable amplitude (loudness) is evidently modulated primarily by muscles of the abdominal wall, and more subtle modulation is apparently effected by vibrations of the lateral tympanic membranes. Contraction of the syringeal muscles evidently shapes the syrinx into a conformation that produces vocalization but that may exert little actual modulatory effect.

Indeed, Gaunt, Gaunt, and Casey proposed that the dove's syrinx, rather than functioning as an oscillating membrane, may operate in the manner of a simple whistle in which the tympanic membranes bow inwardly, constricting the air passage in such a way as to form a narrowed slot. Thereby a whistled sound is generated that is essentially harmonic-free, unless such harmonics are produced by associated resonators. Clearly, comparable work on the vocalizations of owls should be done to see if similar vocal controls might exist in this group, whose hooting calls are quite variable in their harmonic development (see Figures 75 and 76). Inasmuch as A. Miller (1934) stated that he observed vibration of the internal tympaniform membranes of his great horned owl syrinx preparation during sound production, this "whistle hypothesis" may prove inapplicable to owls, but the hypothesis should be kept in mind when further studies of owl vocalizations are undertaken. As Gaunt and Gaunt (1985) have stated, it is possible that within closely related groups, different species, or even different sexes of the same species, might utilize radically different vocal techniques to produce similar sounds.

A few final comments about owl vocalizations are in order. It has been noted that the larger species of owls have lower pitched vocalizations than do the smaller ones. Such low-pitched calls, with their long wavelengths, are more effective at penetrating vegetation and thus carrying farther than are high-pitched sounds, which are more vulnerable to being absorbed by the environment and thus less likely to carry long distances. There is very likely to be some positive relationship between a species' average territorial size and the concentration of low-pitched frequencies in its territorial calls, although of course larger owl species are inherently more prone to have both larger territories and lower pitched calls than are smaller ones. This may, however, help to explain the fact that although male owls are smaller than females they typically have noticeably lower pitched voices, contrary to general acoustical expectations. In the North American species of *Strix, Asio, Bubo,* and *Otus* studied by A. Miller (1934) the males nevertheless have appreciably larger syringes than do females, although the common barn owl is an exception to this pattern in that the female's syrinx measures about 11 percent larger than the male's. In association with this, the screech of the female common barn owl is noticeably "huskier" than that of the male (Bunn, Warburton, and Wilson, 1982).

The flammulated owl is notable for the male's unusually low-pitched voice relative to similar-sized North American owls such as screech-owls. No direct comparisons of average territorial sizes in the relatively omnivorous screech-owls and the mostly insectivorous flammulated owl (see Table 4) appear to be available to test the possibility that its remarkably low advertisement call may somehow be related to unusually large territorial and home requirements. However, in the flammulated owl these seem to average about 14 hectares in area and about 400 meters in maximum diameter, according to Linkhart, Reynolds, and Ryder (1999). These authors characterized the species as having an unusually large home range/territory in relation to its body mass, at least as compared with insectivorous birds of other orders. Apparently this entire area is territorially defended by the male. By comparison, in some ideal riparian habitats of Arizona the territories of the western screech-owl are often only about 50 meters apart (R. Johnson, Haight, and Simpson, 1979), or perhaps total about 0.25 hectares, assuming nonoverlapping and rounded territories, with resultant extrapolated densities of up to about 20 pairs per linear kilometer. Interestingly, the European scops owl (which is highly insectivorous and generally believed to be a close relative of the flammulated owl) has a relatively high-pitched voice and sometimes has also been found nesting at densities of several pairs per hectare (Cramp, 1985).

In sum, he [the scritch-owl] is the very monster of the night, neither crying nor singing out clear, but uttering a certain groan of doleful meaning.

—Pliny, *Natural History*

31

Alone and warming his five wits,
The white owl in the belfry sits.

—Alfred, Lord Tennyson,
"The White Owl"

The behavior of owls, which are elusive, solitary, and primarily nocturnal birds, is still rather poorly understood. Having to a large degree abandoned diurnal activities, the owls have evolved social signals (primarily acoustic, secondarily tactile) that work well under conditions of darkness and have correspondingly largely eliminated visual signals such as bright colors, reducing these to general shapes and patterns that work well under conditions of reduced light. Thus it is likely that the "horns" or "ears" of owls are visual signals that operate as effectively in dim light silhouette conditions as during bright daylight. Likewise, the contrasting yellow iris color found in most owls, as well as the white throat markings exposed by many during calling, are easily seen under low-light conditions.

Nearly all the species of owls having dark brown iris coloration, virtually invisible in the dark, are highly nocturnal owls. Examples in North America include the flammulated owl, barred owl, and common barn owl, all of which are essentially nocturnal species that also have penetrating voices and reduced ear tufts or none at all, and lack a contrasting throat pattern (Figure 13). Many species of *Strix* and *Tyto* fall into this category, although the semidiurnal great gray owl has contrasting yellow eyes set within a framework of concentric gray and whitish rings, and a conspicuous black and white "moustache." Interestingly, the diurnal pygmy owls (*Glaucidium* spp.) not only have bright yellow irises but also have false eye patterns on their napes, which presumably serve as a fake threat signal and thus perhaps help deter possible predators approaching from the rear (see Figures 41 and 45). Many owls with yellow iris coloration have distinctive black rings around the perimeters of their eyes, which provide strong contrast to the yellow, and many owls have contrasting whitish "eyebrows" immediately above the eyes that markedly affect their "expressions." Finally, the downy young of many owl species often have highly distinctive facial patterns (Figure 14). These tend to be better developed in such precocious and often open-site-nesting owls as the snowy owl and hawk-owl, whose young tend to leave their nests relatively early. In such cavity-nesting species as the barred owl and screech-owls, the young remain in the nest for a longer period and presumably require less distinctive juvenile characteristics associated with parental recognition (Scherzinger, 1971a).

Figure 13. Front views of North American owls, showing apparent size relationships of (*A*) great gray, (*B*) great horned, (*C*) snowy, (*D*) barred (and spotted), (*E*) northern hawk-owl, (*F*) barn owl, (*G*) long-eared, (*H*) short-eared, (*I*) boreal, (*J*) burrowing, (*K*) northern saw-whet, (*L*) screech-owls, (*M*) flammulated, (*N*) pygmy-owls, and (*O*) elf owl. Organized by decreasing average length from top of head to tip of tail. In part after J. Sparks and Soper (1970).

Egocentric Behavior

In the category of egocentric behavior is included all the self-directed kinds of behavior related to an individual's survival and well-being, such as ingestive (eating and drinking) behavior, behavior associated with elimination of indigestible materials (pellet regurgitation and defecation), thermoregulatory behavior, and such behaviors associated with feather and body surface care as, for example, preening, oiling, and dusting. Also included are such comfort activities as stretching, shaking, scratching, and the like. Egocentric behaviors blend almost imperceptibly into quasi-social activities when, for example, two or more birds are attracted to the same location in search of food, a suitable roosting site, or a particular nesting habitat.

Owls groom and preen their feathers in a manner similar to that of other birds, gently running their larger body feathers through the slightly opened bill to remove foreign matter and occasionally reaching back to obtain oil from the uropygial gland at the base of the tail, which is then carefully spread over the feathers. Scratching (see Figure 63) is done by moving the feet directly to the head, rather than past the lowered wing as many birds typically do (Clark, 1975). The facial feathers are groomed with the bird's talons, and presumably the specially adapted comblike middle claws of the barn owl are used

for delicate cleaning of the airy feathers of the facial disk. This claw is also used by the bird for scratching and is cleaned regularly(Bunn, Warburton, and Wilson, 1982). Extended preening behavior usually occurs when the bird seemingly has nothing else to do. At such times the bird may also perform extended sun-bathing behavior, orienting the body and head in such a way as to intercept the maximum amount of sunlight and often opening its bill and closing its eyes (Nero, 1980; Clark, 1975).

During preening activities owls often shake their bodies, perhaps to shed loose feathers, and they may also shuffle their wings repeatedly. Wing stretching (either both wings stretched simultaneously, or one wing and the corresponding leg stretched together) is also a common activity (see Figures 49 and 63). Although some hand-raised owls have been observed to perform bathing behavior when presented with water, this is not a universal reaction (Sumner, 1933). Sleeping or dozing may also occur periodically during various times, and the typical manner of sleeping in most owls simply involves a slight drooping of the bill and head on the breast. However, at least in some small owls (perhaps because of their relatively larger heads), the neck may be twisted so that the head is resting on the back and the bill is tucked into the scapular feathers while the birds are sleeping.

When swallowing food, surprisingly large ma-

Figure 14. Adult and natal plumages of North American owls, including (*A*) great gray, (*B*) long-eared, (*C*) short-eared, (*D*) burrowing, (*E*) barred, (*F*) snowy, (*G*) northern hawk-owl, (*H*) boreal, (*I*) northern saw-whet, and (*J*) eastern screech-owl. In part after Glutz von Blotzheim and Bauer (1980), plus author's observations.

terials are sometimes ingested. Even rather small owlets seem to be able to swallow an astonishing number of entire small mice one after another, the bird invariably swallowing the rodents head-first and taking the next almost as soon as the tail of the one before has disappeared down its throat. Although owls swallow their mammalian prey headfirst, in the case of snakes a tail-first swallowing has sometimes been observed (Sumner, 1933). The feeding behavior and digestive process of owls is apparently complicated by the fact that they lack a crop. In addition they have relatively less efficient gastric digestion of bony materials than do, for example, hawks. As a result, their regurgitated pellets contain a higher proportion of bones than those of hawks, although their digestive abilities of soft foods is approximately the same (Mikkola, 1983).

Not surprisingly, the size of the pellets regurgitated by the bird is generally a reflection of its overall body size. A correlation of body weight and average pellet size, based on the summed measurements of length and maximum diameter, indicates a generally linear relationship except for the largest owls, which produce somewhat smaller pellets than would be expected. However, this relationship is only a general one, and it is impossible to identify the pellets of similar-sized owls based on their measurements alone. Pellets produced by individual owls also show seasonal variation, averaging larger in spring and autumn (Mikkola, 1983). Disturbance may also cause the birds to regurgitate their pellets prematurely, in which case they may be larger and softer than normally (K. McKeever, pers. comm.).

Prey-catching behavior of owls is of particular interest, given their remarkable ability to kill their prey under conditions of extremely low light. One of the first studies concerned with the behavioral aspects of prey capture in the dark was that of Payne (1962) who used infrared film to document the method of prey catching by common barn owls under dark conditions. He found that when capturing mice in the dark, the owl approached the prey while flapping the wings quite strongly and continuously all the way to the mouse, swinging its feet back and forth like a pendulum. When it was immediately over the mouse it brought its feet forward, raised its wing, threw its head strongly back, and thus brought its widely spread talons into the same path that its head had been taking a moment previously (Figure 15). The eyes are typically closed as the feet strike the prey.

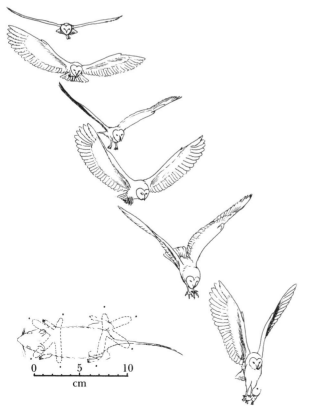

Figure 15. Stages in prey catching by barn owl under dark conditions. After Payne (1962) and J. Sparks and Soper (1970). Inset drawing shows talon spread relative to outline of an average-sized mouse (*Peromyscus*).

0   5   10
cm

Figure 16. Stages in visual prey catching by boreal owl. After Norberg (1970). Talon-spread diagram from specimen, shown relative to outline of an average-sized mouse (*Peromyscus*).

By comparison, Payne observed that when catching prey under lighted conditions, barn owls glided nearly the entire way from their perch to the prey, holding the feet well back until the final instant before impact, when the wings would be raised, the head drawn back, and the feet pushed forward. Under both light and dark conditions the bird oriented its head directly toward the prey prior to taking flight. Similar observations on visual hunting techniques were made by Norberg (1970) for the Tengmalm's (boreal) owl, using wild rather than captive birds and normal photographic flash techniques (Figure 16). He observed that most birds looked for prey from rather low perches (ave. 1.7 meters) and made relatively short (ave. 17 meters) flights to obtain their prey.

When preparing to take flight toward prey the owl would turn its front toward the mouse, look sharply at it, and orient the facial disk directly toward it. Occasionally the bird would make vertical or lateral head movements prior to taking flight or lower its head so that the beak was nearly at the level of its feet. When approaching prey it made shallow wingbeats during the first phase of the flight and then glided silently downward for about the final meter prior to the pounce. When the owl was about to strike, the wings and tail were spread, the feet brought forward and maximally extended, the talons maximally extended, and the eyes closed. Often the

prey would apparently be paralyzed by the impact, but the bird would immediately kill it by grasping its head or neck with the beak. After impact the bird would typically remain with wings and tail spread and lowered to the ground, often looking about with rapid head movements. Then after 20–30 seconds or more, the prey would be carried away, held in one foot. Although Norberg apparently did not observe any prey being carried by the bill, such behavior sometimes occurs in owls, especially when the prey is relatively small, and in some species seems to be the preferred method.

One final method of prey capture in owls should be mentioned, that of the snow-plunging behavior of great gray owls. This behavior has been described by Nero (1980) and Mikkola (1983); Figure 17 is based on photos in the latter source. In this behavior, the bird evidently localizes rodents running under a layer of snow by auditory clues, usually from a listening post within about 20 meters away. When the prey has been localized, the bird flies toward and above its target, often hovering briefly overhead before diving. As it dives downward the bird sometimes drops headfirst almost vertically. Normally it extends its legs and spreads its talons immediately before impact; nonetheless it often hits headfirst, immersing its head in the snow before bringing the legs forward and capturing the prey with its talons.

Plunge holes extending 25 centimeters below

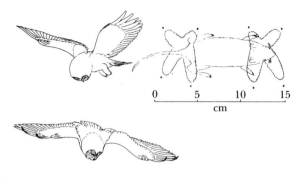

Figure 17. Snow plunging after prey by great gray owl. After photos by E. Kemila, in Mikkola (1983). Talon-spread diagram from specimen, shown relative to outline of an average-sized mouse (*Peromyscus*).

the snow surface have been observed in snow as deep as 76 centimeters. With their very long legs it is possible that great gray owls can capture prey as deep as 45 centimeters beneath the snow. The prey may be swallowed whole before taking flight again, a matter taking only a few seconds, or perhaps it may be carried back to a nesting mate. Similar plunge-diving has been observed in northern hawk-owls and boreal owls, but it is evidently much less common in these species than in the great gray owl (Nero, 1980). Other prey-capture methods, such as ground stalking, fishing by wading, and catching prey in midair, have also been observed in some owls.

The impressive evidence of the great gray owl's hearing ability should not lead one to conclude that it never hunts visually. Nero reported an observation of a bird flying about 200 meters to catch a small mammal on the snow, apparently having seen it from that distance. When visually assessing the distance of an observed object, owls commonly move their heads up and down, or from side to side, apparently to increase visual parallax and associated stereoscopic abilities for estimating distance. This curious head-bobbing behavior is sometimes evident even in owls only a few weeks old (Figure 18). An equally curious behavior may sometimes be seen in owls twisting their neck to a degree that the head seems almost upside down. This is done when the owl is trying to

focus its gaze on an object well above its normal field of vision, such as a bird flying overhead. The owl's sharpest point of visual focus is associated with its temporal fovea of the retina, which lies above the midpoint of the eye and thus allows for sharp imaging of objects directly in front of and somewhat below eye level. Therefore, when an owl must see objects above eye level critically, it rotates its head in such a way as to bring the projected image of the retina into a position corresponding with the location of the temporal fovea, which results in an "upside-down" head alignment (Figure 18).

Social Behavior

Social behaviors in animals include a very wide range of interindividual communications, both within and between species. They include such generalized social responses as social flocking or roosting behavior, as well as much more individualized and complex interactions such as courtship, aggression, and parental behaviors. Regardless of their complexity, social interactions involve some level of communication, or the transmission and interpretation of social signals. These signals can be transmitted in any of several sensory channels, which in owls are most likely to include visual, acoustic, and tactile modes of communication.

Most, perhaps all, owls show distinctive postures when they are alarmed and when threatened. The typical alarm posture of perched owls is one that emphasizes their remarkable capacity for remaining immobile and blending into their environment (Figure 19). This apparently concealing posture, appropriately called the *Tarnstellung* in German, is one in which the owl typically stands upright, often against a vertical tree trunk if it is available, with the wing nearer the object of fear drawn sideways and upward toward the bill, often hiding most or all of the relatively pale and often conspicuous underpart coloration. The eyes are often almost entirely shut so as to form slanted slits, even in species having dark-colored eyes. However, in some species such as the great gray owl and long-eared owl the eyes remain fully open, and they may even be blinked in the case of the long-eared owl (Bondrup-Nielsen, 1983). Additionally, if any ear tufts are present they are erected maximally, and the forehead feathers as well as the "eyebrows" are usually spread, often tending to make these areas less conspicuously contrasting. This feather realignment can cause a pair of dark lines to pass down from the ear tufts past or through the nearly closed eyes, making them considerably less conspicuous than is normally the case. In species such as the scops owl and screech-owls the flattened forehead and eyebrow feathers have a color pattern remarkably similar to that of lichen-covered tree bark, producing an extremely effective facial camouflage. Finally, the rictal bristles forming the "moustache" are sometimes pushed forward in such a way as to hide or nearly hide the beak (Figure

19). Bondrup-Nielsen (1983) has argued that since this posture is of apparently ambivalent motivation rather than being strictly fear-motivated, it is best simply referred to as the "erect posture."

When cornered and facing an opponent, or when protecting a nest, the posture assumed by owls is entirely different. Here, instead of presenting the minimum surface area to view, exactly the opposite response occurs, with the feathers of the head and body maximally fluffed, the tail often spread, and the wings both spread and raised above the back or variably drooping (Figure 19). This remarkable posture is somewhat similar to the two-wing stretching behavior of owls (see Figures 49 and 63), and perhaps represents a ritualized derivative of it. In this posture the bird may hiss menacingly, clatter its beak, and in some species sway back and forth in a snakelike hypnotic rhythm while directly facing its opponent.

That this is an innate response is indicated by the fact that I have seen nestlings of eastern screech-owls and barred owls perform this posture just like adults on seeing a peregrine falcon (*Falco peregrinus*) for the first time, even though at that time their wings were only partially grown and their bodies were still mostly down-covered. Similar intense defensive responses have been reported in hand-raised great horned and barn owls upon initial exposure to snakes (Sumner, 1933). In the common barn owl this defensive posture is typically associated with seemingly almost unending cycles of head swaying alternated with head shaking, accompanied by puffing or beak-

Figure 18. Upside-down peering at overhead object, by short-eared owl (after photo by J. B. Foster in Peterson, 1963), and vertical head bobbing by nestling barred owl (after photos by author).

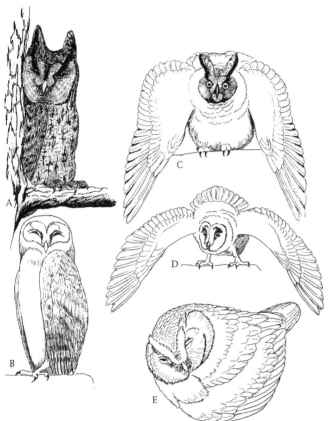

Figure 19. Concealing postures of (*A*) eastern screech-owl and (*B*) barn owl. Also shown are defensive wing-spreading displays by (*C*) long-eared owl and (*D*) barn owl, and (*E*) incubating posture of female eastern screech-owl posture. After various photographic sources.

snapping (apparently actually tongue-clicking) sounds. A similar posture to this is assumed by adult barn owls and by their fledged owlets when threatening one another, and this display is probably more offensive than defensive in motivation (Bunn, Warburton, and Wilson, 1982). This same species may also become prostrate and motionless when it is actually picked up by a human, but such "playing dead" behavior does not seem to be typical of North American owls generally.

The European pygmy-owl, however, does sometimes perform apparent displacement sleeping when it is approached by humans, perhaps as a related kind of defensive behavior. According to Scherzinger (1971b) there is no relationship between the presence of "artificial eyes" on the nape of pygmy-owls and their relative protection from predation, which makes one wonder what other possible social functions those distinctive visual signals might possibly have. Interestingly, it has been reported that European pygmy-owls are more prone to let regurgitated pellets, broken eggshells, and droppings accumulate below their nest hole openings than are somewhat larger species of hole-nesting owls such as boreal (Tengmalm's) owls. Apparently this is because the very small entrances of the cavities used by these owls

are too tiny to allow the entrance of such major predators as martens (*Martes martes*), thereby providing a degree of nesting safety and consequently reducing selective pressures for keeping the nest entrance as difficult to locate as possible (Sonerud, 1985).

An important social behavior for many, perhaps most, owls is mutual preening (allopreening). Even small owlets sometimes perform this behavior, so it is not necessarily a kind of courtship, but it is certainly engaged in primarily by paired birds and almost certainly is an extremely important pair-forming and pair-bonding type of activity. In the common barn owl mutual preening occurs regularly between pair members throughout winter, the female usually approaching the male while uttering squeaks or whistles and preening him all over, but especially around the face and the back of the head and neck. The preened bird responds "with apparent pleasure" by uttering twittering noises and chirrups. Often the two birds doze for a time following such a period of preening (Bunn, Warburton, and Wilson, 1982). In the great gray owl allopreening is one of the strongest behavior patterns evident during pair bonding. It can be readily elicited by humans from adult wild owls of both sexes, as well as from subadults, by approaching the bird with the top of one's head (hairy, not bald!) toward its face, which stimulates the preening response, even in badly injured birds (Nero, 1980). Mutual preening in this species probably serves to reduce aggressiveness between individuals and may provide for sexual recognition and pair-bond maintenance. Forsman and Wight (1979) have suggested that although alloopreening is known to occur in numerous owl species, its function is still uncertain, but it may represent ritualized aggressive biting behavior, and in their opinion it probably is not important as a means of achieving either sexual or individual recognition.

Reduction of danger to both sexes, but especially to the smaller males, is probably an important aspect of pair-bonding behavior in owls, and it has been suggested that one advantage of the reversed sexual dimorphism typical of most owls is that it allows female dominance to be established at the time of pair formation with a minimum of dangerous aggressive encounters between the two birds, which is obviously advantageous to both sexes (S. Smith, 1982). This possible explanation for reversed sexual dimorphism in raptors has been advocated by Mueller and Meyer (1985), using data from Eurasian hawks and eagles. They also observed high levels of reversed sexual dimorphism among falconiform species in which the female was highly involved in territorial defense, which provides a second obvious behavioral advantage to larger female size among raptors. Mueller (1986) extended this explanation for reversed sexual dimorphism to owls, using data from both North American and Eurasian species. Gerhardt and Gerhardt (1997) supported the view that the facilitation of female dominance in pair formation and pair maintenance offers a better explanation for reversed sexual dimorphism in temperate and tropical owls than any of the other hypotheses so far advanced.

Probably for most owls, a major part of courtship signaling and territorial advertisement consists of calling behavior. This usually occurs during evening and night hours, even in the relatively diurnal species such as the burrowing owl, although in high-latitude breeders it regularly occurs during daylight. These advertisement calls are usually initiated by the male, and are predominantly performed by them, either from perches scattered around the male's territory or as "song-flights" from above it, as in the common barn owl (Bunn, Warburton, and Wilson, 1982). Unmated females are probably attracted to such calls and often have distinctive answering calls by which they may gain entrance into the male's territory and begin to establish individualized contact. Vocal duetting is not an uncommon performance by presumably paired owls and occurs, for example, in such diverse groups as some fish owls (*Ketupa*) and scops and screech-owls (*Otus* spp.), as well as in tawny, long-eared, little, and spectacled (*Pulsatrix perspicillata*) owls (Everett, 1977; Cramp, 1985). Although monogamy is certainly the normal mating pattern in owls, scattered instances of polygamy (bigyny) have been reported in the snowy owl (Watson, 1957), Tengmalm's (boreal) owl (Solheim, 1983), northern hawk-owl (Sonerud et al., 1987a) and eastern screech-owl (Gehlbach, 1994a).

Courtship displays in owls often involve aerial activities. In the short-eared owl a major courtship display is an aerial wing-clapping display (Clark, 1975). Like the snowy owl, this open-country species occupies territories that have few if any available elevated perches from which to hoot, and flight displays are an ecologically appropriate type of advertisement behavior. Wing clapping is performed by the male short-eared owl while circling in the vicinity of his territory and occurs before pair formation, thus serving to advertise his availability for mating as well as his being a territory-holder. Courtship calling may occur during this aerial phase of display, but it was also observed by Clark in a perched male shortly before copulation. Courtship feeding may immediately precede copulation, and in cases observed by Clark the female flew to the prey-carrying male, whereupon the male performed a food-begging display (opening his wings and fluttering them while presenting the prey in his mouth to the female). Similar passing on of prey by the male to the female occurs at the nest, during both the incubation and brood-rearing periods (Figure 20). Somewhat similar precopulatory behavior involving food presentation has been observed in the snowy owl (P. Taylor, 1973). Courtship feeding has also been observed in many other owls, such as the common barn owl (see Figure 23), and is probably a regular component of owl pair-forming and precopulatory behavior.

In the common barn owl a wing-clapping display also occurs, but it is much rarer and less loud than is the case in species of *Asio* (such as long-eared and short-eared) where it has been observed. The "moth flight," an aerial display marked by shallow wingbeats, also occurs in the common barn owl, as does a repeated "in-and-out" flight during which the male apparently attempts to entice the female into a nesting site. Copulation in the common barn owl is apparently not invariably associated with prey presentation by the male. It often occurs without apparent prior display when the female begins to "snore" quickly and lowers her body. Thereupon the male mounts her, balancing with his spread wings and

Figure 20. Food passing by male long-eared owl to brooding female (*above*), and feeding of young by brooding female snowy owl (*below*). After photos in Everett (1977) and Mikkola (1983).

holding her nape with his beak feathers. Tongue-clicking, bill-fencing, and cheek-rubbing activities are all regular and important parts of courtship in common barn owls, and are also commonly performed by owlets once they are old enough to become relatively active (Bunn, Warburton, and Wilson, 1982).

Nest building does not occur as such in most owls (although nest excavation does occur in some), but simple scrapes, which are later lined (possibly fortuitously) with a few stalks of stubble, are produced by short-eared owls and snowy owls. Clark (1975) believed that the female short-eared owl does at least some of any nest construction performed by this species. Actual nest-building behavior (as opposed to mere twig-shuffling behavior) by the great gray owl was not observed by Nero (1980), and earlier published reports of it were regarded by him as unproven.

Incubation behavior begins with the laying of the first eggs in most owls, which results in staggered hatching times for the young. In the common barn owl, eggs are laid at 2- to 3-day intervals; thus the eggs normally hatch at intervals of about 2 days, meaning that in a large clutch of 6–7 eggs there may be more than a 2-week span in the hatching dates of the owlets (Bunn, Warburton, and Wilson, 1982). Certainly in the owl species that have been closely studied, only the female is known to incubate; various reports of male owls assisting in incubation thus need confirmation. Incubation lasts an average of about 4 weeks in owls, ranging from as little as 21–22 days in the elf owl to perhaps as long as 34–35 days in the great horned owl.

During the incubation period the male provides his mate with food for her to consume while sitting, but as soon as hatching has occurred the female passes on much of the food provided by the male to her brood (Figure 20). After hatching, the eggshells may be eaten by the female, carried away and dropped some distance from the nest, or sometimes simply pushed into a corner of the nesting site. Brooding of the young while they are still quite small is quite intense, with the female gathering the huddled owlets around and under her, virtually enclosing them in her breast feathers and drooping wings. But gradually the female simply stands beside the older owlets, sometimes partially hiding them from view by wing drooping (Figure 21). Eventually the young become old enough that they can safely be left for short periods, while both the members of the pair begin to gather prey. The role of females in prey catching for the young is seemingly rather variable. In the common barn owl parental brooding gradually ceases when the eldest owlet is about 3–4 weeks old and the youngest about 13–20 days (Bunn, Warburton, and Wilson, 1982). Actual fledging requires about 50–55 days.

In the other North American owls the fledging time varies from as little as 27–28 days in the elf, saw-whet, and pygmy-owls to as long as 63–70 days in the great horned owl. The young of hole- and cavity-nesting owls spend virtually their entire prefledging period safely hidden within the confines of the nesting site, whereas those of such exposed nesting species as snowy, short-eared, long-eared, great horned, and great gray owls often begin to leave their nests for varying lengths

Figure 21. Brooding by great gray owl (*above*) and short-eared owl (*below*). After photos in Mikkola (1983).

of time when they are only about halfway through their fledging period. In the case of the tree nesters, the young owlets soon begin "branching," which consists of clambering about on branches and tree trunks, often even adeptly climbing nearly vertical surfaces while they are still flightless. Similarly the young of snowy owls are able to climb over ground obstacles when only about 20 days old, and young short-eared owls may hide in tall grass some distance from the nest while food is brought to them or dropped from above by their parents (Mikkola, 1983).

While the owlets are quite young the female common barn owl cleans the nesting area by eating their feces, although regurgitated pellets are allowed to accumulate. In the long-eared owl, where feces are usually voided over the side of the nest or dropped through its bottom, such nest hygiene may be lacking (Bunn, Warburton, and Wilson, 1982). Initially the young are fed rather small prey, or torn-up portions of larger prey, but as they grow they are increasingly provided with entire carcasses of larger animals. Observations by Bunn, Warburton, and Wilson indicate that sibling owlets rarely steal food from one another, and indeed the older ones may at times attempt to feed their younger siblings, suggesting that sibling killing and associated cannibalism in owls is probably much rarer than is generally imagined. In the view of these authors, cannibalism, at least in common barn owls, is just one more by-product of undue stress, whether from severe weather conditions, prey shortage, or nest disturbance. Nevertheless, food competition must play an important role in influencing overall reproductive success.

The owl thinks all her young ones beauties.

—T. Fuller, *Gnomologia*

*There shall the great owl make her nest, and
lay, and hatch, and gather under her shading;
there shall the vultures also be gathered, every
one to her mate.*

—Isaiah 34:15

The reproductive period of owls tends to be highly prolonged; few
birds begin their nesting activities as early as do owls in the more
temperate parts of North America, and few terrestrial birds of
comparable size spend so long in hatching and rearing their
young. It is not unusual for great horned owls to begin their egg-
laying activities during January in the southern United States, and
even as far north as Alberta there are egg records for as early as
the latter part of February. Most of these eggs (which have a
month-long incubation period) will have hatched by April, but the
long prefledging period of 9–10 weeks may mean that owls are
still tending nests of unfledged young well into the summer. The
common barn owl has an even more prolonged breeding season,
with egg records for such states as South Carolina and Georgia
extending from March to December, and it is probable that this
species occasionally raises two broods a year in at least some parts
of its North American range. Under captive conditions a pair has
been known to produce as many as five clutches in a 12-month period, from which as many as four broods have been hatched successfully, thus virtually breeding throughout the year without in-
terruption (Betsy Hancock, pers. comm.).

For purposes of this brief overview, we can conveniently divide
the owl's year into three components, the prenesting period, the
nesting period, and the postnesting period.

Prenesting Biology and Population Densities

For some owls it is common for the birds to group into assemblages that roost in communal locations, which seems an unex-
pected kind of behavior for raptors. Presumably these are groups
of birds drawn together to a common suitable roosting site, rather
than true congregations of birds held together by social attraction
to one another. As a result, the number of birds using a given site
tends to vary greatly from time to time, perhaps depending upon
the suitability of the site and the relative abundance of prey and
hunting habitat around the roost. Sites used by short-eared owls
studied by Clark (1975) had several features in common, including
inconspicuous shelter from the weather, proximity to hunting areas, and relative freedom from human disturbance. Similar characteristics seem to apply to long-eared owl winter roosts (D. Smith,
1981; Bosakowski, 1984). Wintering territories for a pair may frequently become extended or modified into breeding territories, at

least in fairly sedentary species such as the short-eared owl. Owls typically establish nesting territories large enough for them to both breed and hunt within, and probably most territorial aggression occurs during the immediate prenesting and early portions of the nesting season (Clark, 1975).

It is apparently typical for raptors, including hawks and owls, to reoccupy the same nesting territories in consecutive years. Thus it is typical for both members of a pair, or at least one of the members, to return annually to the same nesting areas, often to the very same nesting site, until death or some environmental change causes a disruption in the pattern (J. Craighead and Craighead, 1956). The Craigheads observed that one pair of great horned owls occupied the same nesting territory for 7 successive years, and another pair held a territory for 8 years. D. Smith and Murphy (1974) also observed a high level of territorial reoccupation during successive years by great horned owls during a 4-year study, as did Reynolds and Linkhart (1987a) in a 5-year Colorado study of flammulated owls, which is notable in view of the fact that at least in Colorado this is a migratory species. Such fidelity to a particular breeding area probably is important to birds such as raptors in learning the best hunting areas and in avoiding unnecessary overlap of territories or home ranges with other raptors in the same general region.

The Craigheads also have observed a notable consistency in raptor composition and density from year to year, supporting the idea that the raptor population as a whole tends to interact as a single unit. Seemingly the other raptors adjust to minor changes occurring in the nesting patterns of such species as the great horned owl, which, as a year-round resident and early nester, tends to establish the general spatial pattern assumed by the other, generally smaller species. The nesting densities observed by the Craigheads ranged from 0.7 pairs of raptors per square kilometer in some cultivated areas to 1.5 pairs per square kilometer in semiwilderness areas. Substantially lower raptor population levels (averaging 0.4 birds per square kilometer) and somewhat greater year-to-year variability in breeding populations were determined by D. Smith and Murphy (1973) for a semidesert area in Utah, although fully as many raptor species were present there as in the Craigheads' study areas.

Some estimated densities of natural owl populations as reported by these and other researchers are summarized in Table 4. Because of greatly varied census techniques and sizes of areas sampled, these data are probably not fully comparable with one another, but they do provide a general sense of the relatively low biomass represented by owls in most habitats. Frederick Gehlbach (pers. comm.) has informed me that 7-year averages of breeding pair densities per square kilometer in Cave Creek Canyon, Arizona, were 6.4 for whiskered screech-owl, 1.4 for western screech-owl, 2.5 for flammulated, 2.3 for northern pygmy, and 3.6 for elf owl. Only rarely (such as during rodent plague years) do natural densities of larger species of owls exceed a bird per square kilometer.

Some time before the nesting season it is of course necessary that pair bonds be established or reestablished. For permanent resident species it is unlikely that these bonds are ever ruptured during the course of the year, and probably only such migratory owls as the burrowing owl, flammulated owl, and elf owl have to spend any great amount of time locating prior mates or establishing new pair bonds once they return to nesting areas. Pair-bonding mechanisms, which vary, are described in the individual species accounts in Part 2.

Nesting Period and Annual Productivity

The beginning of the nesting season may conveniently be defined for owls as the time that a nest site is established. For many owls this typically involves the takeover of a previously used nest of a hawk or crow, or some other prebuilt nest. For others it requires finding a suitable nesting cavity in a hollow tree. No North American owls actually construct their nests, although common barn owls are known to excavate stream-bank or road-cut cavities in some areas of the western states, and the short-eared owl has been reported to gather materials for its nest scrape on the ground. Among North American owls it is likely that all species become sufficiently mature as to form pair bonds before they are a year old, even though actual nesting may not occur while the birds are still yearlings.

The Craigheads observed several cases of paired great horned owls that established territories and made no nesting attempts in a particular year but did nest the following year, suggesting that first-year owls of at least this species may sometimes require two years to become effective breeders. The Craigheads observed that great horned owls in Michigan began selection of nesting territories as early as late January (about a month before the earliest record of egg laying), compared with late February for eastern screech-owls (about two months before initial laying). In Wyoming, territorial establishment in these two species occurred from late February to early March and from late March to early May, respec-

Table 5

Some Density, Home Range, and Territory Estimates for North American Owls

Species	Average Density (birds/sq km)	Average Home Range (hectares)	Average Territory (hectares)	Reference
Barn owl	0.08	—	—	Sharrock, 1976
	0.42	ca 250	—	Bunn, Warburton, and Wilson, 1982
	0.01–0.24	0–60	—	Glutz von Blotzheim and Bauer, 1980
	—	294–953	—	4 studies in I Taylor, 1994
Flammulated	4.8[1]	—	ca 3–6	Marshall, 1939
	3.11	4.11	4.1	Reynolds and Linkhart, 1987a
E. screech-owl	0.3	—	—	J. Craighead and Craighead, 1956
	—	9–108	—	D. Smith and Gilbert, 1984
	2.3	—	—	Lynch and Smith, 1984
	—	6–30	—	Gehlbach, 1994a
W. screech-owl	0.19	—	—	J. Craighead and Craighead, 1956
	—	—	ca 0.09	R. Johnson et al., 1981
Great horned	0.15–0.26 (various areas)	212	—	J. Craighead and Craighead, 1956
	0.05–0.15 (various years)	23	—	D. Smith and Murphy, 1973
	—	2499	—	Fuller, 1979
	—	725	248–483	Rohner, 1997
Snowy	—	—	260	Watson, 1957
	0.03–0.3	—	—	Manning, Höhn, and Macpherson, 1956
	—	—	286–653	Wiklund and Stigh, 1986
N. hawk-owl	0.04	—	—	Hagen, 1956
Elf	—	1.0	—	Gamel, 1997
	0.6–10	—	—	Henry and Gehlbach, 1999
Burrowing	—	—	0.8	Thomsen, 1971
	0.04	82.9	—	D. Smith and Murphy, 1973
	—	4–5.9	—	R. Grant, 1965
	—	241	—	Haug and Oliphant, 1990
Great gray	0.06	259–405	260	J. Craighead and Craighead, 1956
	—	—	45	Brunton and Pittaway, 1971
	1.5 and 3.4 (2 areas)	—	—	Bull and Henjum, 1987
Barred	0.02	—	—	R. Stewart and Robbins, 1958
	—	231	231	Nicholls and Fuller, 1987
	0.06	—	—	J. Craighead and Craighead, 1956
	—	655	—	Fuller, 1979
	0.07–0.3	—	—	Bosakowski, Speiser, and Benzinger, 1987
Spotted	0.037	—	—	Barrowclough and Coats, 1985
	—	182	—	Gould, 1974
	—	1480	—	Forsman, Meslow, and Wight, 1984
Long-eared	0.02–0.1 (various areas)[2]	55	—	J. Craighead and Craighead, 1956
	0.19–6.6 (various areas)	—	—	Glutz von Blotzheim and Bauer, 1980
Short-eared	—	—	74–121	Clark, 1975
	4.2–6.6[3]	—	18–137	Lockie, 1955
	2.3–5.4	—	20.2	Pitelka, Tomich, and Trichel, 1955b
	0.02–13 (various areas)	—	—	Glutz von Blotzheim and Bauer, 1980

Continued on next page

Table 5 continued

Species	Average Density (birds/sq km)	Average Home Range (hectares)	Average Territory (hectares)	Reference
Boreal (Tengmalm's)	0.2–2.6 (various areas)	—	—	Glutz von Blotzheim and Bauer, 1980
	—	1182–1451	—	Hayward, Hayward, and Garton, 1993
	—	100–500	—	Bondrup-Nielsen, 1978
N. saw-whet	—	114	—	Forbes and Warner, 1974
	0.05	—	—	Hardin and Evans, 1977
	0.002	—	—	Simpson, 1972
	—	8–129[4]	—	Hayward, 1983

[1]Estimates of singing territorial males only.

[2]Excluding some very small (less than 1.0 square kilometer) study areas.

[3]Probably maximum density (associated with high prey levels).

[4]Using 75–95% contour estimate method.

tively. However, in both areas the termination of nesting occurred at very nearly the same time, indicating a telescoping effect of nesting into the shorter climatically available period typical of Wyoming. In Utah, D. Smith and Murphy (1973) found that breeding in the collective raptor population occurred over an eight-month period, with each of the 8–11 breeding species initiating its nesting activity at slightly differing times (and each species also tending to hunt at different times of the day, in different habitats, or concentrating on different prey).

The Craigheads defined their term "nesting range" in the usual sense of "home range," namely an area over which a pair moves in performing the activities associated with its nesting cycle, arguing that the term "home range" is inapplicable to migratory species of birds. However, the owls in their study were all sedentary forms, thus their range data are listed under "home range" in Table 5, together with other estimates of home ranges reported in the literature for various owls and by various workers. "Nesting territories" include those parts of the home range or nesting range that are actively defended, and as such are usually much smaller areas than the total home ranges.

D. Smith and Murphy (1973), in a study similar to that of the Craigheads', found little evidence of actual territorial defense as well as minimum intraspecific overlapping of home ranges, suggesting that the term "maximum home range" best describes the pattern of breeding-raptor spacing observed by them. Various estimates of "territory" sizes as given in the literature (or esti-

mated from data in the literature) are also presented in Table 5; in some cases these might perhaps better be regarded as home-range estimates (for example, the barn owl "home range" estimate from Bunn, Warburton, and Wilson [1982] is based on five birds' "hunting territories"). However, territorial defense of the entire home range does occur in some species such as the barred owl. Nicholls and Fuller (1987) reported that the home range of barred owls in Minnesota was the same as the birds' territorial boundaries. Their criteria of territoriality in barred owls included (1) little evidence of significant home-range overlap among nonpaired birds, (2) a large home-range overlap of paired birds, (3) similar home-range limits in successive years, (4) "inherited" home-range limits between generations, and (5) territorial advertising by vocalizations throughout the home range. However, species such as screech-owls, the flammulated owl, pygmy-owls, and the elf owl actively defend only the multiple nest–food storage–roost cavities and an intermediate radius proportional to tree density (Gehlbach, 1994a, 1995b; Gehlbach and Gehlbach, 2000).

In general, home-range and territory estimates suggest what is intuitively obvious, that primarily insectivorous species such as the elf and burrowing owls should normally have substantially smaller territories than those that are dependent upon vertebrates for their prey. Additionally, in at least some areas territorial behavior may place limits on owl breeding densities. The tawny owl has been reported to have a well-defined territory of about 13 hectares in prime woodland habitat, compared with about 20 hectares in mixed woodland and open ground. Changes in the food supply were not found to alter numbers of breeding owls in this area, but instead altered annual productivity rates and thus regulated populations on a year-to-year basis (Southern, 1970). On the other hand, the short-eared owl is relatively nomadic and able to exploit variations in food supplies in an area by moving in and adjusting its territorial size according to temporal changes in prey abundance (Clark, 1975; Lockie, 1955).

A breeding range or territory may simply be a continuation of a pair's previously occupied winter range or may be reestablished yearly up until early spring. These breeding territories of owls are advertised by calling, which is mostly performed by males and may begin as early as the fall prior to breeding, when older birds reestablish their territorial boundaries and newcomers try to establish new ones (Southern, 1970). Advertisement calling has the possible double function of announcing territorial ownership and attracting new mates or sexually stimulating existing mates.

All owls lay eggs that are completely white, probably because they tend to be well hidden from above and in most cases the nests can be effectively defended by the parent birds. Owl eggs are also relatively round in shape, reflecting the fact that many species tend to nest in holes (round eggs occupy less space for their volume than do other configurations), although a few species (such as snowy and short-eared owls) that nest on flat nest sites or in fairly open situations lay more oval-shaped eggs. Scherzinger (1971a) considered white eggs, long incubation and post-hatching developmental periods, and the absence of visual begging signals to be relatively primitive owl traits, all typical of hole-nesting owls. Owl species that breed in open situations have young that exhibit an earlier development of walking, climbing, and food-tearing behavior, leave their nests sooner than do hole-nesting forms, and show a better development of down in conjunction with greater thermoregulatory needs than occurs in hole-nesting species.

In common with all birds, although larger species of owls lay larger eggs, the relative energy investment in egg production is much greater for smaller species of owls than for larger ones. This trend is especially apparent when entire clutches are considered. Thus a great horned owl lays eggs that are each equal to about 3.6 percent of the adult female weight, and an average clutch size of about 2.5 eggs represents only about 9 percent of the female's weight. By comparison, the elf owl lays an egg representing about 18.3 percent of the

female's weight, and a modal clutch of 3 eggs represents more than half the adult female's weight. The flammulated owl also lays a notably large egg in proportion to the adult female's weight (about 17 percent), and assuming an average clutch of 3.3 eggs, about 55 percent of her weight would be represented by a complete clutch. The Eurasian pygmy-owl has an average clutch size of 5.5 eggs (Mikkola, 1983), with an egg weight representing about 11 percent of the female's weight; thus an average clutch represents approximately 60 percent of her weight. By comparison, the similar-sized northern pygmy-owl produces an egg equivalent to 12 percent of her body weight, and an apparently typical clutch size of 3.2 eggs would represent less than 40 percent of her body weight.

Clutch sizes of owls vary markedly between species and even within species in some cases. Some average clutch-size data are shown in Table 6 for nearly all species of North American owls. Among these owl species, no obvious correlations are apparent between average clutch-size variations and such obvious potentially associated variables as adult body size, average latitude of breeding, primary prey species, and the like. At least in the case of the European *Strix* species, larger average clutch sizes and larger ranges in clutch sizes are reportedly typical of stenophagous owls (such as the great gray owl) having a restricted number of prey species, as compared with more broadly prey-adapted (euryphagous) species such as Ural and tawny owls, according to Mikkola (1983). Apparently the clutch sizes of such narrowly prey-adapted specialists as the great gray owl are prone to vary locally or annually in accordance with the abundance of a few key prey species. This might help account for the generally large but also highly variable clutch sizes typical of such other boreal forms as the northern hawk-owl, short-eared owl, and snowy owl, and the consistently low average clutch size of foraging generalists such as the great horned owl. In one museum study (Murray, 1976) it was found that in most of seven species of North American owls average clutch sizes increase slightly with latitude. However, this is not a universal trend, and a few partially insectivorous species (burrowing and screech-owls) exhibited some latitudinal trend reversals in certain areas. A latitudinal increase in clutch size from south to north is evident in the data for the eastern screech-owl summarized by VanCamp and Henny (1975). A few minor east-west trends in clutch size are also apparent in some owl species, judging from these studies.

As an indication of the variations in clutch sizes typical of owls, a summary of such data is presented for various European owl species (Table

Table 6

Some Clutch Size and Reproductive Success Estimates for North American Owls

Species	Sample Size	Average Clutch Size	Average Brood Size	Average Young Fledged	Percent Reproductive Success	Reference
Barn owl	14	4.2	1.7	1.3	31	D. Smith and Frost, 1974
	91	4.9	2.69	1.84	38	Otteni, Bolen, and Cottam, 1972
	24	5.46	2.31	2.08	38	Reese, 1972
	325	—	—	—	44.6	Marti, 1992
	1639	4.99	—	—	—	Mikkola, 1983
Flammulated	43 nests, 7 broods	3.3	2.9	—	—	Various sources[1]
	11 nests, 26 broods	2.7	2.4	—	—	Reynolds and Linkhart, 1987a
E. screech-owl	91	4.43	4.16[2]	2.55[2]	58[2]	VanCamp and Henny, 1975
	300	3–4.56	—	—	—	VanCamp and Henny, 1975
W. screech-owl	435	3.42–4.0	—	—	—	Murray, 1976
Great horned	19	2.26	1.74	1.56	69	Olendorff (cited in D. R. Johnson 1978)
	22	2.82	2.63[2]	2.0	76	D. Smith and Murphy, 1973
	930	2.44	—	—	—	Murray, 1976
	371 broods	—	—	1.61	—	Henny, 1972
Snowy						
Finland	66	7.74	—	—	—	Mikkola, 1983
Shetlands	49	5.4	4.8[2]	2.5[2]	47	Cramp, 1985
N. hawk-owl	135	6.31	—	—	—	Mikkola, 1983
Elf	90	2.98	2.39[2]	2.26[2]	70	Ligon, 1968
	54	3.04	—	—	—	Various sources[1]
N. pygmy-owl	18	3.22	—	—	—	Various sources[1]
F. pygmy-owl	35	3.28	—	—	—	Various sources[1]
Burrowing	—	—	4.7	3.3	—	Butts, 1973
	439	6.48	—	—	—	Murray, 1976
	15	—	5.2[2]	4.3[2]	—	Martin, 1973a
	18	—	3.9[2]	3.2[2]	—	Thomsen, 1971
Barred	315	2.41	—	—	—	Murray, 1976
Spotted	46 broods	—	2.13	1.8[3]	85	Forsman, Meslow, and Wight, 1984
	5377 females	—	—	0.6	32.8	Burnham, Anderson, and White 1996
Great gray						
Finland	241	4.4	—	—	—	Mikkola, 1983
Finland	42	3.6	3.3[2]	2.4	69	Mikkola, 1981
Oregon	67	—	—	2.3	78	Bull, Henjum, and Rohweder, 1989b
Long-eared	393	4.49	—	—	—	Murray, 1976
Britain	287	4.15	1.2[2]	0.8[2]	20	Glue, 1977a
Europe	413 nests, 329 broods	4.9	3.0	—	—	Mikkola, 1983
Short-eared	21	6.33	1.86[2]	0.8+[2]	131+	Pitelka, Tomich, and Trichel, 1955b
	186	5.61	—	—	—	Murray, 1976
Europe	121	7.3	—	—	—	Mikkola, 1983
Germany	17	7.1	2.58[2]	1.94[2]	27	Holzinger, Mickley, and Schilhansl, 1973
Boreal (Tengmalm's)						
Finland	701	5.35	4.57[2]	3.64[2]	54	Korpimäki, 1981
Finland	110	5.5	—	—	—	Mikkola, 1983
N. saw-whet	156	4.28	—	—	—	Murray, 1976
	9	5.9	4.4	2.4	42.5	Cannings, 1987

[1]Mostly museum data from Western Foundation of Vertebrate Zoology.

[2]Values calculated from authors' data; may exclude possible effects of complete clutch/brood loss or of renesting.

[3]Unweighted average of 5 years' data.

Table 7

Variations in Clutch Sizes of Representative European Owls

Species	Sample Size	Clutch Size Category															Average
		1	2	3	4	5	6	7	8	9	10	11	12	13	14	15+	
Eagle	481	10	226	180	57	5	3	—	—	—	—	—	—	—	—	—	2.6
Ural	98	8	18	40	23	6	3	—	—	—	—	—	—	—	—	—	3.1
Little	391	3	26	106	159	75	16	6	—	—	—	—	—	—	—	—	3.9
Pygmy-owl	49	—	—	4	13	12	7	7	4	1	1	—	—	—	—	—	5.4
Barn owl	146	—	3	13	22	37	35	17	5	4	5	1	1	1	—	2	5.7
N. hawk-owl	135	—	—	5	16	31	27	23	17	8	3	4	—	1	—	—	6.3
Short-eared	121	—	2	—	10	8	19	25	27	18	8	2	1	1	—	—	7.3
Snowy	66	—	—	—	—	9	14	10	12	11	3	1	3	2	1	—	7.7

Note: Adapted from data presented by Mikkola (1983); species arranged by increasing average clutch size.

7). What is notable in this summary is the remarkable variation around the mean clutch size that is apparent in most species, suggesting that selection has favored the evolution in owls of a relatively variable clutch size rather than a fixed one as is typical, for example, of virtually all shorebirds. A possible explanation for the advantage of such variability becomes evident when British data for a single species, the barn owl, are examined (Table 8). Using data provided by Bunn, Warburton, and Wilson (1982), the reproductive results of nests with clutch sizes ranging from 1 to 9 eggs may be compared. Although the mean clutch size in this sample was about 4.7 eggs, the most productive clutch size (that producing the most surviving young per nest) was of 7 eggs, and the most efficient clutch size (that producing the highest percentage of fledged young relative to eggs laid) was only 2 eggs.

The modal clutch size of 5 eggs might thus be regarded as a kind of selective compromise between these extremes. In such clutches the female's productivity is relatively high, but there is less wastage of her energies on eggs or chicks that do not live to be fledged young than would be the case if a larger clutch had been produced. It is further likely that the flexibility in owl clutch-size production enables individual females to produce larger clutches in those years or situations in which they are in prime physiological condition (such as during good prey years) than during years of relative prey scarcity and associated poor breeding condition. Thus the snowy owl may produce surprisingly large clutches of up to about a dozen or so eggs in good prey-years and perhaps not breed at all in years of prey scarcity. In Europe such species as the great gray owl and tawny owl are also reported to vary their clutch sizes ac-

Table 8

Variations in Clutch Sizes and Productivity in British Barn Owls

	Sample Size	Clutch Size Category								Average
		2	3	4	5	6	7	8	9	
Number of clutches	115	3	12	36	40	15	6	2	1	—
Total eggs produced	543	6	36	144	200	90	42	16	9	4.72
Number of young fledged	256	4	19	85	87	38	19	4	0	—
Percent eggs fledged	—	66.7	52.7	52.1	43.5	43.3	45.2	25	0	47.10
Percent nests fledging all eggs	—	33.3	33.3	27.8	12.5	6.7	0	0	0	20.00
Average young fledged per nest	—	1.33	1.58	2.36	2.17	2.58	3.16	2.0	0	2.23

Note: After Tables 11–13, in Bunn, Warburton, and Wilson, 1982.

cording to prey abundance. The latter species is known to vary its average clutch size by about 25 percent during prey cycles, and the former species is reported to lay as many as two replacement clutches of eggs (following egg removal by humans) during good vole years (Mikkola, 1983). Likewise, a study in England indicated that the average clutch size of the tawny owl was higher (2.9 eggs) during years when prey was abundant, and renesting sometimes occurred following nest loss in such years. In years of low prey availability there was extensive nonbreeding, and clutch sizes averaged substantially lower (2.0 eggs). There were higher levels of egg and chick losses, and high mortality rates of the young occurred during their first fall and winter, especially from starvation, during years of low prey availability (Southern, 1970).

Regardless of initial clutch-size variations, it is apparent from Table 6 that the reported average reproductive success rates of North American owls vary greatly, ranging from surprisingly low (under 15 percent) to remarkably high (more than 80 percent), the former reported for a northern rodent specialist with a large average clutch size and a fairly exposed nest site and the latter belonging to an arthropod generalist having a relatively low average clutch size and a fairly secure nesting site. Thus, as with clutch size, there is perhaps no typical rate of reproductive success that can be attributed to owls. The Craigheads (1956) observed an overall nesting success rate (percent of initiated nests resulting in hatched young) of 66 percent for 161 active raptor nests and an average fledging success rate (percent of hatched young fledging) of 62 percent. D. Smith and Murphy (1973) reported annual hatching success rates (percent of all eggs laid that hatch successfully) from 139 nests of from 75.6 to 82.5 percent and annual fledging success rates of from 53.4 to 61.6 percent for all raptors studied. One of the major apparent influences on reproductive success in such studies is the level of human interference to which the birds are exposed, often causing nest desertion.

In a resurvey of the Craigheads' original Wyoming study areas, F. Craighead and Mindell (1981) found that the 1975 breeding population there had declined 30 percent since 1947. They also found that average clutch sizes and average size of fledged broods were smaller, and that only 40 percent of the raptor pairs studied fledged any young, compared with 88 percent in 1947. They considered that increased levels of human activity in the Jackson Hole area were a probable major factor in causing these reductions in productivity.

Postnesting Molt and Dispersal

It is perhaps convenient to define the end of the breeding season as being that time when the onset of the annual molt occurs, for by then it is certain that one or both birds will have gone out of breeding condition, and further breeding activity is unlikely. In common barn owls this molting period lasts about three months, and their wing-molt pattern differs from that of many owls (which shed their primaries in the usual manner, from the inside outward) in that it begins in the middle of the primaries and proceeds in both directions. In all owls the primary molt occurs slowly enough that flight effectiveness is not noticeably impaired. The tail molt in barn owls is unusually slow and is irregular in sequence (Bunn, Warburton, and Wilson, 1982). By comparison, in some small owls (including at least some species of *Athene, Glaucidium,* and *Otus*) all the tail feathers molt almost simultaneously, for reasons that are presumably adaptive but not yet apparent (Mayr and Mayr, 1954). However, the small flammulated owl apparently lacks a simultaneous tail molt in spite of some assertions to the contrary (Reynolds and Linkhart, 1987b). In the burrowing owl the tail molt may actually vary from being essentially simultaneous to relatively gradual in different individuals, and probably the rate of tail molt does not significantly influence the bird's efficiency at insect catching in this species (Courser, 1972).

If nothing conclusive can be said of average annual recruitment rates in owls, as measured by variations in rates of annual productivity, can any generalizations at least be made of the counterpart vital statistic, annual mortality rates? Mortality rates of a variety of owl species are not available, and representative figures for various North American owls are summarized in Table 9. Here, a much higher level of consistency seems to prevail than for recruitment rates. Nestling mortality is certainly highly variable and is probably the primary contributor to the high variations in reproductive success statistics. However, first-year postfledging mortality rates for juveniles (those occurring from about the time of fledging, which is usually assumed to be equivalent to the time the young are old enough to band) are rather consistently in the range of 50–70 percent.

The Craigheads (1956) estimated in their study that immature raptors exhibited an approximate annual postfledging mortality rate of 88 percent, compared with an estimated adult annual mortality rate of only 12 percent. They emphasized that both these figures represented only approximations, but the data summarized in Table 9 also indicate that a much higher rate of

Table 9

Some Estimated Mortality Rates of North American Owls

Species	Nestlings	Juveniles[1]	1–2 yr.	2+ yr.	Reference
					Mortality rate (%)
Barn owl	—	65	43	37	Barrowclough and Coats, 1985 (average of several studies)
	—	68	51	43	Glutz von Blotzheim and Bauer, 1980
	—	75	ca 40	—	Frylestam, 1972
E. screech-owl	39	69	39	—	VanCamp and Henny, 1975
	—	69	41	25–33	Ricklefs, 1983[2]
Great horned	—	—	46	31	P. Stewart, 1969
	15	58	44	28	Houston, 1978; Adamcik, Todd, and Keith, 1978
	—	32	24	15	Houston, Smith, and Rohner, 1998
Burrowing	20.5	70	—	—	Thomsen, 1971
	—	—	81	32–41	Millsap and Bear, 1992
Spotted	—	—	81	15	Barrowclough and Coats, 1985
Great gray	—	46	8–29 (adults)	—	Bull and Henjum, 1987
Long-eared	—	52	31	31	Glutz von Blotzheim and Bauer, 1980
Short-eared	56	—	—	—	Pitelka, Tomich, and Trichel, 1955b
Boreal (Tengmalm's)	80	—	—	—	Korpimäki, 1981
	—	—	—	46 (adults)	Hayward, 1989

[1]Postfledging mortality to end of first year.

[2]Based on VanCamp and Henny's (1975) data.

mortality apparently occurs in first-year raptors than in adults. As has been suggested by the Craigheads and by later observations by Southern (1970), starvation presumably accounts for a substantial part of these losses, especially after the young are left to shift for themselves by their parents. However, young and inexperienced birds are also probably much more prone to accidental deaths such as those caused by collisions with traffic and various inanimate objects, as indicated by banding data for the great horned owl (P. Stewart, 1969), three species of European owls (Table 10), and a high incidence of young birds among road-killed great gray owls (Nero and Copland, 1981). Young and unwary birds probably also suffer a relatively high mortality as a result of illegal shooting by hunters (D. Smith and Murphy, 1973).

Adults and young often remain within their nesting ranges until the onset of fall migration in the case of migratory species or until the adults no longer feed their young in the case of resident populations. Until they stop feeding their young, the adult owls often range progressively farther from their nest sites, as the brood develops and their appetites increase (J. Craighead and Craighead, 1956). In some parts of their range (espe-

cially those having long available nesting seasons), common barn owls may begin a second brood, with the eggs of the second clutch frequently being laid in the original nest, sometimes even appearing before the last owlets of the first brood have departed. Feeding of these young by the female may occur up to within a few days of her laying the first egg of the second clutch. There are a few reports of third or even fourth broods being produced within a calendar year by common barn owls (usually captive birds), but such cases are probably extremely rare under natural conditions, at least in temperate latitudes.

As the adult barn owls become occupied with their next brood, those of the first gradually begin to roost elsewhere and to wander about, either drifting out of their home territory or sometimes actually being forced out by growing neglect or even aggression on the part of their parents. As the owlets begin to leave their nests, or even beforehand, their parents may begin to avoid them by roosting away from the nest site, sometimes to points quite distant in the territory. By the time the young are 14 weeks old their parents may show overt aggressive behavior toward the owlets, especially on the part of adult females.

Table 10

Causes of Death of Barn, Eagle, and Tawny Owls in Europe

	Barn Owl[1]		Eagle Owl[2]	Tawny Owl[3]
	First-Year Birds	Age Unspecified	(All Ages)	(First-Year)
Total sample	289	59	387	579
Unnatural Causes				
Collisions	124 (42.9%)	14 (23.7%)	46 (11.9%)	32.9%
Traffic	103	11	46	20.5%
Wires	7	1	—	5.5%
Other	14	2	—	6.9%
Electrocuted	—	—	82 (21.2%)	—
Shot/poisoned	5 (1.7%)	4 (6.8%)	62 (16.0%)	tr.
Trapped	23 (7.9%)	4 (6.8%)	25 (6.5%)	17.8%
(in traps, buildings, etc.)				
Natural Causes				
Sickness or injury	—	1 (1.7%)	38 (9.8%)	—
Starvation	—	5 (8.4%)	20 (5.2%)	1.4%
Drowned	9 (3.7%)	1 (1.7%)	5 (1.3%)	4.1%
Miscellaneous	11 (3.8%)	3 (5.1%)	18 (4.6%)	8.2%
Unknown Causes	117 (40.5%)	27 (45.8%)	91 (23.5%)	35.6%

[1]Adapted from Tables 18 and 39b in Bunn, Warburton, and Wilson, (1982).

[2]Adapted from Table 11 in Mikkola (1983).

[3]Adapted from data of Saurola as summarized by Mikkola (1983).

As they begin to fend for themselves, the young birds may spend a good deal of time catching easy prey such as insects or performing apparent play behavior while perhaps practicing prey-catching techniques (Bunn, Warburton, and Wilson, 1982).

During the young birds' first year of life, substantial dispersal typically occurs in owls, even among nonmigratory species. This dispersal tendency has been extensively documented for common barn owls (Bunn, Warburton, and Wilson, 1982), eastern screech-owls (VanCamp and Henny, 1975; Gehlbach, 1994a), great horned owls (Houston, 1978), and spotted owls (Gutiérrez and Carey, 1985; G. Miller and Meslow, 1985), plus several European owls (Mikkola, 1983). These studies suggest that it is not uncommon for young owls to move out of their natal areas to a distance of 75–150 kilometers, and rarely up to several hundred kilometers. They thus must pass through totally unfamiliar habitats and probably thereby are exposed to considerably increased risk of competition from established residents, as well as to greater probabilities of death through accidents, starvation, or even predation by larger raptors. Once the birds establish breeding territories they tend to become much more sedentary, except of course for the relatively few migratory

species of North American owls. Gehlbach (1994a) found no difference in dispersal tendencies between the sexes of eastern screech-owls.

Little is known of the migration routes or navigational mechanisms of the highly migratory owls, such as the elf and flammulated owls. In the case of the latter species, migratory habitats chosen in the fall may differ from those of the spring, mainly in that higher elevations are used in fall than in spring. This is apparently because large nocturnal insects such as moths are available to provide an ample food supply during fall, allowing the birds to exploit relatively high-altitude habitats during migration that are similar to those in which they breed (Balda, McKnight, and Johnson, 1975). Probably the elf owl also has evolved migration times and routes that are associated with the relative seasonal availability of insects, which are its primary prey. The northern saw-whet owl is also migratory, with definite north-south routes and predictable migration schedules, but females are additionally inclined to be nomadic.

In addition to these highly migratory owls, the boreal (Tengmalm's) owl in Europe is apparently partially migratory, with the males tending to be residential in breeding areas but the females and young birds relatively migratory, although the species has also been reported to be periodically nomadic or irruptive, according to relative prey availability (Mikkola, 1983). Many of the other northerly breeding owls with semiresidential ranges make periodic invasions into areas well to the south of their normal breeding and wintering ranges. These species include typical arctic or boreal species such as the snowy, northern hawk-owl, great gray, and to some extent the more broadly distributed great horned, long-eared, and short-eared owls. For species such as the snowy and perhaps also short-eared owls, these invasions are associated with yearly variations in various microtine rodent populations such as voles (*Microtus*) and lemmings (*Lemmus*), which often peak and crash at about 3–4 year intervals. At least in Europe the periodic invasions of great gray owls are also known to be associated with changes in *Microtus* populations, which oscillate at fairly regular 3–4 year intervals (Mikkola,

1983). In North America the causes of the periodic and irregular southward invasions of this species are less certain, and it has been suggested that such factors as unusually high owl populations, icy substrate crusts, or unusually deep snow levels may contribute to such invasions (Nero, 1980). On the other hand, great horned owl invasions in northern North America are likely to be influenced by annual variations in populations of snowshoe hares (*Lepus americana*), whose numbers tend to fluctuate over relatively long time periods of about 8–10 years (Keith, 1963). During the winter of 2000–2001 a major movement of great horned owls, great gray owls, and northern hawk-owls into southern Canada and the northern United States coincided.

> *I fain would know what man ever found a scritch-owl's nest and met with any of their eggs.*
>
> —Pliny, *Natural History*

> Noctua *the owl is called a* Noctua *because it*
> *flies about by night (*nox*). It cannot see by day,*
> *because its sight is weakened by the rising*
> *splendor of the sun.*

> —From a 12th-century bestiary,
> T. H. White, *The Book of Beasts*

Owl Myths of the Old World

The association of owls and humans is an archaic one, reaching
back to the very dawn of human history. The winged, bird-footed
Mesopotamian goddess Lilith was the goddess of death, and she
was depicted on a Sumerian tablet of 2300–2000 B.C.E. wearing a
headdress of horns, having taloned feet, and being flanked by owls.
Lilith also appears in the Old Testament Hebrew as an owl or female witch, both of which controlled the power of live and death.
It is thus possible, inasmuch as beliefs emanating from Crete and
the Middle East were certainly important in influencing early
Greek religion, that Lilith was the germinal basis for the later
Athenian goddess of wisdom and warfare, Athene, who was symbolically associated with an aegis (shield) and with owls (the little
owl being very common in the vicinity of Athens). Alternatively, it
has been suggested that Athene was originally a pre-Hellenic rock-
goddess from Anatolia (now western Turkey). She became a symbolic mountain-mother for the Acropolis, and the owls living in its
rocky crevices naturally became associated with Athene as a living
manifestation of her presence (E. Armstrong, 1970). The poetic
epithet Pallas was later added to her name.

Regardless of her origins, in early Greek culture the goddess
Pallas Athene became closely associated with the owls of the
Acropolis, perhaps in part because of the nocturnal (and especially
the lunar) association of owls and the corresponding associations
between female fertility goddesses and the cycles of the moon. The
Greeks believed the owl to be a transmuted form of the daughter
of Nukteus ("She of the night") who, upon falling in love with her
father, was in danger of being put to death by him. However,
Athene took pity on Nukteus and changed her into an owl (*Noctua*),
which always fled from the daylight. Athene perhaps closely associated herself with owls because, like them, she could reputedly see
in the darkness. In her earlier Hellenic form Athene was considered largely as a goddess of storm and lightning (the term Pallas is
perhaps derived from a Greek word meaning "to strike"). Homer
described her as *glaucopsis,* or flashing-eyed, perhaps thus reflecting these associations with lightning. She was also believed to
delight in three fear-inspiring creatures: the dragon, the owl, and
the Athenian people themselves (de Gubernatis, 1872). However,

the aegis-protected Athene gradually became a warrior goddess of great power, who helped the Athenians win many important battles. When the Athenians won the battle of Marathon against the Persians in 490 B.C.E. they believed that Athene led them from overhead while assuming the form of an owl (Rowland, 1978). Agathokles of Syracuse reputedly used tethered owls to help defeat the Carthaginians in 310 B.C.E.; when released, the owls settled on his warriors' helmets and shields and thereby increased the confidence of his men. Owls eventually became so closely associated with Athene and Athens that the expression "taking owls to Athens" described a useless activity or gift, and "there goes an owl" was a way of predicting success or victory. However, even to the Greeks, owls sometimes foretold death, as one did to Pyrrhos of Epeiros by landing on his spear (E. Armstrong, 1970).

The Romans assimilated many beliefs from the Greek and Middle Eastern cultures, and owls became associated with their goddess of prophetic wisdom, Minerva. The Hindus had regarded Manus as the first man and the father of all men. He also became the first of the dead and as such was associated with the moon and the kingdom of the dead. Minerva eventually became conjoined with Manus in a goddess role similar to that of Athene and with the owl rather than the moon as her symbol (de Gubernatis, 1872). The prophetic qualities of owls in Rome became especially strong, particularly their associations with imminent death. Virgil stated that the hooting of an owl foretold the suicide of Dido; Pliny reported that great fear and confusion were brought about when an owl entered the forum; and Horace specifically associated owls with witchcraft. As a result, the Romans sometimes used owl representations to combat the evil eye, and the feathers or internal parts of owls became widely used as magical potions or as pharmaceutical components. For example, the ashes of an owl's feet were reported by Pliny to provide an antidote to serpent venom. During the same era an owl's heart placed on the breast of a sleeping woman would supposedly force her to disclose her most closely held secrets. The influence of Pliny and his Roman contemporaries as to European attitudes toward owls persisted for many centuries. The consumption of owl eyes or their cooked or powdered eggs, or even the burning of owl feathers, were widely thought to have medical or magical effects at least well into the Middle Ages. Until quite recently the nailing up of a dead owl or its wings has been widely believed in some parts of Europe to help ward off such dangers as pestilence, lightning, and hail. Similar pseudomedical applica-

tions and occultist views of owls are still widespread in India (Kumar, 1984).

In China the owl was regarded as a god-figure associated with thunder and lightning, and owl-like ornaments were often placed on the corners of roofs to help protect the building from fire. Similarly, the Ainu of northern Japan placed carved eagle owls on their houses to protect their families from famine or pestilence. The symbolic role of owls in China is remarkably similar to that of some western European countries. For example, the owl was the emblem of a royal clan of Chinese masters of the thunderbolt and of the regulators of seasons. Additionally, when a person was soon to die in a Chinese village, the voice of the owl, telling the residents to begin digging a grave, could always be heard. Some Orientals also believed that the owl carried away the dead person's soul (E. Armstrong, 1970).

Similarly, in Sicily it was generally believed that the "horned owl" (probably scops owl) would sing around the house of a sick person for three days before his death, and in such countries as Italy, Russia, Germany, and Hungary owls have continued to be regarded as representing deathly omens (de Gubernatis, 1872). Throughout most of Eurasia owls have long been believed to serve as familiars for witches. This is especially true of the barn owl, a highly nocturnal owl notable for its white ghostly appearance and its blood-curdling screams. In various cultures it has been called the ghost owl, demon owl, phantom owl (Germany), and owl of evil omen (France). Barn owls sometimes even glow in the dark, apparently because luminous bacteria and fungi become attached to their feathers in their roosting trees. The birds may then emit a strange nocturnal glow as they course silently over otherwise darkened fields and marshlands, a sight likely to inspire terror in even stout-hearted observers (Hoyo, Elliott, and Sargatal, 1999).

With these kinds of emotional associations, it should come as no surprise that the early Christian church seized upon the owl as a symbol of evil and of demonic possession. One commonly held early Christian view was that owls symbolized the Jews, who had rejected Christ. On Roman carvings of the early Christian era Jews were thus often represented as owls, typically shown as being tormented by doves or sparrows, which of course represented the righteous Christians. The owl was probably an especially convenient target of the early Church since it is associated with darkness, is notable for its haunting calls and nocturnal predatory powers, and was considered by the clergy as representing a seeker after vain knowledge but unable to perceive the Truth. In later medieval carvings and illustrations owls were often shown in association with apes, which were regarded as the worst of all beasts and were usually identified with the devil himself. The somewhat simian appearance of owls, especially the barn owl, probably only increased the strength of this association, for just as the devil cunningly trapped unwary human souls, the ape and owl sometimes lurked together in the shadow of the Tree of Knowledge, with the ape using the owl to attract and capture small and innocent birds (Rowland, 1978).

It is well known that a variety of birds will approach and harass owls, and in some parts of Europe songbirds have traditionally been lured to their deaths by attracting them to a tethered owl, the birds becoming trapped when they land on sticks that have been covered with birdlime and placed around the owl. A comparable well-known enmity between owls and crows occurs and can be traced in legend and folklore back through Aristotle's writings to the earliest Hindu manuscripts, such as the *Mahabharata*. This manuscript details the war between the dark night (symbolized by the crow) and the luminous moon (the owl), and the associated tendency for owls to kill crows at night, while during the daytime crows consistently harass and mob owls (de Gubernatis, 1872). To this day tethered owls (or artificial owl decoys) are regularly used as an effective means of attracting crows within the range of hunters' guns.

Owl Myths of the New World

Just as in the Old World, there has long been a strong association in North America between owls and death or the supernatural. Remarkable parallels exist between the beliefs of many North American Indian tribes and Oriental beliefs, which might suggest that the two cultures are linked by ancient oral traditions. For example, in many if not most tribes of North American Indians owls are closely associated with impending death, and an owl often serves as a soul-bearer or vehicle by which the spirit of the dead person is transported to a life beyond. The Kwakiutl Indians of the Pacific Northwest considered the owl to represent both the deceased individual and his newly freed soul. In the case of the Ojibway (Chippewa), the bridge over which the spirit of the dead had to pass was called the owl-bridge (J. Sparks and Soper, 1970). For the members of the Oto-Missouri tribe, the hooting of an owl provides a clear and undeniable message of death, a belief held even today among older members of the tribe, as revealed by the following transcript of a narrative by a tribal elder (Waters, 1983):

The owl is the one that gives the death warning. The owl that's got the horns they are the ones that warn you. You can hear them way in the distance and they give that kind of humming you hear. And it will be a while but you might get a bad message that means death. Hear them in the distance, it never fails, never fails, death is close. So, that's what they're here for, "Look out, look out, danger is coming."

And, in a variation of this story by another elder:

They say that [the screech-owl] was not wanted among the rest of the owls. And there are four classes of them. But the screech owl according to them is not welcome, they don't want him around. So he fought them, as small as he is, he fought them. And he told them that if he's put out of the group he will cause famine, famine will strike. And that he's going to be the one that is going to bring a bad omen to the world. And it seems that way. And they kind of go around in a flock and then again you will see just one. And whenever you hear a screech owl or the others there is going to be bad news, death is going to take place.

In a Pima song, the deathly fear caused by hearing an owl calling in the distance is suggested by the following freely translated song (F. Russell, 1904–1905):

There came a gray owl at sunset,
There came a gray owl at sunset
Hooting softly around me,
He brought terror to my heart.

In a variant translation a second smaller owl, perhaps the burrowing owl (which sometimes produces distinctive clicking calls), also appears as an agent of death (Holmgren, 1988):

Gray Owl came to me at sunset,
Came just as the last rays faded,
Sang to me of death's dark journey,
Sang as I froze in terror,
As I cried out for the sunlight
While I saw the last rays vanish,
While Small Gray Owl stood beside me
Clicking, clicking his gourd rattle.

Such terror is understandable, as the Pima believed that owls served as special guides, first alerting the person who was about to die then leading him into the dark and frightening world of the dead. Sometimes an owl feather was placed in the hand of a dying person, to help the owl decide who might be ready for the frightening journey. The Kwakiutl tribe believed that owls were the souls of people and that to kill one would result in the death of the person to whom the soul belonged. The Mojave Indians of Arizona believed that after death a person's soul was first transmogrified into an owl, then into a water beetle, and finally transformed into pure air (Weinstein, 1989). The Kiowas similarly believed that after a medicine man died, he was transformed into an owl. In a somewhat similar way, in Australian aboriginal folklore the souls of men were thought to pass into bats and those of women into owls (Freethy, 1992).

Because of the ability of owls to see in the dark, they brought with them unique powers that could at times be put to special use. For example, among the Cherokees the eyes of children were bathed in water containing the feathers of owls, in order that they might be able to remain awake all night. The Creek medicine men kept an owl skin among their sacred amulets and regarded the bird as symbolic of wisdom. Similarly, in the beautiful and symbolically rich Hako ritual of the Pawnees, part of the ceremonial pipe was decorated with owl feathers. This was done because in a vision an owl came to a holy man and instructed him as follows (Fletcher, 1900–1901):

> Put me upon the feathered stem, for I have power to help the Children. The night season is mine. I wake when others sleep. I can see in the darkness and discern coming danger. The human race must be able to care for its young during the night. The warrior must be alert and ready to protect his home against prowlers in the dark. I have the power to help the people so that they may not forget their young in sleep. I have power to help the people to be watchful against enemies while darkness is on the earth. I have power to help the people keep awake and perform these ceremonies in the night as well as the day.

In many tribes owls play an important role in their creation myths, when some of the owls' most typical attributes were obtained or during which owls helped to determine the alternation of day and night, sometimes after losing a contest. Thus the Cherokees explained the nocturnal associations of owls (Mooney, 1995):

> When the animals and plants were first made—we do not know by whom—they were told to watch and keep awake for seven nights, just as young men now fast and keep awake when they pray to their medicine. They tried to do this, and nearly all were awake through the first night, but the next night several dropped off to sleep, and the third night others were asleep, and then others, until on the seventh night, of all the animals only the owl, the panther and one or two more were still awake. To these were given the power to see and to go about in the dark, and to make prey of the birds and animals which must sleep at night.

The Menominee explained the alternation between daytime and nighttime in the following myth (Hoffman, 1892–1893):

> One time as Wabus [the rabbit] was traveling through the forest, he came to a clearing on the bank of a river, where he saw, perched on a twig, Totoba, the saw-whet owl. The light was obscure, and the rabbit could not see very well, so he said to the saw-whet owl, "Why do you want it so dark? I do not like it, so I will cause it to be daylight." Then the saw-whet owl said, "If you are powerful enough, so do. Let us try our powers, and whoever succeeds may decide as he prefers."
>
> Then the rabbit and the owl called together all the birds and the beasts to witness the contest, and when they had assembled the two informed them what was to occur. Some of the birds and beasts wanted the rabbit to succeed, that it might be light; others wished the saw-whet to win the contest, that it might remain dark.
>
> Then both the rabbit and the saw-whet began, the former repeating rapidly the words "wabon, wa-bon" [light, light], while the owl kept repeating "uni-tap-qkot, uni-tap-qkot" [dark, dark]. Should one of them make a mistake and repeat his opponent's word, the erring one would lose. Finally the owl accidentally repeated after the rabbit the word "wa-bon," when he lost and surrendered the contest. The rabbit then decided it should be light, but he granted that night should have a chance, for the benefit of the vanquished.

A similar genesis tale is told by the Jicarilla Apache. They thought that originally animals as well as people could talk, and all living things were below in the underworld, where it was dark. The humans and the animals that live by day wanted more light, but the nighttime animals such as the bears, the mountain lion, and the owl all wanted darkness. After an argument, they decided to settle the question by playing a thimble and button game (guessing if a thimble holds a hidden button), with the outcome to determine whether day or night would prevail. The magpie and quail, who have fine eyesight and love the light, watched the game intently and were able to see the button inside, through the thin wood of the stick that served as a thimble. They then told the humans where the button was, and so the humans won the first round. At this, the morning star appeared and the black bear ran away to hide in the darkness. They played a second time, and again the humans won. Now the sky grew brighter in the east, and the brown bear ran away to hide. They played a third time, and again the humans won. Now it grew still brighter, and the mountain lion crept away toward the remaining darkness. Finally they played a fourth time, and again the humans won. Now the sun actually appeared in the east, and the owl flew away to hide forever from the daylight (Erdoes and Ortiz, 1984).

According to the Oto-Missouri, children were sometimes magically transformed into owls against their will, the story perhaps being used as an effective means of convincing youngsters not to stray off into the woods while playing (Waters, 1983):

> They say that one time the owls were pretty much next to human beings. When the children were out playing they would tell them not to go into the timber, told them to stay away from it. But this little fella he got away from the rest of the children and he got lost. And the owls found him and they took him. And each one of them I don't know but ever so many of them, they each took some feathers off of themselves. And then they fixed some kind of a clay and they put it on this little boy. And they pasted him up and stuck all the feathers on him and then they taught him to sing a song. And the owls, they heard the people calling for the boy but he couldn't answer. So they just all scattered out looking for him. Finally once when it was still they heard the child crying and crying, and they stopped to listen to him. And he said, "I'm that little boy that got lost from the crowd and the owls found me. And I don't know how I'm ever going to get back to you people, but keep on searching. They even taught me how to sing the owl's songs."

In another transformation tale, the Montagnais Indians of Quebec reportedly called the boreal owl (probably actually the locally breeding northern saw-whet owl) "phillip-pie-tschch," or water-dripping owl, based on their belief that it was once the largest owl in the world and was very proud of its great voice. It even tried to imitate the noise of a waterfall and drown out its roar. Because of its inordinate pride, the Great Spirit humiliated the bird by transforming it into a tiny owl and changing its song to one that sounds something like slowly dripping water (Comeau, 1923).

The Pueblo Indians are great observers of nature, and these people have names for at least seven kinds of owls. The most important of these are the screech-owl, the great horned owl, and the burrowing owl. Although in the central Rio Grande pueblos owls tend to be related to witchcraft (probably because these villages have been most strongly affected by Spanish cultural influences), in the more western Hopi and Zuni pueblos owls are noted for both their good and evil effects, with the fertility motif perhaps uppermost (Tyler, 1979).

The Zunis call the burrowing owl the "priest of the prairie dogs" and knew these owls to live amicably with prairie dogs, rattlesnakes, and horned lizards. The prairie dogs are especially friendly toward the owls, considering them to be birds of great gravity and sanctity. Thus owls never disturb the councils or ceremonies of the prairie dogs, but instead remain at a respectful distance whenever their dances are occurring. Part of the prairie dogs' respect for burrowing owls comes from a time when the usually meager summer rains instead became floods, washing away all their favorite foods. The prairie dogs asked the owl for advice and to help them. Thereupon the owl captured a darkling beetle and forced it to disgorge its vile smell into a bag. The owl then began to strike the bag with a stick, releasing the terrible stench, and each time he did so the storm clouds scudded farther away, until the sky was perfectly clear. Then the prairie dogs all came out of their holes to give praise and thanks to their benefactor, the burrowing owl (Tyler, 1979).

The Hopi Indians identify the burrowing owl with their god of the dead, who is called Masau'u. Masau'u is also the deity of the night, just as owls are guardians of the darkness. However, this same god is guardian of fires and tends to all underground things. This role includes regulating the germination of seeds, thus overall the owl has a positive image (Tyler and Phillips, 1978). In the Chippewa tradition the hooting of the owl portended bad weather and a north wind, whereas the warm south wind was generated by a butterfly.

In the Hidatsa tradition of the northern plains (Dakota) Indians, "big owl" or "speckled owl" (probably the great horned or snowy owl) was a keeper-of-game spirit who among other things watched over and controlled the all-important buffalo, which the owl kept corralled for part of the year inside a great butte. Big owl's companion and assistant in such buffalo-herding activities was "little owl" (probably the burrowing owl). The burrowing owl was also a protective spirit for warriors, and at times when a Hidatsa warrior would sally forth to attack an enemy the owl would fly above him, letting the others of the tribe know that he was safe, because this bird was his personal protective god. Members of the dog society of the Hidatsa tribe always wore owl feathers, because of the protective value and guardian role of owls (Tyler and Phillips, 1978). How different is this view of owls from that of many other northern and Pacific Coast tribes, where to hear the owl call your name in the night is to know that you are being summoned to join your ancestors. Likewise among the Zapotec of southern Mexico it is believed that barn owls are charged with giving notice when a person is about to die and collecting his or her soul.

Perhaps we need a new, quasi-mythic view of

our North American owls—one that not only recognizes the birds for their obvious aesthetic beauty and traditional mystery but also takes into account the fact that they represent an evolutionary pinnacle that we can comprehend no better than we might imagine living in some other physical world or with a more perfect sensory awareness of our own world. Unless we adopt that view, and in so doing protect not only the owls but also their habitats, the birds that once warned individuals of their coming demise can only foretell our bleak collective human fate.

> *It was the owl that shrieked, the fatal bellman*
> *Which gives the stern'st goodnight.*
>
> —Shakespeare, *Macbeth*

PART TWO

Natural Histories of North American Owls

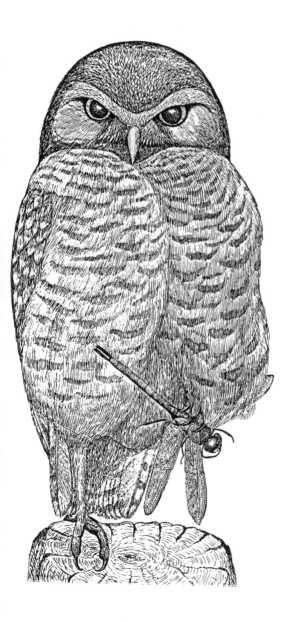

FAMILY TYTONIDAE

Only a single member of this family exists in North America, the widely distributed barn owl. The many distinct osteological features of *Tyto* have been recognized for more than a century (Beddard, 1888), but the characteristics and affinities of *Phodilus,* the other genus usually included in the family, have been much less well documented, and its taxonomic position remains somewhat equivocal. The osteological studies by Ford (1967) have remedied this situation. He found a predominance of characteristics aligning *Phodilus* with *Tyto,* as well as some similarities to the Strigidae. The latter include such things as the pattern of notching on the posterior edge of the sternum, relatively larger orbits for the eyes, and a relatively broader and flatter skull than is found in *Tyto.*

Interestingly, the Old World bay owls of the genus *Phodilus* lack dermal ear flaps, or opercula, but their ear openings (auditory meatuses) are asymmetrical in that the left one is situated higher on the head than the right one, as also occurs in *Tyto.* Additionally, the structures of the stapes, or columella, are very similar in both genera (Feduccia and Ferree, 1978). Finally, bay owls are similar to barn owls in their threat and defensive behavior (Wells, 1986). In all, it would seem that *Phodilus* probably belongs within the family Tytonidae, but it possesses several characteristics that tend to bridge the two owl families, and so deserves subfamilial separation. A general comparison of traits typical of the Tytonidae (based on the conditions found in *Tyto*) and of the Strigidae is shown in Table 11.

Beyond these traits, one might also mention that barn owls lack the characteristic hooting calls usually associated with owls, are highly nocturnal, and are rarely seen by people except when flushed from nests or daytime roosting sites. They are also distinctly tropical to subtropical in distribution, their ranges rarely extending more then 40 degrees north or south of the equator. There are never ear tufts present; the eyes are fairly small and distinctly oval in shape; and although a squarish operculum is present in front of it, the actual opening of the ear canal (the auditory meatus) is extremely small.

Table 11

Comparative Traits of Tytonidae and Strigidae

Trait	Tytonidae	Strigidae
Inner (2nd) toe	Same length as 3rd	Much shorter than 3rd
Middle (3rd) toe	Pectinated	Not pectinated
Legs	Relatively long	Relatively short (in most)
Facial disk	Heart-shaped	Rounded
Eyes	Relatively small	Relatively large
Interorbital septum	Thick	Thin or perforated
Auricular area	Smaller than eye	Enlarged (esp. Striginae)
Preaural flap	Present	In some (Striginae)

Continued on next page

Table 11 continued

Trait	Tytonidae	Strigidae
Postaural flap	Absent	In some (Striginae)
Skull and beak	Relatively long	Relatively short
Sternum edge	Not deeply notched	With 4 deep notches
Sternum	Lacking manubrium	Manubrium present
Furcula (wishbone)	Fused to sternum	Separate from sternum
Tarsal feathering	Directed upward	Directed downward
Natal plumage	Reduced, unpatterned	Often patterned
Secondaries (number)	15	11–18
Primaries (length)	10th longer than 8th	10th shorter than 8th
Primary emargination	Lacking on all	Variably developed
Primary molt	Bidirectional from 6th	Variable, often distally
Rectrices (number)	12	Usually 12, rarely 10
Longest rectrices	Outermost pair	Middle pair
Head-swaying threat	Present	Lacking
Sex defending nest	Male	Both

Barn Owl   *Tyto alba* (Scopoli) 1769

Other Vernacular Names:
American barn owl; golden owl; monkey-faced owl

North American Range (Adapted from AOU, 1983.)

Resident in North America from southwestern British Columbia, western Washington, Oregon, northern Utah, southern Wyoming, Nebraska, Iowa (rarely north to North Dakota and southern Minnesota), southern Wisconsin, southern Michigan, southern Ontario, New York, southern Vermont, and Massachusetts south through the United States and Middle and South America to Tierra del Fuego. Northernmost populations in North America are partially migratory, with some birds reaching southern Mexico. Wanders casually north to southern Alberta, southern Saskatchewan, southern Manitoba, northern Minnesota, southern Quebec, New Brunswick, Newfoundland, and Nova Scotia. Local in the West Indies (Cuba, Hispaniola). Other races occur widely in the Old World. (See Figure 22.)

North and Central American Subspecies (Adapted from AOU, 1957, and J. Peters, 1940, with some recent additions.)

T. a pratincola (Bonaparte). Occurs in North America as described above, south to eastern Guatemala and probably eastern Nicaragua.

T. a. lucayana (Riley). Resident in the Bahama Islands.

T. a. furcata (Temminck). Resident in Cuba.

T. a. niveicauda Parkes and Phillips (1978). Resident on the Isle of Pines.

T. a. glaucops (Kaup). Resident in Hispaniola and the Tortuga Islands.

T. a. guatemalae (Ridgway). Resident in western Guatemala, El Salvador, western Nicaragua, and Panama to the Canal Zone. Presumably also in mainland Honduras, although both the validity of *guatemalae* and its geographic range are still uncertain (Parkes and Phillips, 1978).

T. a. bondi Parkes and Phillips (1978). Resident on the Bay Islands, off the Caribbean coast of Honduras.

Measurements

Wing (of *pratincola*), males 314–346 mm (ave. of 18, 328.6), females 320–360 mm (ave. of 18, 336.9); tail, males 126–152.5 mm (ave. of 18, 138.1), females 127–157.5 mm (ave. of 18, 141.1)

(Ridgway, 1914). The eggs average 43.1 × 33 mm (Bent, 1938).

Weights

Earhart and Johnson (1970) reported the average weight of 16 males as 442.2 g (range 382–580), and that of 21 females as 490 g (range 299–580). Mikkola (1983) reported the average of 17 males and 55 females of the Eurasian population as 312 and 362 g, respectively. The estimated male:female mass ratio is 1:1.11–1.16. The estimated egg weight is 24.4 g, and the proportional egg-to-female mass ratio is 5.0 percent.

Identification

In the field. Though not often seen, barn owls are easily recognized by their nearly pure white underparts and the distinctive heart-shaped facial disk surrounding dark eyes. The typical call is a loud screaming *shrreeeee* uttered in flight, which is variably hissing, somewhat gargling-like, or tremulous, and usually drawn out to last about 2 seconds. There are also a wide variety of other calls, none of which resembles the hooting sounds usually attributed to owls. Highly nocturnal, and rarely observed during the day unless flushed from a roost or nest.

In the hand. This is the only North American owl with a heart-shaped facial disk, and the only one in which the claw of the middle toe is comblike. Older birds can be tentatively sexed by plumage color (females average darker, especially on their underparts, and have more large flecks on their outer primaries), bill color (females have dusky bill edging), and middle talon flange width (females average wider). Nestlings are entirely immaculate white in their first natal plumage; a second natal down follows 12–14 days after hatching that is longer, thicker, and buffish creamy. Following the loss of natal down (which may persist on lower underparts for several months) immatures are almost identical to adults in plumage (Mikkola, 1983). Juveniles have flight feathers and rectrices that are uniform in shape, wear and pattern, and tend to be narrower than in adults. The dark bars on these feathers are relatively large and have shorter distances between them than occur in older birds. The flange on the middle toe talon is narrow and smooth or only slightly serrated (Pyle, 1997b). The sequence of wing molt is useful in ageing barn owls for several years. The first wing molt typically replaces only primary

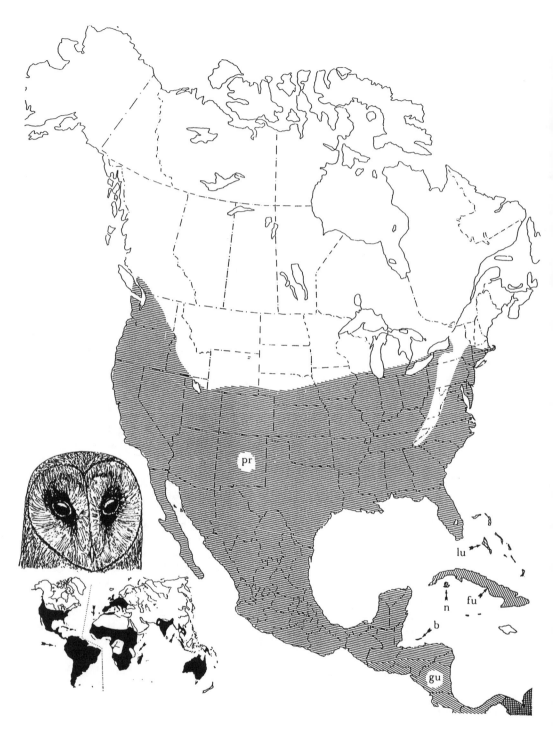

Figure 22. North American distribution of the barn owl, showing residential ranges of races *bondi* (b), *furcata* (fu), *guatemalae* (gu), *lucayana* (lu), *niveicauda* (n), and *pratincola* (pr). Indicated racial limits of *guatemalae* are only approximate. World distribution shown in inset, with arrows indicating isolated insular populations.

#6; the second one replaces 2–4 primaries either side of the sixth primary; and the third molt, the rest of the primaries. Subsequent molts, still centered on primary #6, replace the primaries every other year (Hoyo, Elliott, and Sargatal, 1999).

Vocalizations

Sound production in the barn owl is extremely diverse, as is to be expected in a highly nocturnal species; Bunn, Warburton, and Wilson (1982) described 15 distinct calls as well as tongue clicking and wing clapping modes of sound production. Seven of the vocalizations consist of screaming or screeching calls, of which the screech, often uttered in flight, is perhaps best known. This rather eerie and unpleasant vocalization functions as a "song" in that it serves to proclaim territory, attract unmated females, and sexually stimulate the pair. Both sexes utter it, the female's note generally being huskier. A series of mellow screeches by the male, or "purring," is used to attract his mate, and a similar wailing call, of lower pitch than the screech, is a probable female call. Warning screams are used as an alarm signal, and other screams are used as anxiety, distress, or mobbing signals.

Hissing sounds include sustained and brief defensive hisses, single hisses occurring during courtship or mate-recognition situations, and "snoring," a wheezing or almost whistling hiss that varies greatly but is persistently repeated. It is primarily a call uttered by females and young, mainly during the breeding season. It often is stimulated by hunger, but in females also is uttered during copulation. A variety of chirrups, twitters, squeaks, and similar brief notes occur when young or adults are quarreling or when otherwise excited, as when mates are greeting or one is being preened by the other. During copulation the male utters a more staccato, squeaking call, and a fast, chattering twitter is used during food presentation by adults.

Nonvocal tongue clicking (or bill snapping) often accompanies defensive hissing, but may occur during courtship or serve as an intimidation signal. Wing clapping, a single loud clap sometimes followed by a softer one, is produced during courtship as the male hovers in front of the female, apparently on the upstroke. Except perhaps for this sound, none of the signals is clearly sex-limited, and most intergrade with one another, making a total count of discrete calls essentially impossible. Further, none of the calls are typical owl-like hoots, although defensive hisses sometimes grade into hooting-like sounds.

Bühler and Epple (1980) performed a similar vocal analysis and estimated that the barn owl has a repertoire of 18 different calls. These fall into five functional signal categories. Territorial calls consist of screeches, purring notes, and screams. Defensive signals include hissing calls and bill-snapping (tongue-clicking) behavior. Begging and feeding calls include snoring and chittering notes. Social contact calls consist of a variety of twittering and snoring sounds, including copulation calls of both sexes. Finally, various calls of half-grown nestlings were impossible for the authors to assign any particular function. Their analysis did not suggest any obvious vocal homologies between barn owls and those of typical strigid owls.

Habitats and Ecology

The original habitats of barn owls may have been quite different from those now utilized; Bunn, Warburton, and Wilson (1982) believed that barn owls were probably originally cliff-haunting birds in Britain, where their light plumage coloration closely matched the chalky and limestone backgrounds. However, now the birds are largely associated with countrysides having an abundance of open fields and hedgerows for hunting, and with numerous old buildings (or large hollow trees) used for breeding sites. Generally, low-lying areas of arable land near coasts, which have a mild winter climate and abundant foods, and young forestry plantations with rich supplies of voles in the associated tall grasses, support large populations. Areas of severe cold weather, and with little vegetation, are shunned.

In California, Bloom (1979) reported that abundant populations in the Sacramento Valley are found where grasslands, riparian vegetation, marshes, and oak-sycamore woodlands persist, but have virtually disappeared where the valley has become intensively cultivated. In coastal southern California the birds are variably common but are declining in the face of increasing habitat loss; in most arid areas they are scarce, but more common where marshlands or pastures occur adjacent to arid lands. In favorable hunting habitats nesting densities appear to be limited only by available nest sites. Where prey and nesting sites allow, pairs can coexist with greatly overlapping home ranges and may defend very small territories of only up to about 10 meters in diameter around the nest itself. Thus, D. Smith and Frost (1974) observed a colony of barn owls in Utah (numbering 28–38 birds) that nested in an abandoned steel mill and hunted in the surrounding vicinity, up to 3 kilometers or more from the nesting or roosting site. Bunn, Warburton, and Wilson (1982) suggested that in favorable habitats an area of about 2.5 square kilometers is adequate to support an adult barn owl, even when rodents are at fairly low ebbs in their population cycle.

A study by Fast and Ambrose (1976), using a single owl, suggested that it had a prey preference for *Microtus* over *Peromyscus* (41 vs. 17 captures), and for hunting in open fields rather than woods-like habitats (44 vs. 13 successful hunting trips). Various studies (Knight and Jackman, 1984; Rudolph, 1978) suggest that a variety of factors, including behavioral differences such as timing and methods of hunting, and differences in size and identity of preferred prey, may help to reduce competition between the barn owl and the great horned owl, although they certainly exhibit substantial habitat overlap and at least local overlap in food-niche characteristics. Great horned owls may also be important predators of barn owls in some areas, thus affecting their local distribution and abundance. A causal analysis of more than 1100 barn owl deaths in Britain showed that predation there is a minor mortality factor, perhaps caused mainly by dogs and cats. Collisions or other accidents caused more than half the fatalities, while natural causes (especially starvation) accounted for nearly a third (Newton, Wylie, and Dale, 1997).

Movements

A considerable amount of information that has been collected on local and long-distance movements of barn owls in Britain and Europe has been summarized by Bunn, Warburton, and Wilson (1982). These movements can conveniently be classified as postfledging dispersal movements of young birds, later movements of older birds, and movements brought on by unusually cold winters.

Early postfledging movements, or those that occur up to about 3 months after the birds have been banded in the nest, suggest a progressive movement away from the nest, with most birds remaining within about 20 kilometers of the nest but a few attaining dispersals of 100 kilometers or more. Dutch and British banding returns for the first 12 months after banding suggest that about 2 percent (Britain) to 10 percent (Holland) of the birds had moved at least 100 kilometers during that period. Of those in the Dutch sample that traveled more than 300 kilometers, most moved in a southwestern direction (toward Spain) relative to the point of banding. Generally similar long-distance trends have been observed in German and North American (P. Stewart, 1952) studies.

In addition to such regular juvenile dispersal, in some years relatively massive barn owl move-

ments occur, which probably are linked to a combination of high local barn owl densities and falling rates of prey (small rodent availability). Most data suggest that birds engaging in these large-scale movements are primarily young. In a Dutch sample, about 6 percent of the birds banded as adults were subsequently recovered more than 300 kilometers away, but none of the more sedentary British sample were found more than 200 kilometers distant (Bunn, Warburton, and Wilson, 1982). In North America there are a number of cases of primarily northerly-nesting adults traveling southward during autumn for distances of more than 300 kilometers and rarely moving more than 900 kilometers (P. Stewart, 1952; Soucy, 1980, 1985). There are also a few cases of comparable long-distance movements in more southerly-nesting (Texas) birds (Bolen, 1978).

Estimates of home ranges by adults during the breeding season have been performed by radio-tracking in various areas. In three North American studies, males were estimated to have mean ranges of 953.3 ha (eight males), 308 ha (two males), and 418 ha (two males). The corresponding mean ranges for the same numbers of females were 752.1 ha, 294 ha, and 369 ha (I. Taylor, 1994). There was high individual variability in these estimates, but the mean ranges of females averaged consistently smaller than those of males. However, female home ranges tend to be similar to those of their mates as to size and shape. Home ranges during the breeding season appear to be defended little if at all, since those of adjacent pairs often overlap. However, the area immediately around the nest is strongly defended. Work by I. Taylor (1994) in Scotland indicated that home ranges there averaged about 3.2 km^2 during the breeding season, about equivalent to a 1-kilometer radius around the nest. In winter the birds ranged up to 5 kilometers from the nest. The highest breeding success occurred in areas where a substantial amount of woodland edge was present within a kilometer of the nest, and especially where grassy strips immediately bordered these woodland habitats and provided easy hunting opportunities.

Foods and Foraging Behavior

Over the barn owl's nearly worldwide range a vast number of studies of barn owl foods have been performed, using regurgitated pellet analysis (Bunn, Warburton, and Wilson, 1982). These studies collectively indicate that the species has no innate food preferences, but rather feeds on those animals that are small enough to be easily killed and are susceptible to predation by their ecologies, periodicities of activities, and the like, namely those occurring in open habitats during nighttime hours. These are mainly rodents, especially microtine rodents such as voles (Cricetidae, especially *Microtus* spp.), with shrews (Soricidae) most commonly serving as a secondary group of prey, the frequencies of these two prey categories often varying in a reciprocal fashion. In Britain the short-tailed vole (*M. agrestis*) and common shrew (*Sorex araneus*) are not only the two major prey species but also the two most abundant small mammal species in the open habitats that are favored for hunting. In some areas of Europe the Muridae, especially house mice (*Mus musculus*) and rats (*Rattus* spp.), are important components of the diet, especially where nesting occurs around human habitations. However, moles, rabbits, mustelids, and bats are generally rarely or only locally exploited in Europe, and birds seem to be taken when regular mammalian prey becomes scarce or where they can be easily captured, as for example sparrows (*Passer* spp.) or European starlings (*Sturnus vulgaris*) at colonial roost sites. Amphibians are usually taken in small numbers, and even fewer reptiles and fishes have been reported as prey. Invertebrates (insects, earthworms) comprise an essentially insignificant part of the species' diet (Bunn, Warburton, and Wilson, 1982; Mikkola, 1983). A variety of European studies suggest that mice (Murinae) and small voles (Microtinae) collectively comprise about 60–90 percent of the food intake on a percentage-live-weight basis, with shrews contributing 6–33 percent, and larger mammals about 1–33 percent. Birds usually represent less than 5 percent, but can reach as high as about 15 percent (Cramp, 1985).

In North America a large number of local studies of barn owl foods have been undertaken by pellet analysis (e.g., Marti, 1969; Fitch, 1947; Maser and Brodie, 1966). Marti (1992) has summarized this information for five separate regions, with total samples of more than 100,000 pellets analyzed. The birds tend to prey mainly on a single general food source. This was *Microtus* in three of the five areas listed, where it comprised 60.8–90.6 percent of total prey items. In another region (Louisiana) rice rats (*Sigmodon*) made up 50.2 percent of the prey total. Wallace (1948) found that barn owls in Michigan concentrated on a common species of vole (*Microtus pennsylvanicus*) and a large shrew (*Blarina brevicauda*). In several studies, deer or white-footed mice (*Peromyscus* spp.) are as important as voles in barn owl diets. Additionally many other rodents such as pocket gophers (*Thomomys* spp.), ground squirrels (*Citellus* spp.), pocket mice (*Perognathus* spp.), and kangaroo rats (*Dipodomys* spp.) are locally significant

prey. Bent (1938) stated that nearly every available species of mouse and rat is consumed, plus shrews, moles, and some larger mammals (rabbits, muskrats, skunks), as well as various birds, frogs, and a few insects. The average prey size of barn owls in four different studies ranged from 27 to 123 grams, or generally lighter than the prey of coexisting great horned owls (Knight and Jackman, 1984).

In a world summary of more than 50 barn owl food analyses tabulated by I. Taylor (1994), 15 North American studies were included. Worldwide, small mammals of up to 100 grams (usually up to 50–60 g) made up the majority of prey. In most studies rodents comprised more than three-fourths of the total prey items, and only a few species of prey comprised the majority of the diet. Among the North American studies, small mammals comprised 84.4 to 100 percent of all prey items found, and rodents alone comprised 66.6–100 percent of these items. Average prey weights ranged from 28 to 76 grams, with the majority of prey species averaging between 28 and 42 grams. The number of species comprising at least 80 percent of the total observed diet ranged from 1 to 8, with only 2 or 3 species typically represented in this percentage. By simple prey numbers, *Microtus* voles were the most frequent food item in eight studies. The *Microtus* species of North America averaged about 40 grams or about twice the mass typical of widespread *Peromyscus* and *Perognathus* species. *Perognathus* mice were most frequent in three and *Sigmodon* rats in two. The most important prey species, based on collective prey biomass, were *Microtus* voles in eight studies, *Sigmodon* rats (averaging about 75–100 g) in three studies, and *Thomomys* pocket gophers (averaging about 100 g) in two. In these study areas the total number of small mammal species present ranged from 7 to 18, suggesting that barn owls are quite selective in their prey choice, usually concentrating on the fatter and slower microtine voles rather than the more agile mice, which also typically have better hearing and eyesight. Further, the owls appear to become more generalists predators when prey densities are low and greater specialists on one or a few prey species when prey densities are high.

Hunting is typically done by extended flights over rather open terrain, the flights often beginning about dusk but in some situations before dusk, perhaps to take advantage of diurnal or crepuscular rodent activity and to allow for better visual searching. Foraging is done solitarily, the birds evidently consistently following favored routes, but probably not flying with the wind, in order to avoid flying too fast to locate prey. A low wing-loading allows for slow flight speeds without stalling, sudden turns, and an ability to carry off fairly heavy prey. The owls probably feed primarily during three general periods, the first at about dusk, the second around midnight, and the third around dawn (Bunn, Warburton, and Wilson, 1982). In some areas the presence of great horned owls, which are significant predators on barn owls, may restrict the activity periods of the latter to the hours of darkness (Rudolph, 1978). An average daily food intake of about 100–150 grams is probably typical for wild adult birds, although there may be substantial seasonal differences in this figure, and captive birds probably consume only about half this amount (Bunn, Warburton, and Wilson, 1982).

Hunting is done primarily by ear, at least at night, although vision may be used for avoiding large obstacles and general searching. This owl's hearing is among the best known for any animal. Hearing is especially precise at sound frequencies concentrated between 6 and 9 kHz, and for sounds presenting a wide range of frequencies. A high capability for differentiating sounds that differ only slightly in their component frequencies is also present, and this ability may be important in prey recognition. Very high frequencies tend to be filtered out by the stiff feathers of the facial ruffs, and these ruffs also evidently aid in judging the elevation of sound sources. The highest accuracy of sound-source localization is attained when sounds come from directly in front of and at nearly the same level as the bird; thus aerial hunting is usually done at low elevations (Konishi, 1983; I. Taylor, 1994). Apparently a hunting height of about three meters is close to ideal, whether hunting in flight or from a perch.

Social Behavior

At least in Britain, where the birds are fairly sedentary, barn owls often remain on their territories throughout the year, and at least some pairs remain together for extended periods. The birds are essentially monogamous, although at least one confirmed case of a wild male pairing bigamously with females and raising broods with both has been found in England. Some pairs remain at their nest site throughout the year, roosting together and performing mutual preening and other activities that probably help to maintain the pair bond. When one member of the pair succumbs, the remaining bird (especially the male) may remain at the nest site until it is joined by another mate, which sometimes occurs during the same breeding season. On the other hand, a series of barn owls may use the same nesting site every year for up to 30 years, and in some cases up to 70 years (Bunn, Warburton, and

Wilson, 1982). The longevity of wild barn owls, which only rather infrequently attain ages of 5 or more years, would suggest that persistent habitation of nest sites for many years is the result of successive pair usage. Unlike with typical owls, territorial (mainly nest site) defense is apparently performed only by the male. Additionally, mate choice by females is apparently not directly linked to courtship feeding, as copulatory behavior typically begins seasonally well prior to the start of courtship feeding (Epple, 1985).

True courtship begins in late February in England and is marked by screeching song flights of males as they patrol their territories and search for prey to present to their mates. Sometimes pairs may be seen in flight together, and sexual chases are frequent, with the male following the female while both birds scream loudly. One male display occurring during this time is the "moth flight," during which the male hovers in front of the female at her head level, exposing his white underparts. A second display is the "in-and-out flight," during which the male repeatedly flies in and out of the nest site, apparently trying to entice the female into it. This flight is marked by shallow, rapid wingbeats and repeated calling. A female responds to her mate by uttering juvenile-like snoring calls that stimulate the male to present food to her. This call is acoustically similar to the food-begging calls of nestlings. Copulation usually occurs at possible nest sites and often follows food presentation by the male (Figure 23). Treading typically is preceded by the female snoring quickly and softly. She then lowers her body, whereupon the male quickly mounts, maintaining balance with his wings and holding her nape feathers in his bill. Upon dismounting the male often begins to doze, while the female preens him, especially his head and underparts. Although copulation is probably most prevalent during the egg-laying period, it has been observed as late in the breeding cycle as when the oldest chick was 29 days old (Bunn, Warburton, and Wilson, 1982). Often copulations occur before the male leaves for his first foraging trip at night, and is repeated several times when he returns with prey or even without it (I. Taylor, 1994). Sometimes a second female may establish a nest site near an established pair; I. Taylor (1994) recorded bigamous matings that resulted in actual clutches among 7 of 419 nesting attempts in Scotland. In each case one of the females received much less food than the other and was less reproductively successful. In the Columbus, Ohio, *Post Dispatch* of May 17, 2001, a barn owl was reported to have fathered at least 13 chicks with three different females, and helped feed the young in all three nests.

Figure 23. Barn owl behavior, including courtship feeding (*A*) and copulatory sequence (*B–D*). After drawings in Glutz von Blotzheim and Bauer (1980).

Nesting sites in Britain are most frequently in barns, in holes of hollow trees, and in other holes in walls, towers, roofs, chimneys, and the like. Natural rock cavities, such as in cliffs, mines, quarries, and so on, are only infrequently used, and gullies or road cuts are evidently almost never used. However, in western North America these are common nest sites; and in areas such as the sandhills of Nebraska, where the substrate is soft enough, a good deal of actual excavation may be done by the birds, using the feet. The nests are rarely more than 10 meters aboveground in the case of tree nests, averaging about 5 m, and are usually in a cleft or cavity of the main trunk (Bunn, Warburton, and Wilson, 1982).

Breeding Biology

Information on clutch sizes in barn owls is presented in Tables 7 and 8, and in general, clutches are highly variable, ranging from 2 to 11 eggs. Clutch sizes in this species varies from year to year and place to place, probably in close correlation with food supplies. In various North American studies they have ranged from 4.9 to 7.2 eggs (Andrusiak and Cheng, 1997), and the fact that not only are replacement clutches common but second broods may be produced in favorable years makes this owl species one of the most flexible of all North American owls in its reproductive potential.

The eggs are laid at two- to three-day intervals, with incubation beginning with the laying of the first egg, as in all owls. Thus as much as a three-week difference in hatching times is possible between the youngest and oldest hatchlings. However, fledging success drops off sharply in nests with clutches of 5 or more eggs, and in general it appears to be closely associated with the relative abundance of prey during the chick-raising period. In southern Texas the hatching success of eggs averaged 54.9 percent (2.7 chicks per nest, average clutch of 4.9 eggs) over a seven-year period, while an average of 2.5 young per nest were raised in years of prey abundance, compared with 1.0 young per nest during years of prey scarcity (Otteni, Bolen, and Cottam, 1972). In six North American studies the average number of young fledged per nest ranged from 2.0 to 5.1 (Andrusiak and Cheng, 1997).

The young fledge at ages of about 56–62 days and soon begin to venture away from the nest site. As that occurs the adults begin to roost away from the nest, apparently to avoid the attentions of their young. Usually courtship begins again when the first brood is about seven weeks old. The female may even begin to lay a second clutch before the youngest owlets of the first brood have fledged. The eggs of the second clutch may be laid in the same nest as the first, sometimes while the last owlets are still present, though some hens choose new sites. The total length of a single breeding cycle is about four months, so that two broods per year are easily possible in areas with long summers. Schulz and Yasuda (1985) found that 56 percent of the birds using nest boxes in a California study had two clutches, the average observed clutch size being 6 eggs and the hatching success 72 percent. A few rare instances of three broods per year have been reported, these typically being associated with captive birds (Bunn, Warburton, and Wilson, 1982). One pair using a nest box in Illinois hatched five clutches and fledged young from four of these nestings during a period of 23 months (Walk, Esker, and Simpson, 1999). Even more remarkably, one captive barn owl trio at the Raptor Rehabilitation Center in Lincoln, Nebraska, produced five clutches of eggs in a 12-month period, four of which resulted in reared young, while the other attempt was aborted (because the eggs froze) before incubation began (Betsy Hancock, pers. comm.). This trio consisted of one male and two females (both of which participated in egg laying and parental care). Cooperative biandry in captivity has also been observed (Epple, 1985). When defending its nest, an adult barn owl not only lowers and spreads its wings in the usual manner of owls, but also hisses, shakes its head, and rhythmically moves the head up and down. It may also lunge at the intruder or even squirt feces at it. However, when picked up, it is likely to lie still and "play dead" rather than struggle.

It is clear that by virtue of its highly flexible clutch size, as well as its potentially prolonged breeding season, the barn owl is highly adapted to maximizing its annual productivity in favorable years or situations. Colvin and Hegdal (1985) reported that yearly differences in nest site use and annual productivity in New Jersey were related to relative grassland and *Microtus* availability, while Schulz and Yasuda (1985) correlated nesting success variations in California with the relative quality of the nest site.

In Scotland, 99.3 percent of 137 males, and 95.1 percent of 150 females remained on the same nest site from one year to the next, representing a high level of site fidelity that perhaps results in an improved knowledge of the local prey base over time (I. Taylor, 1994). Most movements to new sites occurred after the loss of a mate and typically were simply to a nearby nest site where the birds paired with others who had also lost their mates during the nonbreeding season. Fewer data of this type are available for North America, but the limited information suggests a similar behavior pattern may exist here. Young birds move well away from their nest site soon after fledging, with females tending to disperse farther than males. Perhaps because of this juvenile dispersal, few if any matings seem to occur between parent and offspring, and only rarely do siblings mate. Since first-year mortality rates average about 65–75 percent, only about a third of the fledged young reach sexual maturity (I. Taylor, 1994). Second-year mortality rates are slightly lower (40–60 percent), and third-year rates still somewhat lower (30–40 percent), so probably only about 10–12 percent of fledged birds survive to the end of their third year, and thus breed for at least two seasons.

In his world review of barn owl productivity, I. Taylor (1994) tabulated results from 11 studies, of which 4 were from North America. Worldwide, estimated annual production per pair ranged from 1.6 to about 7 fledged young, and in North America these ranged from 1.9 to 4.0. These American estimates were all based on annual single-brooding rather than double-brooding; the latter appears be widespread in those parts of the world where the seasonal duration and prey base permits it. In two of the American studies the percentages of young fledged relative to the average initial clutch sizes were 39 percent (88 clutches) and 43 percent (146 clutches). Maximum known longevity under natural conditions is 15 years and 5 months (Patuxent Wildlife Research Center banding data).

Evolutionary Relationships and Conservation Status

The evolutionary relationships between *Tyto* and *Phodilus* are discussed above (under "Family Tytonidae") and need no additional attention. The general population status of this species in North America appears to be unfavorable, particularly near the northern end of its range in the Midwest, where agricultural practices have had negative effects on it (Colvin, 1985). There is an apparent downward but not statistically significant national population trend for barn owls, based on annual Breeding Bird Surveys done between 1966 and 1993 (Price, Droege, and Price, 1995). A more recent and updated summary of these same data (http://www.mbr-pwrc.usgs.gov/bbs/bbs.html) suggests that a 1.7 percent annual population decline occurred from 1966 to 2000, which is also not statistically significant. This apparent if not yet proven North American decline is parallel to one seemingly occurring in Britain and Europe, which started in the 1930s or 1940s and accelerated in the 1960s and 1970s (I. Taylor, 1994). The U.S. federal authorities have not yet listed the barn owl as a species warranting special attention.

During the 1999 and 2000 Christmas Bird

Counts (CBC) the species was detected on an average of 273 counts, or about 15 percent of all U.S. and Canadian counts. In terms of localities where the species was detected, the barn owl was the fifth most widely reported of the 18 North American owl species then encountered. The average total number counted among the 100 counts having the highest species totals was 628 birds, or about 7 percent of all owls reported in the tabulated counts for the U.S. and Canada during those years. The maximum number ever reported on a single CBC was 109, at Sacramento, California, in 1981. The Canadian population has recently been estimated at 250–750 pairs (Kirk and Hyslop, 1998), and is considered stable or fluctuating but vulnerable. The barn owl is unusually vulnerable to cold winter temperatures, which has important implications for its conservation and survival in Canada (Andrusiak and Cheng, 1997). In Mexico this species has been reported from all 31 states plus the Federal District and from several adjoining islands (Enriquez-Rocha, Rangel-Salazar, and Holt, 1993). This species would thus appear to be the most widely distributed of all Mexican owls.

Family Strigidae

This family includes all the North American owls except the barn owl, the major distinctions of which are discussed above. Worldwide, there are about 190 species of Strigidae, as compared with fewer than 20 species of Tytonidae. The Strigidae have traditionally been separated into two subfamilies, the Buboninae and the Striginae, whose primary characteristics are listed in Table 12. Although this provides a convenient means of separating most (but not all) the strigid owls, recent evidence suggests that this is an artificial classification and that specialized hearing adaptations have evolved more than once in the family. As a result, Ford's (1967) tribal subdivision of the Strigidae (see Chapter 1) probably offers a more realistic organization of the typical owl family, and was adopted by Hoyo, Elliott, and Sargatal (1999) for their world survey of owls.

The Strigidae generally conform to the average person's conception of typical owls, having voices that are often low-pitched and hooting, especially in the larger species, large eyes that are more frontally oriented and appear more fully rounded (less oval) than is the case with barn owls, and sometimes ear tufts or "horns" that provide a distinctive "owlish" profile. Perhaps these ear tufts, together with contrasting eye colors, provide important social signals under dim light conditions. It is of interest that the owl species having both well-developed ear tufts and bright yellow eyes tend to be crepuscular species; owls with dark brown eyes are seemingly all highly nocturnal species, and with few exceptions these mostly nocturnal species tend to have poorly developed ear tufts.

Table 12

Comparative Traits of Buboninae and Striginae

Trait	Buboninae	Striginae
Size of ear opening	Less than one-half size of skull	At least one-half size of skull
Dermal ear flaps	Lacking	Present
Ligamentous bridge across ear conch	Lacking	Sometimes present
Facial disk	Centered below eye; well developed	Centered at eye; often poorly developed
Ear asymmetry (size or position)	Lacking or poorly developed (*Bubo*)	Often present; may include skull shape

Flammulated Owl   *Otus flammeolus* (Kaup) 1853

Other Vernacular Names:
flammulated screech-owl; flammulated scops owl (when considered conspecific with
O. scops).

Range (Adapted from AOU, 1983.)

Breeds locally from southern and southeastern
British Columbia, north-central Washington,
eastern Oregon, Idaho, western Montana, and
northern Colorado south to southern California,
southern Arizona, southern New Mexico, and
western Texas; also in Coahuila, Nuevo León, the
state of Mexico, and Veracruz. Winters from central Mexico (Jalisco) south to the highlands of
Guatemala and El Salvador, casually north to
southern California. (See Figure 24.)

North American Subspecies

None recognized here. Marshall (1967) regarded
idahoensis (Merriam) as a synonym. The supposed
race *rarus* from Guatemala was described on the
basis of a wintering specimen and thus is an invalid form (Marshall, 1968). Several other races
(*borealis* from British Columbia, *frontalis* from Colorado, and *meridionalis* from Mexico) have been
described by Hekstra (1982) but have not been
verified. He regards *rarus* as the form breeding
from western Washington to southern California.
McCallum (1994b) did not recognize any subspecies.

Measurements

Wing, male 128–138 mm (ave. of 12, 132.9), females 128.5–144 mm (ave. of 14, 135.2); tail,
males 58–63.5 mm (ave. of 12, 59.7), females 58–
67 mm (ave. of 14, 62.2) (Ridgway, 1914). The
eggs average 29.1 × 25.5 mm (Bent, 1938).

Weights

Earhart and Johnson (1970) reported the average
weight of 56 males as 53.9 g (range 45–63), and
that of 9 females as 57.2 g (range 51–63). The
estimated male:female mass ratio is 1:1.06.
N. Johnson and Russell (1962) listed 11 males as
averaging 55.9 g (range 48.8–66.1) and 2 females
as 60.3 and 78.2 g. Reynolds and Linkhart (1987b)
found that 9 males captured during the incubation period averaged 57.1 g, but after hatching
the average weight of 13 males was 53.4 g.
Weights of females varied greatly through the
nesting cycle. The estimated egg weight is 9.8 g.
The estimated proportional egg-to-female proportional mass ratio is about 17 percent, one of
the highest such ratios among North American
owls.

Identification

In the field. If visible, the owl's tiny size and the
dark brown eyes are distinctive and found in no
other small owl. The "ears" are short, compared
to those of screech-owls. The male's song is a
single- or double-noted and very low-pitched hoot
that is sometimes preceded by one or two even
lower-pitched notes, and usually persistently repeated at about 1–10 second intervals. As in
screech-owls, two plumage morphs are present, a
cinnamon-toned red morph and a gray morph
that is mostly wood brown. The existence of a distinct red morph has been questioned, with the
more reddish birds considered simply as individual plumage variants (Voous, 1988).

In the hand. This species is easily separated from
the other small "eared" owls by its brown eyes,
legs with densely feathered tarsi but the feathering terminating at about the base of the short (to
10 mm) bare toes, and its relatively long wings
(outermost primary longer than secondaries).
The wing is relatively pointed, and the pale shaft
streaks are broad and often tinged with rufous.
Adult females can be reliably sexed if a brood
patch is present, but measurements overlap
greatly. Nestlings are initially covered with snowy
white down, with pinkish gray bills and feet, and
dark blackish brown iris (Bent, 1938). The upperparts of juveniles are barred with grayish white,
or pale grayish, and dusky, but without any longitudinal streaks; the underparts are dull white or
grayish white, broadly barred with dusky gray or
grayish dusky. The remiges and rectrices of first-
year birds are similar to adults' but more uniform
in coloration and wear, the outer three primaries
gradually developing more tapered tips with wear,
and their coverts usually with small rounded spots
present (Pyle, 1997b).

Vocalizations

The primary advertisement call of the male is an
extremely low-pitched (ca. 440 Hz), mellow hoot
or *boo* note that is produced at regular (usually
about 2- to 4-second) intervals. These series are
often preceded by one or two preliminary even
lower-pitched "grace" notes (Marshall, 1968).
When an intruding male is detected on the territory, the call is a more rapid three-syllable *boop-
boop-boop*, with the last note accented, or at closer
range, a hoarse but quieter version of this same
call. A soft location call, *boop-boop-hoo*, is uttered

Figure 24. Distribution of the flammulated owl, showing breeding (cross-hatched) and migratory or wintering (stippled) ranges.

by the male when returning with food to his mate. A similar "mating" or courtship call is a two-syllable *boo-boot* call, the second strongly emphasized and usually dropping slightly in pitch at the end, and the calls usually about 1.5 seconds apart. According to Weyden (1975), this call has its highest pitch in the range of 450–500 Hz, making it one of the lowest pitched of all North American owl vocalizations. Although it has been suggested that duetting may be absent in this species, duetting does occur during pair contact calling and territorial singing (Frederick Gehlbach, pers. comm.). The female has a much higher-pitched quavering, whining call, which was described by Reynolds and Linkhart (1985) as sounding like raspy meows and used when soliciting food. Both sexes, when alarmed or apprehensive, produce harsh barking notes sounding something like the warning call of an elf owl. Screeches and shrieks may be uttered if a nesting bird is closely approached by humans or possible predators (Reynolds and Linkhart, 1998); screeching calls by flammulated owls are rarer than in typical screech-owls but do occur when under severe disturbance (Frederick Gehlbach, pers. comm.).

During courtship both sexes utter clucking or chittering noises. Paired males sing nightly during the pairing and incubation periods but less often during the brood-rearing period and thereafter. Unpaired males sing throughout the spring and summer (Reynolds and Linkhart, 1987b); such calls may be heard up to several hundred meters away under ideal conditions. When near the nest and feeding young, adults produce two-syllable mewing calls; as begging calls, young produce rasping or wheezing notes every 5–10 seconds, which can sometimes be heard for up to about 100 meters (Cannings et al., 1978; Cannings and Cannings, 1982).

Habitats and Ecology

In his survey of the distribution and habitat needs of this species in California, Winter (1979) reported that it is mainly limited to the higher parts of the coniferous forest zone of California, where besides ponderosa pine (*Pinus ponderosa*) and Jeffrey pine (*P. jeffreyi*) the principal forest species include sugar pine (*P. lambertiana*), Douglas fir (*Pseudotsuga menziesii*), white fir (*Abies concolor*), incense cedar (*Libocedrus decurrens*), and black oak (*Quercus kelloggii*). However, the owl's breeding range is most closely associated with ponderosa and Jeffrey pines, where the summers are warm (average maxima 80–93 degrees F) and dry (annual precipitation about 60–200 cm) and insect life is abundant. The belt of these pines ranges from 390 to 1800 meters altitude in the

north, 650–2100 meters in the central areas, and 800–2900 meters in the southern part of California. Above this pine belt the flammulated owl is less common, but the species has been taken up to above 3000 meters in the lodgepole pine (*P. contorta*) and red fir (*Abies magnifica*) belt, where limited breeding probably occurs. In northwestern California the birds are most often associated with black oak and ponderosa pine, especially where these occur in dense clumps of tall, mature trees (Marcot and Hill, 1980).

Studies in northeast Oregon (Goggans, 1985) indicate that the birds favor middle montane zone ponderosa pine–Douglas fir forests of rather open nature; of 20 nests, 74 percent were in such forests, and 76 percent were on sites having less than 50 percent canopy coverage, even though these sites, respectively, constituted only 50 percent and 22 percent of the study area. Similarly, of 37 roosts, 54 percent were in mixed coniferous forests, a habitat type covering more than 32 percent of the combined home ranges. Ponderosa pines were used for 67 percent of the roosting sites, although they constituted only 17 percent of the trees present. Of 352 radio-tracked locations, 77 percent were in ponderosa pines and 67 percent were forest-edge communities, which, respectively, comprised 22 percent and 26 percent of the home ranges. At the northern edge of its range in the Okanagan Valley of British Columbia the birds occupy dry, submontane stands of mixed-aged Douglas fir forests between 550 to 1200 meters elevation and having some old Douglas fir and ponderosa pines present. There the density of breeding birds is quite low, with about 0.4–0.7 calling males per 40 hectares (Howie and Ritcey, 1987).

In Colorado the species is largely associated with successional aspen (*Populus tremuloides*) communities and coniferous (especially mature ponderosa pine) forests occurring from about 1900 to at least 3200 m. Breeding birds are also limited to habitats having available natural or preexcavated nesting cavities, usually those made by northern flickers (*Colaptes auritus*) and equally large or even larger woodpeckers (Bailey and Niedrach, 1965; Webb, 1982b). Among records collected for the *Colorado Breeding Bird Atlas* (Kingery, 1998), ponderosa pine and aspen communities accounted for 40 and 28 percent of the records, respectively. Because of their rather soft wood, aspens are notable for their attractiveness as nesting sites for woodpeckers, and ponderosa pine forests tend to be open rather than densely spaced, allowing room for effectively catching insects. Returning males seem particularly prone to establish territories in old-growth stands of ponderosa pine mixed with Douglas fir, which evidently provides ideal foraging habitat characteristics (Linkhart, Reynolds, and Ryder, 1999).

In Arizona the species is not found in pure stands of ponderosa pine or in cut-over forests, but instead apparently requires the presence of some undergrowth or intermixture of oaks. It also occurs in aspen forests and at least locally is common in fir and spruce forests. Where the oaks or pines are large and dense at the lower edge of the so-called transition (warm-temperate forest) zone, it enters the "upper Sonoran" (hot desert) zone, where it comes into local contact with the elf owl (Phillips, Marshall, and Monson, 1964). In Arizona the birds are closely associated with ponderosa pine and mixed conifer forests (Reynolds and Linkhart, 1998). Old-growth forests are preferred over younger stands, probably because of the larger numbers of dead and dying trees, but some use of second-growth and open, logged stands has also been reported.

In northern Utah, where ponderosa pine does not grow, flammulated owls nest in montane deciduous forests dominated by quaking aspen (*Populus tremuloides*) (Marti, 1997b). They have also been found nesting in mixed deciduous and fir forests in Idaho (Powers et al., 1996). A marginal population in New Mexico breeds in a transition zone between ponderosa pine forest and piñon-juniper woodland, and at lower elevations (under 2300 m) than occur elsewhere in the species' range. (McCallum, Gehlbach, and Webb, 1995).

Density estimates are not numerous, but Marshall's (1939) estimates of 24 males in an area of about two square miles suggest an approximate density of about 5.7 males per square kilometer. Winter (1979) found a density of about 5.3 males per square kilometer in similar California habitat. He suggested that the species may be described as loosely colonial, congregating in small, rather dense and discrete breeding populations, but with other areas of seemingly optimum habitat having no birds present. Reynolds and Linkhart (1985, 1987b) reported that 4–6 nesting attempts occurred during each of five years on a 452-hectare area in Colorado, representing an average density of 0.8–1.3 nests per square kilometer. The same area had 2–3 additional territorial but apparently nonbreeding males present each year. McCallum, Gehlbach, and Webb (1995) reported a higher breeding density of 2.9 nests per square kilometer in New Mexico.

Movements

This is generally believed to be one of the most migratory of all North American owls, although remarkably little is known of the details of this

migration (Phillips, 1942). Other than a January record from the San Bernardino Mountains in 1885, only two additional substantiated winter records are known for the United States (Winter, 1979). N. Johnson (1963) suggested that the data, while not disproving a partial or complete migration, might also be interpreted to mean that the species is a permanent resident on or near breeding areas in the western United States and Mexico. However, this possibility has not been supported by later observers (Banks, 1964), and Balda, McKnight, and Johnson (1975) concluded that the species is indeed migratory, at least in the northern parts of its range. Based mainly on data from Arizona and New Mexico, there is apparently a relatively rapid movement northward in spring and a more leisurely fall passage southward. Their spring migrations appear to occur at lower elevations than the fall movements, which is probably related to relative arthropod abundance (especially large nocturnal insects) at these various altitudes during the two time periods. Marshall (1968) judged that wintering might be concentrated in the mountains peripheral to the southern Mexican Plateau. Birds breeding in southern Mexico are likely to be resident, whereas those in northern Mexico are probably migratory.

Local movements of flammulated owls were studied in Colorado by Linkhart (1984) and Linkhart, Reynolds, and Ryder (1999), who found that the home ranges of seven nesting pairs varied from 8.5 to 24 hectares, averaging 14.1 hectares. The sizes of the home ranges appeared to be determined by the degree of patchiness of the overstory trees and the age of the overstory, whereas range shape appeared to be determined by topography and the home ranges of neighboring conspecific birds, which had little if any apparent overlap and thus resembled territories. Territorial song posts of males were mostly associated with mature, open stands of mixed ponderosa pine and Douglas fir. Within each home range were up to four intensive foraging areas, which averaged 0.5 hectares individually and 1.0 collectively. The centers of these foraging areas were usually less than 140 meters from the nest, and in most cases (six of seven) the nest site was within an intensive foraging area. The mean territory (home range) diameter was estimated at 424 meters (Reynolds and Linkhart, 1987a).

Foods and Foraging Behavior

Marshall (1957) examined stomachs of 27 flammulated owls from pine-oak woodland habitats and found that all the prey were arthropods. There were primarily medium-sized insects of types likely to be caught in the air or among foliage, with some also probably taken from the ground or large branches. In diminishing frequency of occurrence they included beetles, moths, caterpillars, crickets, other insects, centipedes, spiders, scorpions, and other arachnids. In earlier papers he reported on smaller samples from California and Oregon, which also had such nocturnally active insects as moths and nocturnal crickets in preponderance.

Ross (1969) reviewed the available information on the foods of this species, which is clearly almost exclusively insectivorous. He examined the stomachs of 46 specimens from various parts of the range, which had clearly selected moths (both larvae and adults), beetles, and grasshoppers, plus various other primarily flying insects and some noninsect arthropods such as spiders, scorpions, centipedes, and millipedes. The lengths of the prey ranged from 6 to 55 millimeters, but most were at least 15 millimeters long.

Goggans (1985) reported that the diets of birds studied in Oregon included eight arthropod orders, but 72 percent of the total were orthopterans, nearly all of which were associated with grasslands. In another Oregon nesting study, one pellet containing the remains of a *Clethrionomys* vole was found, and the feathers of a dark-eyed junco (*Junco hyemalis*) were found at one nest, providing virtually the only evidence of vertebrate consumption by this species.

McCallum (1994b) summarized the available information on this species' foods, based on five published studies. Among insect groups, orthopterans (grasshoppers and relatives), lepidopterans (moths and butterflies) and coleopterans (beetles) were present in all five; and collectively comprised from 75 to 100 percent of the food items encountered. Other insect groups were present in very small quantities. Arachnids were present in four of the analyses; centipedes were also present in four and millipedes in three, but all these groups were of minor quantitative significance. Several prey-catching methods are used, especially hawk-gleaning (landing on a tree trunk to glean resting insects), hover-gleaning (catching insects from surfaces while hovering), drop-pouncing, and least frequently, hawking (catching insects in flight) (Reynolds and Linkhart, 1987b, 1998).

Flammulated owls are distinctly nocturnal in their foraging behavior, although Marshall (1957) thought that the birds forage mostly at dusk and dawn and are less active at night. Linkhart and Reynolds (1985) reported that the frequency with which food was delivered by males to the nest was highest immediately after darkness had fallen. However, foraging continues periodically through-

out the night at reduced intensity, according to Linkhart (1984). He found that most insects were captured by gleaning among the needles of conifer crowns or tree trunks, with flying insects hawked occasionally or captured on the ground during short flights from tree crowns. When adults deliver food to their nests or fledged young, they do so one prey at a time, the rate of feeding increasing from about 3.5 trips per hour during incubation to 9.8 trips per hour during the nestling stage of brood-rearing (Reynolds and Linkhart, 1987b).

Social Behavior

Although little is known of the details of courtship in this elusive and nocturnally active species, some knowledge of its general social patterns is starting to emerge. Thus Reynolds and Linkhart (1985, 1987a) established, by marking and studying for five years a population of birds in central Colorado, that males arrive on their nesting areas first and always reoccupy their previous year's territories. Of 22 banded birds, 59 percent returned to nest on the study area the following year. Returning females arrived later and settled into their previous territories as well, provided that those territories were already occupied by males. If not, they moved into the territories of neighboring and unpaired males. Of 12 pairs, 10 nested together for only one year; one remained together two years, and one for three years. One female nested in three different but adjoining territories during the study, and there was an average dispersal distance of 474 meters of five females from their previous year's territory. Evidently there was an initial competition for territories among males and later competition among females for obtaining established territorial males as mates. Marshall (1939) observed distinct territorial behavior among males of this species, judging that the territories were relatively small, usually under 300 yards (275 m) in diameter. He also (1957) reported that in one year there were 18 territorial males in a distance of about 3.6 kilometers surveyed between Sunnyside and the head of Sylvania Canyon of the Huachuca Mountains, Arizona, or one territory about every 200 meters of linear distance. Flammulated owls in New Mexico were found by Arsenault (1999) to nest in aggregations, with the nests separated by as little as 150 meters but with no evidence of extra-pair fertilizations, judging from DNA fingerprinting of 17 broods.

Linkhart (1984) observed that males sing territorially from throughout their home ranges, suggesting that territorial and home range limits are essentially identical. When singing, the males usually stood against the tree trunk but sometimes sang from the lower crowns of trees. Singing was most frequent during incubation (June in central Colorado), but males also sang after hatching, usually later in the night after the broods had been fed. Territorial encounters at the edges of territories were common in Linkhart's study and sometimes involved actual fighting. Linkhart and Reynolds (1997) found that over a 15-year period 14 territories were typically present within a study area of 452 hectares. From 3 to 6 of these territories were occupied by breeding pairs, and 3 to 7 by unpaired males. Unpaired males occupied territories an average of 3.9 years, and breeding pairs an average of 5.1 years, in both cases not necessarily consecutively. Most territories remained fairly fixed in area over the entire study period. Those territories most consistently occupied by breeding pairs had the highest percentage of old-growth (200–400 years) pine-fir forests. These same territories also produced the greatest number of owlets.

Copulation occurs away from the nest, with the male approaching silently or while uttering faint hoots. The female utters food-solicitation notes and is fed by the male prior to copulation. Mutual preening may follow copulation, and there is some evidence that extra-pair copulations not only occur but also may be part of the male's mating strategy (Reynolds and Linkhart, 1990a; McCallum, 1994b). Remating with the previous year's mate is typical if both members return to the same territory; males whose mates do not return tend to remain on their previous territory, whereas females whose mates fail to appear move to adjacent territories. No case of "divorce" from a previous mate has been detected. One banded female lived for at least seven years, and one male survived at least eight years (Reynolds and Linkhart, 1998). The annual rate of return of adults to prior territories was 8/17 for males and 10/19 for females (Reynolds and Linkhart, 1987a), suggestive of an annual survival rate of at least 50 percent, assuming some surviving birds move to new areas. Known maximum longevities noted above also suggest a fairly high survival rate. Typically, although adults exhibit territorial fidelity in subsequent years, they only infrequently use the same nest site repeatedly (McCallum, 1994b). Some representative postures of adult flammulated owls are shown in Figure 25.

Breeding Biology

Relatively few egg records are available for this species. Bent (1938) lists 11 Colorado records as extending from June 2 to 27, but a larger sample of 27 clutches (including Bent's, and mostly from the Western Foundation of Vertebrate Zoology) runs from May 17 to July 7, with 14 between June 5 and 20. The mean date of clutch completion for 14 nests in a Colorado study area was June 7 (range May 29–June 14) (Reynolds and Linkhart,

Figure 25. Comparison of (*A*) concealment posture of whiskered screech-owl and (*B*) mild alarm or curious posture of flammulated owl. Also shown (*C*) and (*D*) are aggressive-defensive postures of adult whiskered screech-owl (*C*) and flammulated owl (*D*) near nest. After photos by Art Wolfe.

1987b). Eight Arizona and New Mexico records (including 5 cited by Bent, 1938) are from April 18 to June 11. Three Idaho and Utah egg records are from April 25 to July 1. Bull and Anderson (1978) noted that incubated eggs have been found in Oregon from June 8 to July 3, and nestlings observed to August 2. The eggs in one British Columbia nest were probably laid about July 1, with hatching about July 24–26 (Cannings and Cannings, 1982), but a newly fledged young has also been seen in mid-July (Cannings et al., 1978). Mexican nest records are few, but in northern Mexico eggs may be incubated as early as April, with some hatching occurring by May (McCallum, 1994).

Clutch-size data for this species are quite limited, but Reynolds and Linkhart (1987b) reported that 11 Colorado clutches ranged from 2 to 3 eggs, with an average of 2.7. Other clutch samples from New Mexico, Oregon, and British Columbia have means of 2.28 to 2.7 eggs (McCallum, 1994b). Of 23 clutches reported from throughout the species' range, the mean was 3.0 eggs and the range 2–4 eggs (various sources, mostly from Western Foundation of Vertebrate Zoology, and excluding some misidentified clutches). The modal clutch size is of 3 eggs; only a few verified 4-egg clutches are known. It has been reported that females mated to inexperienced males typically lay only 2 eggs, but those with experienced mates lay 3 (McCallum, 1994b). Female age may also be a factor operating here.

Of 28 nest records known to me, 18 were in aspens, 6 were in ponderosa pines, and the rest in other trees including unspecified pines. Of 30 nests, 12 were in woodpecker holes, 12 were in deciduous tree cavities (some of which were probably also woodpecker holes), and 5 were in conifer snags or cavities. Of 27 nests, the heights of the openings were from 2.4 to 12 meters aboveground, averaging 6.1 m. All of 4 nests described by Bull and Anderson (1978) were in ponderosa pines, 3 of them dead, and all were old woodpecker holes, including those of northern flickers and pileated woodpeckers (*Dryocopus pileatus*). All the nest cavities found by Reynolds and Linkhart (1984) had minimum entrance diameters of 4–10 centimeters, and were from 18 to 40 centimeters deep. A study of 33 nest sites in Oregon by Bull, Wright, and Henjum (1990) found that there the birds prefer large-diameter dead trees, with openings at least as large as those of northern flickers and especially pileated woodpeckers, and those located in conifer stands on ridges and upper slopes having eastern or southern slopes. The average cavity height was 12 m, and the mean tree diameter at breast height was 72 cm. A sample of 17 sites from New Mexico (McCallum and Gehl-

bach, 1988) were in trees of lesser mean diameter (46.1 cm), lower mean cavity heights (4.9 m), and had mean cavity entrance diameters of 5.9 cm. These authors found that the birds preferred sites having low shrub densities, high canopy heights, and mature stands of Colorado piñon pine (*Pinus edulis*). In New Mexico, the birds favored the holes in ponderosa pines made by northern flickers, whereas western bluebirds most often chose to nest in the somewhat smaller cavities of acorn woodpeckers (*Melanerpes formicivorous*), most often located in oaks (Arsenault, 1999).

Incubation requires 22–24 days, and apparently begins after the second egg is laid. Probably 2 days elapse between the laying of at least the later eggs of a clutch. The females do all the incubating and brooding, while males are the sole providers of food until late in the nestling period. During this time the males mainly capture small moths (Noctuidae), supplemented by other insects, spiders, and other arthropods. A total of 26 nesting attempts produced a mean of 2.4 young per brood. There was no known mortality during the prefledging period among nests observed, but some young were taken shortly after fledging, probably by a sharp-shinned hawk (*Accipiter striatus*) (Reynolds and Linkhart, 1987b). Birds known to be nesting for the first time together had a slightly smaller average brood size (0.25 fewer young) than did pairs having one member known to have nested previously (Reynolds and Linkhart, 1987a).

By 10 days after hatching the soft gray and horizontally banded juvenile plumage was nearly completed, and the remiges were about three-fifths grown. The young fledged 22–25 nights after hatching in Reynolds and Linkhart's Colorado study area, and similar fledging periods of 21–23 days have been reported elsewhere (Cannings and Cannings, 1982). At the time of fledging (late July in Colorado) the five observed broods were divided within 3 nights of fledging (the first such report of brood division in owls), with one subgroup being attended by each pair member. These subgroups dispersed in different directions and apparently had no subsequent contact. After 34–40 nights following fledging the young were no longer provided with food, and at that time (late August) the young began to disperse from natal areas. Adults left the study area by mid-October (Linkhart and Reynolds, 1985).

Hatching success rates (at least one egg hatching per clutch) have been reported as 96.5 percent in 11 New Mexico clutches and 89 percent in 26 Colorado clutches. Estimates of average numbers of fledglings produced for all documented nesting efforts in various locations range from 1.3 to 2.3 young per nesting effort, and a mean of 2.7 fledglings produced among 10 successful nests (McCallum 1994b). Similar figures were presented by McCallum, Gehlbach, and Webb (1995). Summarizing his own and several other studies, Marti (1997b) estimated that mean clutch sizes have ranged from 2.3 to 2.7 eggs, mean number of nestlings in hatched nests from 2.2 to 2.4 chicks, and a mean of 1.8–2.7 fledglings in successful nests. Based on a mean clutch of 2.6 eggs in 27 total nests, and a mean of 2.1 fledglings in 22 successful nests, a breeding success rate of 66 percent would seem typical of Marti's study, a surprisingly high rate of reproductive success for such a tiny owl. Predation by small mammals such as squirrels may cause most nesting losses. Maximum known longevity under natural conditions is 7 years and 11 months (Patuxent Wildlife Research Center banding data).

Evolutionary Relationships and Conservation Status

Marshall (1967) suggested that this species be included in the superspecies *Otus scops,* all four members of which have small body size, tiny feet with naked toes, and some rufous coloration, at least on the wing coverts. The flammulated owl was considered by him to constitute a distinct species within this group, while the three other Old World forms (*scops, sunia,* and *senegalensis*) were considered by him to be conspecific with one another. Weyden (1975) agreed with Marshall that the vocal affinities of the flammulated owl are with the Old World species of "scops owls," rather than with the New World screech-owls, and Hekstra (1982) has provided additional support for this general position. More recent owl taxonomies have generally followed this same arrangement, although DNA data from the cytochrome b gene would place it as a distinctly peripheral member of the New World *Otus* group (Wink and Heidrich, 1999).

The status of this highly elusive species is difficult or impossible to gauge, but probably it is more common over much of its range than is generally appreciated. Thus the first nesting record for Montana was obtained near Missoula in 1986 (Holt and Hillis, 1987), and only six occurrence records for British Columbia existed prior to 1980, although it is now known to be moderately common in a few areas of the Okanagan Valley and the Rocky Mountain trench (Howie and Ritcey, 1987). However, the U.S. Forest Service has classified the flammulated owl as a sensitive species in several of its management regions, and it has been listed as a vulnerable species in Canada. The Canadian population has recently been estimated at 1500 pairs (Kirk and Hyslop, 1998) and is considered vulnerable but possibly stable. The species has not been detected in recent Christmas Bird Counts.

Plate 1. Common Barn Owl. Watercolor by Louis Agassiz Fuertes. Courtesy of the Cornell Lab of Ornithology, Ithaca, N.Y.

Plate 2. Eastern Screech-Owl (red and gray morphs). Watercolor by Louis Agassiz Fuertes. Courtesy of the Cornell Lab of Ornithology, Ithaca, N.Y.

Plate 3. Great Horned Owl. Watercolor by Louis Agassiz Fuertes. Courtesy of the Cornell Lab of Ornithology, Ithaca, N.Y.

Plate 4. Snowy Owl. Watercolor by Louis Agassiz Fuertes. Courtesy of the Cornell Lab of Ornithology, Ithaca, N.Y.

Plate 5. Burrowing Owl. Watercolor by Louis Agassiz Fuertes. Courtesy of the Cornell Lab of
Ornithology, Ithaca, N.Y.

Plate 6. Barred Owl. Watercolor by Louis Agassiz Fuertes. Courtesy of the Cornell Lab of Ornithology, Ithaca, N.Y.

Plate 7. Long-eared Owl. Watercolor by Louis Agassiz Fuertes. Courtesy of the Cornell Lab of Ornithology, Ithaca, N.Y.

Plate 8. Short-eared Owl. Watercolor by Louis Agassiz Fuertes. Courtesy of the Cornell Lab of Ornithology, Ithaca, N.Y.

Plate 9. Boreal Owl. Watercolor by Louis Agassiz Fuertes. Courtesy of the Cornell Lab of
Ornithology, Ithaca, N.Y.

Plate 10. Northern Saw-whet Owl. Watercolor by Louis Agassiz Fuertes. Courtesy of the
Cornell Lab of Ornithology, Ithaca, N.Y.

Plate 12. Eastern Screech-Owl (gray morph). Photo by author.

Plate 11 (opposite). Flammulated Owl. Alan G. Nelson Nature Photography.

Plate 14. Whiskered Screech-Owl. Acrylic by Mark E. Marcuson.

Plate 13 (opposite). Western Screech-Owl (gray morph). Photo by Kenneth W. Fink.

Plate 15. Western Screech-Owl (gray morph). Alan G. Nelson Nature Photography.

Plate 16. Western Screech-Owl (gray morph). Photo by Kenneth W. Fink.

Plate 17 (right). Whiskered Screech-Owl. Photo ©
D. A. Rintoul.

Plate 18 (below). Great Horned Owl. Photo by author.

Plate 19. Northern Hawk-Owl. Photo by Heidi Hughes Valega.

Plate 20 (opposite). Northern Pygmy-Owl. Photo by William C. Shuster.

Plate 21 (right). Ferruginous Pygmy-Owls. Photo by author.

Plate 22 (below). Elf Owl. Photo by Kenneth W. Fink.

Plate 23. Northern Pygmy-Owl. Alan G. Nelson Nature Photography.

Plate 24. Spotted Owl. Photo by Kenneth W. Fink.

Plate 25. Barred Owl. Photo by author.

Plate 26. Great Gray Owl. Photo by author.

Plate 27. Long-eared Owl. Photo by author.

Plate 28. Short-eared Owl. Photo by author.

Plate 30. Northern Saw-whet Owl. Photo by author.

Plate 29 (opposite). Great Gray Owl. Photo by Thomas D. Mangelsen, Inc./Images of Nature.

Plate 32. Boreal Owl (juvenile). Photo by Pat and Greg Hayward.

Plate 31. Boreal Owl. Photo by Pat and Greg Hayward.

Plate 33. Northern Saw-whet Owl (pair). Alan G. Nelson Nature Photography.

Plate 34. Vermiculated Screech-Owl (*guatemalae*). Photo by José Luis Rangel and Paula Enríquez.

Plate 36. Spectacled Owl. Photo by Kenneth W. Fink.

Plate 35. Crested Owl. Photo by José Luis Rangel.

Plate 37. Spectacled Owl (juvenile).
Photo by Kenneth W. Fink.

Plate 38. Mottled Owl (adult). Photo
by Rick Gerhardt.

Plate 39. Black-and-White Owl (adult). Photo by Mark Kasprzyk.

Plate 40. Black-and-White Owl (adult). Photo by José Luis Rangel.

Plate 41. Stygian Owl. Photo © Jim Culbertson.

Plate 42. Striped Owl. Photo by Kenneth W. Fink.

Eastern Screech-Owl *Otus asio* (Linnaeus) 1758

Other Vernacular Names:
common screech-owl; Florida screech-owl (*floridanus*); Hasbrouck's screech-owl
(*hasbroucki*); little owl; Rocky Mountain screech-owl (*maxwelliae*); shivering owl; southern
screech-owl (*asio*); Texas screech-owl (*mccallii*).

Range (Adapted from AOU, 1983.)

Resident from southern Saskatchewan, southern Manitoba, northern Minnesota, northern Michigan, southern Ontario, southwestern Quebec, and Maine south through eastern Montana, eastern Wyoming, eastern Colorado, western Kansas, western Oklahoma and west-central Texas, eastern San Luis Potosí, southern Texas, the Gulf coast, and southern Florida. Recorded in summer and probably breeding in southern and central Alberta (north to the Swan Hills). In local sympatric contact with the western screech-owl in northern Mexico, the Big Bend and Edwards Plateau area of Texas, and in southeastern Colorado. Hybridization occurs locally along some of these zones of sympatry, which appear to be of recent origin (Gehlbach, in press). (See Figure 26.)

Subspecies (As recognized by Marshall [1967] and Gehlbach [1995b]. Gehlbach [in press] has redefined several of the races.)

O. a. maxwelliae (Ridgway). Southeastern Saskatchewan, southern Manitoba, eastern Montana, and the Dakotas south to central and southeastern Wyoming (Big Horn Mountains and Snowy Range), western Nebraska, western Kansas, and eastern Colorado. Possibly also breeds in central Alberta. Includes *swenki* Oberholser.

O. a. hasbroucki (Ridgway). Central Kansas to Oklahoma and Texas.

O. a. mccallii (Cassin). Lower Rio Grande to the southern border of Tamaulipas and southeastern San Luis Potosí.

O. a. asio (Linnaeus). Minnesota, peninsular Michigan, southern Quebec, and southern Maine south to Missouri and northern parts of Mississippi, Alabama, and Georgia. Here includes *naevius* Gmelin, but Gehlbach (in press) has reinstated this form to include region shown north of dashed line on the map.

O. a. floridanus (Ridgway). Florida and the Gulf Coast west at least to Louisiana and north to Arkansas.

Measurements

Wing (*of asio*), males 139–151.5 mm (ave. of 12, 144.7), females 144–162.5 mm (ave. of 12, 151.3); tail, males 62.5–73.5 mm (ave. of 12, 66.7), females 67–76.5 mm (ave. of 12, 71.3) (Ridgway, 1914). Gehlbach (in press) provided wing chord and culmen length for the six races of *asio* that he recognized. The collective mean wing chord of 199 males of all subspecies was 155.2 mm, and of 210 females was 160.3 mm. The wing chord trend is clinal, with the northwestern race *maxwelliae* averaging largest (169.6 mm for 27 birds of both sexes) and *floridanus* the smallest, averaging 145.1 mm for 64 birds of both sexes. The eggs of *asio* average 35.5 × 30 mm (Bent, 1938); average size decreases from the northwest to the southeast and varies annually, seasonally, and within clutches (Gehlbach, 1995b).

Weights

The collective mean mass of 27 males of all subspecies measured by Gehlbach (in press) was 137.1 g, and of 33 females was 163.9 g. The estimated male:female mass ratio is 1:1.2. The mass trend is clinal, with the northern race *naevius* averaging largest (196.2 g for 20 birds of both sexes) and *floridanus* the smallest, a single male weighing 111.0 g. Henny and VanCamp (1979) reported the average weight of 31 males of *naevius* as 167 g (range 140–210), and that of 66 females as 194 g (range 150–235g). The mean mass of 22 *hasbroucki* eggs was 18.2 g, or 10.1 percent of mean adult female incubation weight (Gehlbach, 1994a).

Identification

In the field. The primary (advertising or alpha) song of the eastern screech-owl is a whinny that typically starts upward in pitch, falls gradually, and becomes a vibrato or tremolo terminally ("Oh-O-O-O that I had never been bor-r-r-rn" call). Farther west in Texas the tremolo may be very fine and short, and in the Rio Grande valley only the inflected portion is uttered. The secondary song is a long trill of rapid notes at a constant pitch but gradually increasing volume and is apparently present in all populations of the *asio* group (Marshall, 1967). This secondary song, given by a male in response to a female, is common in *asio* and often is uttered as a synchronized duet between the sexes (Weyden, 1975). Some representative postures of adult and nestling eastern screech-owls are shown in Figure 27.

Figure 26. Distribution of the eastern screech-owl, showing residential ranges of races *asio* (as), *floridanus* (fl), *hasbroucki* (ha), *mccallii* (mc), and *maxwelliae* (ma). The range of the presumptive race *naevius* occurs north of the dashed line (Gehlbach, in press). Racial boundaries are approximations only; cross-hatching indicates regions of intergrades or of uncertain racial status. See Figure 28 for historic distribution of rufous morph.

Figure 27. Eastern screech-owl behavior: adults in (*A*) pseudo-sleeping concealment posture and (*B*) actual sleeping posture, juveniles in (*C*) inquisitive and (*D*) aggressive-defensive postures. After photos by author except for *C*, which is after a published photo.

In the hand. Generally separable from the western screech-owl in that the bill is never black, the ear tufts are slightly longer, and the ground color is brighter, the rufous plumage morph being rich in bright brown, buff, and ruddy tones. There are complex clinal variations in morph frequency that suggest climatically related genetic polymorphism, with warm, moist climates evidently favoring survival of the rufous morph, such as in the Ozarks and southern Appalachians (see Figure 28). In Tennessee, for example, the rufous morph is 1.7–3.4 times more common (or comprising 63–75 percent of the total) than the gray morph (Nicholson, 1997), and in Florida the red morph represents about 55 percent of museum specimens (Stevenson and Anderson, 1994). However, the rufous morph comprises less than 25 percent of the Michigan population (Brewer, McPeek, and Adams, 1991), represents only about 7 percent of the birds in central Texas (Gehlbach, 1994a), and in Vermont was reported only twice between 1973 and 1983 (Laughlin and Kibbe, 1985). Percentages of rufous morph by subspecies, *asio*, 35–75 percent; *floridanus* 35 percent; *maxwelliae*, 7 percent; *hasbroucki*, 5 percent; *mccallii*, 0 percent (Pyle, 1997b). Intermediate shades of brown plumage are also usually present, confusing the genetic interpretation of these differences. The dorsal plumage of both morphs is more strongly patterned laterally than linearly, and the recurved

ventral crossbars are as wide as the dark shaft streaks. The presence of a brood patch will identify females during breeding, and a combination of weight and linear measurements may accurately sex nearly 90 percent of adults (D. Smith and Wiemeyer, 1992). Nestlings are initially covered with pure white down, which soon stains to a dirty gray and is replaced by a second downy plumage that is olive to umber above and whitish below, with sepia barring (Bent, 1938). The juvenal upperparts are deep grayish brown, indistinctly and rather broadly barred with dusky; many of the feathers are tipped with dull white. In the rufous morph the grayish or grayish brown markings are distinctly rufescent. Remiges and rectrices of first-year birds are similar in color and wear; the outer three primaries are initially relatively broad toward their tips but gradually become more tapered through wear (Pyle, 1997b).

Vocalizations

Gehlbach (1994a, 1995b) recognized two basic types of songs in the eastern screech-owl and at least six different calls. Of the songs, the "monotonic trill" (also called the bounce, tremolo, secondary song, and warble) is most familiar and consists of a 3–6 second soft trill on the same pitch but with slowly increasing amplitude before stopping suddenly. It is used for pair- and family-bonding; variants also are used for cavity-signaling and in association with copulation. The other song, the "descending trill" (also called whinny or primary song) is shorter in length (up to 2 seconds) and is uttered as a single descending note. It is mainly used during intraspecific territorial defense. The major calls include a "hoot" in mild alarm, a "bark" prior to attack, a "screech" in nest defense, a "rasp" in food-begging by adult females or nestlings, a "chuckle-rattle" in mobbed adults, and a "chuck-rattle" uttered in mild annoyance.

In the New York region, male calling begins in January, with the trilled "mating" song of soft, monotone *who-who-who . . .* notes that become especially prevalent in March and April but decline during May and June. This vocalization lasts a few seconds, and the series dies away abruptly. During July, when the young are fledged, the familiar descending trill or "whinny" call is begun, and this is the regular call uttered by the birds until about January, when the mating song again begins (Hough, 1960). D. Smith, Walsh, and Devine (1987) reported that the whinny is mainly evoked by playbacks of recorded calls during the fall and winter period of territorial establishment, while the "warble" (bouncing) call is mainly evoked during the pairing and breeding season in spring. Males and the incubating mates sing antiphonal duets both day and night, and neighboring males sing synchronously at night. The weather and amount of moonlight affect song intensity, as do the time of day or night and the seasonal progression of breeding (Gehlbach, 1995b).

Cavanagh and Ritchison (1987) analyzed the bounce (monotone trill) and whinny songs of eastern screech-owls. They determined that the former averaged 2.5 seconds in length and had an average of 35.8 notes that were uttered at a nearly constant frequency. The whinny averaged 1.24 seconds in length and had an initial unmodulated portion of about 0.25 seconds followed by a terminal portion lasting about 1 second. Substantial frequency variations occurred during the whinny song, but it averaged higher in pitch throughout than the bounce song. Significant individual variation occurred in some aspects of both songs, suggesting that they may be useful in individual recognition, while other aspects were less variable and may be important for species recognition.

According to Marshall (1967), the *asio* group (*sensu lato*) of screech-owls have the most varied calls of any owls known to him. Besides the primary (territorial) and secondary (duetting or courtship) songs, there are dawn calls that are different from those uttered at dusk, calls by the female from inside the nest cavity when bringing food to her young, excitement calls around the nest, alarm notes in response to great horned owl hoots, various food calls of the young, and explosive barking notes uttered in flight. Males have lower-pitched voices than females and are more likely to trill, which is used for communicating with their mates; females are more prone to hoot and bark in alarm or defense, as when they are defending their family (Gehlbach, 1994a).

The fact that screech-owls will respond vocally to playbacks of their own territorial songs allows for auditory censusing of this species (Nowicki, 1974; R. Johnson et al., 1981; Lynch and Smith, 1984). Responses to playbacks of the bouncing song suggest that it serves a role in aggressive territorial announcement. Additionally owls are able to discriminate between the songs of neighbors and nonneighbors, responding more strongly to those of neighbors (Ritchison and Cavanagh, 1986). The trilled songs are used mainly by males for advertising nest sites and for duetting with a mate, whereas the whinny songs are used almost entirely for territorial advertisement and defense. This latter song is more suited for long-distance communication, as it carries better through forest vegetation (Ritchison et al., 1988). Trilled duets are typically initiated by males, and appear to provide individual recognition information.

Habitats and Ecology

Marshall (1967) described the habitats of the *asio* group (*sensu lato*) of screech-owls as "any open woods," with oaks, cottonwoods, and mesquites especially favored. They also abound in suburban areas having large shade trees, including exotic trees, with cavities suitable for nesting, and will use artificial nest boxes. In a Connecticut study, there was a positive correlation between owl abundance and the percent of natural habitats within urban open space areas, including the percentage of shrubs, old fields, and marshes, as well as with the total habitat diversity and the linear amount of available habitat edge. Areas with a high mix of habitats, including a relatively high amount of undisturbed successional communities and associated edges, tended to support good owl populations (Lynch and Smith, 1984). Dwight Smith (pers. comm.) has observed some variations in roosting habitat choice of the two color morphs, with gray-morph birds tending to roost closely beside the tree trunk, while red-morph ones often roost out among the leafy foliage. Conners (1982) suggested that morph preferences in roosting sites might be related to differences in thermal tolerances.

In another study of owl habitats in suburban areas of Connecticut, D. Smith and Gilbert (1984) used radiotelemetry data to estimate relative habitat usage. Four habitats, including red maple (*Acer rubrum*) woodland, upland woodland, evergreen hedgerows, and edge habitats were used by the owls more often than one would expect from random distribution, and lawns, mixed woodlands, and evergreen woodlands were used less than expected. Although lawns were not found to be a selected habitat type, they were a major component of all monthly home ranges. Use of red maple woodlands and upland woodlands was highest during winter months, as was use of old fields. In a similar study of habitat use, Ellison (1980) reported that local eastern screech-owl distributions were positively associated with habitat edge, running water, wet woodlands, and open weedy areas, but negatively associated with dry upland woods, especially those of softwood (evergreen) species. At the western edges of their range in Wyoming, eastern screech-owls are closely associated with riparian cottonwood stands on the high shortgrass plains and in the Bighorn Basin. There is no known contact with western screech-owls in that state (Dorn and Dorn, 1994). However, in southeastern Colorado there is limited sympatric contact between them in Bent County (Kingery, 1998), and more extensive contact occurs in Texas and Mexico (Gehlbach, in press).

Population density estimates vary greatly for eastern screech-owls. Cink (1975) estimated a spring density of 0.24 owls per square kilometer in Kansas, and Allaire and Landrum (1975) made a summer estimate of 0.12 owls per square kilometer in Kentucky. Lynch and Smith (1984) provided monthly estimates of owls ranging from less than 1 to more than 7 owls per square kilometer, with an overall average of 2.3 for four Connecticut study areas. Nowicki (1974) produced an estimate of 0.4–2.4 owls per square kilometer over a township in Michigan, the latter figure based on estimated amounts of suitable woodland habitat in the total township. Gehlbach (1994a) reported that in Texas the population density averaged 4.4–7.4 pairs per square kilometer in suburbs, but only 0.7–2.2 on a rural study area. Both populations exhibited population cycles that were seemingly correlated largely with differences in reproductive efficiencies of novice and experienced breeders. Differences in predation, food, and weather were also significant influences, but living in suburbs tends to ameliorate these environmental effects and allows a more stable age-class ratio to be maintained.

Movements

In spite of some statements that eastern screech-owls are migratory at the northern end of their range there is no evidence for this, and instead the birds scarcely wander from their natal homes, even in winter. In northern Ohio, near the northern end of the breeding range, young birds disperse in early fall, with about 75 percent of them moving more than 10 kilometers by the following spring, the average being about 32 kilometers. However, among adult birds the vast majority (87 percent) remained within 16 kilometers of the banding site, and none moved more than 64 kilometers. The direction of dispersal is apparently random (VanCamp and Henny, 1975).

Gehlbach (1986) reported that in Texas juvenile screech-owls begin to disperse in late summer and that recoveries of young birds from 6 to 48 months after fledging indicate that they may move up to about 15 kilometers away from natal areas, although most have settled little more than a kilometer away. Males are prone to disperse farther than females, nearly half of which in one study settled within a kilometer of their natal area (Gehlbach, 1994a).

The young of three families of owls that were radio-tagged in Kentucky were found to disperse an average of 1.8 kilometers, beginning their dispersals 45–65 days after fledging (Belthoff and Ritchison, 1986). Roost sites used by the young birds on consecutive nights averaged only about 40 meters apart, and on nearly one-fourth of the nights the entire family (both adults and all juveniles) roosted in the same tree. Trees used most frequently for such roosts were eastern red cedar (*Juniperus virginiana*) and shagbark hickory (*Carya ovata*) (Belthoff, 1986). The home ranges of adult birds were found to vary from 40 to more than 300 hectares, averaging larger during the breeding season than outside it (Sparks, Ritchison, and Belthoff, 1986).

Estimated home ranges in the radiotelemetry study by D. Smith and Gilbert (1984) were quite variable, ranging from 8.8 hectares in December to 107.5 hectares in June, averaging smallest during December–January and again during nesting in April–May. Estimated total home ranges increased during the entire period that the individual birds were studied, and nightly activity ranges varied from about 5 to 12 percent of the total estimated home range. Both members of one tracked pair typically hunted "over only a small part of their total home ranges each night, with the male averaging smaller movements than the female." The total estimated home ranges ("territories") of six tracked owls averaged 0.8 square kilometers, with estimated nightly activity ranges varying from 0.25 to 0.6 square kilometers. Most of the home ranges of these six birds showed substantial overlap. One or more roosting sites were present within the home range, which in the case of three roosts averaged 4.9 meters high, and most frequently (75.5 percent of 151 cavity roosts) faced south. Nearly half of 48 open roosting sites were in bittersweet or in Norway spruce (Gilbert, 1981).

The actual "territory," or defended portion of the home range, of eastern screech-owls is quite small; only the nesting and roosting cavities and areas in the immediate vicinity of the nesting site are defended by males, and neighboring pairs may nest as close as 45 meters apart. In suburban areas their home ranges typically consist of about 4–6 hectares, while rural pairs in central Texas typically range over about 30 hectares (Gehlbach, 1994a). In suburbs, the pairs tend to be regularly dispersed and closely spaced, but distribution in the countryside is scattered and patchy. Suburban males may defend up to three cavities suitable for nesting and roosting within an area of 0.5 hectare and have home ranges of 4.0–13.2 hectares (Gehlbach, 1994a).

Foods and Foraging Behavior

The varied list of prey of the eastern screech-owls was summarized by Bent (1938), after which he concluded that the birds tend to consume the most readily available food sources. Where mice, rats, and similar small mammals are common they tend to utilize them, but the total mammal prey list includes shrews, moles, flying squirrels, chipmunks, and bats. The bird prey list is even longer and includes many songbirds as well as larger nonpasserines such as rock dove, northern bobwhite (*Colinus virginianus*), ruffed grouse (*Bonasa umbellus*), American woodcock (*Scolopax minor*), American kestrel (*Falco sparverius*), and even other screech-owls. Gehlbach (1995b) tallied 18 mammal species (mostly rodents), 83 bird species (85 percent passerines), 16 reptile species (mostly snakes), 12 amphibian species, and 9 species of fish among known food items. Many invertebrates, especially insects, are also eaten. In northern regions more mammals and birds are generally taken, and in the south more invertebrates and ectothermic (cold-blooded) vertebrates. During winter more endothermic (warm-blooded) vertebrates are eaten, whereas in summer a greater proportion of insects and ectothermic vertebrates are consumed. Predation tends to be opportunistic, with prey sizes averaging only 28 grams, and the largest known prey a 4-kilogram domestic fowl. Items up to at least 130 grams may be carried away and cached. Unlike

other snakes, blind snakes (*Leptotyphlops*) are often carried to the nest alive, and although some are killed and fed to the chicks, others survive and eat fly and ant larvae and pupae, where perhaps they help maintain nest hygiene during the fledging period. In any case, nest sites with blind snakes had young that grew more rapidly than those without snakes, suggesting a symbiotic relationship (Gehlbach, 1995b).

A. Allen (1924) found that examples of 24 bird species were fed to one brood of developing young, the total list probably including more than 100 individual birds. Other vertebrate prey reportedly consumed by eastern screech-owls includes snakes, lizards, frogs, toads, salamanders, and small fish, and invertebrate prey includes a large number of insects, many apparently caught in flight. Other invertebrate prey includes snails, crayfish, spiders, scorpions, millipedes, and earthworms (Bent, 1938). Judging from the observations of A. Allen (1924), parent screech-owls evidently hunt throughout the night for their young, from just after dusk until just before dawn, capturing whatever is most easily secured.

Gehlbach (1986) noted that although parent owls often hunt in opposite directions from the nest, once a food source has been located both birds fly to and from it repeatedly. He frequently observed the birds catching insects on the ground or in tree foliage. When feeding nestlings, the male captures fewer and larger prey (such as birds) early in the nestling period, whereas later smaller prey such as lizards and insects are taken during more frequent hunting trips, especially after the female begins to assist the male (Gehlbach, 1994a). Marshall (1967) thought that screech-owls usually forage by taking short flights from trees; thus hunting is aided by fairly open ground around trees or at the edges of groves. He regarded large invertebrates as the usual food of *O. asio* (*sensu lato*), with vertebrates occasionally taken. However, Marshall's observations were made mainly in more southern areas, where insects are fairly abundant all year, and most studies in northern parts of the birds' range indicate that they are highly opportunistic and readily shift to vertebrate foods during winter or at other times when insect prey is less available.

Social Behavior

Although pair bonding is monogamous and apparently lifelong in screech-owls, their rather short lifespans tend to make actual pair-bond lengths fairly brief. Gehlbach (1986) observed that one female had three successive mates in the course of a single breeding season, her two earlier ones being killed by traffic. He estimated an average lifespan of 3.6 years for his central Texas population, with most females living only long enough to breed a single time. During years of abundant food and high populations, a small percentage of males (6 percent in Gehlbach's study) may be concurrently or sequentially polygynous. There is an apparently low incidence (6 percent) of "divorce" occurring during or after unsuccessful nesting efforts. Suburban owls had higher survivorship and longer lives than rural ones, with about 10 percent of a marked population living longer than 5 years (Gehlbach, 1994a). However, there are records of very old wild screech-owls, such as one living for at least 13 years (VanCamp and Henny, 1975), and another wild female lived for more than 14 years (Gehlbach, 1995b).

Gehlbach (1994a) has speculated on the adaptive significance of the reversed sexual dimorphism on screech-owls (females average about 17 percent heavier than their mates), judging that larger females are better able to store energy for breeding and can more effectively defend the nest. The smaller males may be more efficient in catching the relatively abundant small prey and require less to eat to support themselves, and their lighter wing-loading may result in improved hunting and flying efficiency. Upper limits on female size may be controlled by the need for females to be able to enter small cavities such as woodpecker holes, and males likewise probably encounter constraints on minimum body size, such as physiological tolerances to prolonged cold weather or short-term famines.

Pair-bonding behavior evidently consists of mutual calling, usually in synchronized duets, and mutual preening or nibbling of the facial area. Although copulatory behavior is apparently undescribed for the eastern screech-owl, it probably takes the same form as described below for the western species.

Breeding Biology

The breeding season in eastern screech-owls is not very long; 25 egg dates from New York and New England are from April 12 to May 18; 53 records from Pennsylvania and New Jersey are from March 23 to May 19; 37 records from Florida are from March 11 to May 18; and 16 midwestern records (Illinois to Iowa) are from March 29 to May 11. These records are primarily concentrated (numbering at least half of the total sample) between April 4 and 27. Clutch sizes in the eastern half of the United States vary greatly, with ranges of from 1 to 8 eggs reported, but means of from 3.00 eggs (Florida) to 4.56 eggs (Ohio, Indiana, Illinois, and Wisconsin). The average clutch size increases clinally from south to north and from east to west (VanCamp and Henny, 1975). In Texas the mean clutch size declines though the breeding season, with 4-egg clutches typical early in the season, and 3-egg clutches typical of replacement clutches. The amount of winter precipitation, mostly rain, positively influences the size of initial clutches, and older, larger females lay larger clutches than do smaller and younger birds. (Gehlbach, 1994a).

In northern Ohio, owls may be seen together at nest sites as early as the first week of February, and the first eggs there are probably laid in early to mid-March, based on early hatching dates of mid-April. The first 2–3 eggs in a clutch are laid a day apart, with later eggs dropped at increasing intervals. The incubation period is now known to be about 30 days (range 27–34) and the fledging period 28–32 days (Gehlbach, 1995b). The nesting success rate (nests fledging at least one young) among 511 nests in northern Ohio over a 29-year period averaged 86.1 percent, and the average number of young fledged in 440 successful nests was 3.8 (VanCamp and Henny, 1975). Assuming an average clutch of 4.43 eggs, the overall breeding success rate (percent of eggs producing fledged young) was about 73.8 percent if one accepts the estimate of 3.8 young fledged. However, this was considered by the authors to be probably an overestimate owing to sampling error, and a more conservative estimate of 2.63 fledged young per nesting attempt (or about 60 percent breeding success) was suggested. Generally, early nestings produced slightly more fledged young per nest than later ones, perhaps because older birds may nest somewhat earlier than inexperienced ones. Additionally some of the late nests may have actually been renests, with associated smaller average clutch sizes. The incidence of renesting is apparently rather low in this species, in part because of the fairly short span of the nesting period. Gehlbach (1994a) documented clutch sizes for 63 initial nestings and 19 renests. No renests were attempted after the first week of June; later nestings would probably result in starvation of the young and would also place severe energy constraints on premolting adults.

No significant differences in breeding success of the various plumage morph combinations could be detected by VanCamp and Henny, and it was judged that the genetic basis for the morphs is due either to one pair of alleles, with red dominant to gray, and the intermediate phenotype the result of genetic modifiers, or to a series of three alleles, with a graded order of dominance of red over intermediate over gray. Gray birds evidently were able to survive stressful periods of heavy snowfall and low temperatures better than red-morph ones. Gehlbach (1994a) found that in Texas the red morph averaged about 7 percent of the total population, but with too many intermediate birds present to make any firm genetic conclusions beyond the statement that gray morphs appear to be homozygous recessives and red morphs either homozygous dominant or heterozygous. He believed the rufous plumage morph to be climatically associated with regions having mild winters as well as warm and wet or cloudy weather. Even in Gehlbach's small study area the rufous morph was more common in the climatically buffered suburbs. The presence of two morphs in a single region may thus represent an example of balanced polymorphism, with the rufous birds doing better in warm and wet years, and the gray morph better in cold and dry years. The red morph is most cryptic in dim light conditions, and the rufous feathers are more prone to abrasion than are gray ones, as well as perhaps having less insulative effectiveness under cold conditions. Some other screech-owl populations have somewhat rufous plumage variants, but none is so clearly dimorphic as this species.

Based on various mortality factors, VanCamp and Henny (1975) estimated that an annual recruitment rate of 2.22 fledged young per pair is needed to maintain population sizes; in central Texas the rate is closer to 1.8 (Gehlbach, 1994a). As with other owls, the highest mortality rates of nestlings (estimated at 69.5 percent) occur during the first year; thereafter the annual mortality rate is about 34 percent. Being hit by motor vehicles is evidently a major mortality factor both in Ohio and in Texas. VanCamp and Henny estimated that 77–83 percent of first-year screech-owls attempt to breed, and probably all older birds attempt to breed, compared to about 25 percent of first-year great horned owls breeding. Gehlbach (1994a) estimated that 85 percent of yearling males breed, as do 91 percent of first-year females. He also estimated an overall multi-year hatching success of 55.3 percent of all nests studied. A total of 87 broods had 330 nestlings (3.79 chicks per nest), and fledging success was 86 percent in the suburbs as compared with 72 percent in the rural areas. Yearling females averaged 1.7 fledged young per breeding season; older females, 2.4. Fratricide and infanticide accounted for more than half the known chick mortality. Productivity increased with increasing clutch size, but only up to 4-egg clutches. Older juveniles were most often killed by vehicles, with poisoning in second place as a known mortality factor. First-year survival ranged from 30 percent (rural areas) to 36 percent (suburbs) in Gehlbach's (1995b) studies.

Evolutionary Relationships and Conservation Status

The relationships of the North and Middle American screech-owls were discussed at length by Marshall (1967). He believed that 7 species of *Otus* exist in this general region, with *O. asio* and *O. trichopsis* constituting a closely related but specifically distinct species pair. The species *O. asio*, as he visualized it, consists of several geographically isolated "incipient species," three of which are widely ranging and highly variable geographically. One of these is the *asio* group, the eastern screech-owl; a second is the *kennicottii* group, the western screech-owl. Additional groups of still uncertain affinities to *kennicottii* occur in Mexico and Central America. Apparently because of hybridization between members of the *asio* group and the *kennicottii* group in the Big Bend area of the Rio Grande, Marshall stated that these two forms must be considered conspecific. He believed that present-day *trichopsis* was probably derived from an ancestral *kennicottii* form and now coexists with its own parental stock without interbreeding. He also found no evidence for character displacement between whiskered screech-owls and other forms of *Otus* in their areas of sympatry.

In the 1998 edition of the *Check-list of North American Birds*, the AOU's committee on classification and nomenclature designated *asio* and *kennicottii* as allospecies of a superspecies and attributed the mixing of pairs in eastern Colorado (Arkansas River) and southern Texas (Rio Grande) to long-distance dispersal in marginal habitats rather than to significant levels of hybridization. Gehlbach (in press) reported that where the two species approach each other or are sympatric their mean bill-length differences are greater than where they are allopatric farther north. These differences may reflect selection for niche specialization through interspecies competition, as strong isolating mechanisms have not yet developed. Gehlbach suggested that interbreeding may result from a scarcity of mates in sparse populations along range margins. In some areas of contact interspecific vocal mimicry may facilitate such hybridization. He has also noted (pers. comm.) that unpublished cytochrome b data support a species separation of *asio* and *kennicottii*.

At least in northern Ohio, the screech-owl population has fluctuated since the 1940s, with no clear directional trends evident. Similar uncertainty exists for other midwestern states, largely because of censusing difficulties. Use of suburbs by the owls may be balanced by habitat losses involving woodlands and creek bottoms (VanCamp and Henny, 1975). It is possible that there has been a westward expansion of *asio* in southern Canada, where it has reached southern Saskatchewan and probably southern Alberta. The western screech-owl has also moved east in western Texas, and perhaps also in Colorado. There is a statistically nonsignificant annual population trend of +1.8 percent, based on annual Breeding Bird Surveys done between 1966 and 2000 (http://www.mbr-pwrc.usgs.gov/bbs/bbs.html). During the 1999 and 2000 Christmas Bird Counts the species was detected on an average of 708 counts, or about 39 percent of all U.S. and Canadian counts. In terms of all localities where the species was detected, the eastern screech-owl was

the second most widely reported of the 18 North American owl species then encountered. The average total number counted among the 100 counts having the highest species totals was 2625 birds (averaging 26.2 birds per count, or one bird per 6.7 square miles, the highest reported mean density of any owl species). This total represents 27 percent of all owls reported in the tabulated summaries of counts for the United States and Canada during 1999–2000. The maximum number ever reported on a single CBC was 220, at Lynchburg, Virginia, in 1977. The Canadian population has recently been estimated at 10,000–15,000 pairs (Kirk and Hyslop, 1998), and is believed stable.

Western Screech-Owl   *Otus kennicottii* (Elliot) 1867

Other Vernacular Names:
Aiken's screech-owl (*aikeni*); Guadeloupe screech-owl (*suttoni*): Kennicott's screech-owl
(*kennicottii*); California screech-owl (*bendirei*); Pasadena screech-owl (*quercinus*); Yuma
screech-owl (*yumanensis*).

Range (Adapted from AOU, 1983)

Resident from south-coastal and southeastern
Alaska, coastal and southern British Columbia,
northern Idaho, western Montana, Colorado, extreme western Oklahoma, and western Texas (locally extending to east of the Pecos River on the
Edwards Plateau) south to southern Baja California, northern Sinaloa, and across the Mexican
highlands through Chihuahua and Coahuila as
far as the Distrito Federal. (See Figure 28.)

Subspecies (As recognized by Marshall
[1967]. Gehlbach [in press] has reinstated
two of Marshall's synonymized races,
synonymized one, and geographically
redefined some of the others; his treatment
has recently been adopted by Cannings and
Angell [2001].)

O. k. kennicottii (Elliot). Coastal Alaska to coastal
Oregon. Includes *brewsteri* Ridgway.

O. k. bendirei (Brewster). Includes *macfarlanei*
(Brewster) and *quercinus* (Grinnell).Washington
and Idaho south to southern California and
east to the continental divide of Montana and
Wyoming. A single specimen from near Jackson,
Wyoming, is the only such record from that state,
and has been described as approaching *macfarlanei*
(Dorn and Dorn, 1994). This latter race has been
reinstated by Gehlbach (in press), to include
birds from regions north and east of the dashed
line on the distribution map, east to Montana and
Wyoming. The range of *bendirei* was correspondingly restricted by him to California, with *aikeni*
replacing it from the Sierras east, as shown by the
dotted line on the map, and extending from there
east to the Great Plains.

O. k. xantusi (Brewster). Baja California, at least
in the south, but see *cardonensis* below. The taxon
xantusi has also been considered to perhaps be
taxonomically closer to *O. cooperi*, because of reported vocal similarities (AOU, 1998).

O. k. cardonensis (Huey). Baja California Norte.
Considered a synonym of *xantusi* by Gehlbach (in
press).

O. k. aikeni (Brewster). North-central Sonora, eastern California, Nevada, Utah, Arizona, New Mexico, southern Colorado, and extreme western

Oklahoma. Includes *gilmani* Swarth, *inyoensis*
Grinnell, and *cineraceus* (Ridgway).

O. k. suttoni (Moore). Big Bend area of Texas and
Guadalupe Canyon, Arizona, south to the Mexican Plateau.

O. k. yumanensis (Miller and Miller). Colorado
Desert, lower Colorado River, and northwestern
Sonora.

O. k. vinaceus (Brewster). Central Sonora to
Sinaloa.

Measurements

Gehlbach (in press) provided wing chord and culmen length for the eight races of *kennicottii* that
he recognized. The collective mean wing chord of
286 males of all subspecies was 160.0 mm, and of
256 females was 165.9 mm. The wing chord trend
is clinal, with the northernmost race *macfarlanei*
averaging largest (179.1 mm for 58 of both sexes)
and *xantusi* the smallest, averaging 147.5 mm for
113 birds of both sexes. Wing (of nominate *kennicottii*), males 170.5–190.5 mm (ave. of 9, 176.5),
females 170.5–187.5 mm (ave. of 8, 179.2); tail,
males 82–98.5 mm (ave. of 9, 89), females 85.5–
98.5 mm (ave. of 9, 89.2) (Ridgway, 1914). Wing
and body mass measurements of all subspecies
are provided by Cannings and Angell (2001). The
eggs of *kennicottii* average 37.8 × 32 mm (Bent,
1938).

Weights

The collective mean mass of 134 males of all subspecies measured by Gehlbach (in press, and in
Cannings and Angell, 2001) was 125.6 g, and of
71 females was 142.7 g. The estimated male:
female mass ratio is 1:1.14. The mass trend is clinal, with the northern race *kennicottii* averaging
largest (170.1 g for 6 birds of both sexes) and
vinaceus the smallest, averaging 111.1 g for 15
birds of both sexes. Earhart and Johnson (1970)
reported that 14 males and 11 females of *kennicottii* averaged 152 and 186 g, respectively, while 35
males and 18 females of *aikeni* ("*cineraceus*") averaged 111 and 123 g, and 26 males and 10 females
of *bendirei* ("*quercinus*") averaged 134 and 152 g.
The estimated egg weight of *kennicottii* is 20.1 g,
representing an egg-to-female proportional ratio

Figure 28. Distribution of the western screech-owl, showing residential ranges of races *aikeni* (ai), *bendirei* (be), *cardonensis* (ca), *kennicottii* (ke), *suttoni* (su), *vinaceus* (vi), *yumanensis* (yu), and *xantusi* (xa), and of the Pacific screech-owl, *O. cooperi* (co). A taxonomically unresolved population in Oaxaca (*lambi*) (la) evidently intergrades with *cooperi*. Gehlbach (in press) recognizes an additional western screech-owl race (*macfarlanei*) occurring north of the dashed line in Oregon, and extends the range of *aikeni* west to the dotted line in California. Also shown is the historic relative frequency of the eastern screech-owl's rufous morph (adapted from Hasbrouck, 1893a).

of 10.8 percent. One fresh egg weighed 17.3 grams (Cannings and Angell, 2001).

Identification

In the field. Along the Pacific coast the usual song of the western screech-owl consists of a bouncing-ball series of about 12 to 15 notes, ending in a fine roll. A double trill, the second longer and gradually dropping in pitch, is also common. Farther inland the number of trilled notes is shorter, usually 8 or 9 but sometimes as few as 4. Separation in the field from the whiskered screech-owl is feasible only by voice. The western screech-owl's duetting song is a series of short notes on the same pitch that speed up, the series lasting from 2 to 4.5 seconds. By contrast, the whiskered screech-owl utters an evenly spaced series of slower notes or a syncopated series of short and long notes. In those areas where the western and eastern screech-owls may both occur, the eastern screech-owl produces a primary song that is a whinny, and a secondary song that is a long (about 4 seconds) trill of rapid notes at the same pitch but slowly increasing in volume. The western screech-owl also utters a trilled call, but it tends to be short and two-parted, with a distinct pause in the series.

In the hand. Separable from the eastern screech-owl in that the bill is almost always black, with a whitish tip (except in northern examples of *bendirei*, where it is greenish gray); the usual dorsal pattern is one of linear black streaks; and the ventral patterning has thinner crossbars (coarser in the Pacific Northwest). Gray-morph birds are typical through most of the range, with a reddish brown (but not true rufous) morph present only in the Pacific Northwest race (*kennicottii*), where about 7 percent is of this plumage variant (Pyle, 1997b). Although the presence of a brood patch will serve to identify breeding females, at other times multiple criteria such as a combination of weight and linear measurements may be useful in judging sex (A. Miller and Miller, 1951), as indicated for the eastern screech-owl. Plumages of the young are similar to those of *asio* (Bent, 1938; Sumner, 1928).

Vocalizations

A complete comparative study of the vocalizations of eastern and western screech-owls has unfortunately not yet been performed. Additionally, in areas of sympatric contact, interspecific vocal mimicry may occur, perhaps as a means of reducing interspecific competition (Gehlbach, in press). Marshall (1967) characterized the western screech-owl as having territorial (primary) and duetting (secondary) songs of mellow, pure tones and a constant pitch. The primary song is similar to that made by "a ball bounding more and more rapidly over a frozen surface"; the secondary song is a double trill, consisting of a short burst of rapid notes followed by a longer series of the same. In the Cape area of southern Baja California, the vocalizations closely approach those of the Pacific screech-owl of Mexico's western coast.

Henry Hinshaw, cited by Bent (1938), described the double trill call of *aikeni* (*"cineraceus"*) as consisting of two prolonged syllables, with quite an interval between, followed by a rapid utterance of 6–7 notes, which are run together at the end. It has also been described as a short trill followed immediately by a longer one, and as two sets of 6–12 notes, separated by a half-second interval. A young female of *aikeni* raised by C. Aiken was described as having a short *wow* bark when excited, hungry, or demanding attention, similar to a puppy's bark. A gentle and soft *cr-r-oo-oo-oo-oo-oo* was seemingly used as an affectionate greeting, and a similar note resembled the noise made by ducks' wings in flight. This soft, quavering note was described by Bonnot (1922) as an apparent contentment call, and he listed five other calls as typical of *bendirei*.

Habitats and Ecology

A wide range of habitats is used by the several races of this widespread form, varying from the tropical coastal lowlands of the Baja Peninsula and the hot Sonoran desert habitats of Arizona to the humid temperate rain forests of western British Columbia and southern Alaska. Perhaps in general the birds prefer partially open country, dominated by deciduous trees and their associated grayish brown bark coloration, although reddish brown birds occur in the humid portions of the Pacific Northwest in firs and other conifer-dominated forests. In the Southwest and Mexico the birds occur sympatrically with the whiskered screech-owl, evidently without interbreeding. There it occurs in lower elevation open pine-oak-juniper woodlands not occupied by the whiskered screech-owls, which favor higher and more densely wooded habitats. North of the range of the whiskered screech-owl it also occurs in dense oak groves, suggesting that it is restricted ecologically in areas of competitive overlap (Marshall, 1957). Where sympatry occurs with the eastern screech-owl, the western is likely to occur in somewhat higher and drier habitats (Gehlbach, 1995b).

Open deciduous woods, especially riparian hardwoods or arroyos with oaks (*Quercus*) or sycamores (*Plantanus*) present, are favorite habitats, but the birds also locally occur in stands of giant cardon and saguaro (*Carnegiea*) cacti in upland deserts, in Joshua trees (*Yucca brevifolia*), in stands of sycamores, cottonwoods (*Populus*), tamarisk (*Tamarix*), and willows (*Salix*) along rivers, in groves (bosques) of mesquite (*Prosopis*), in open piñon-juniper woodlands or pine forests, and in low, dense Douglas fir (*Pseudotsuga*) forests.

In Idaho, Hayward (1983) found that *macfarlanei* exhibited a relatively narrow habitat niche breadth, being limited to deciduous habitats at low elevations, with a strong preference for deciduous river bottoms. Some use of adjacent bunch grass areas also occurred, presumably for hunting. Nearly half the roosts he found were in deciduous cover, and more than 80 percent of the birds perched next to the tree bole, where their bark-like plumage makes them nearly invisible. Roost heights averaged 4.6 meters aboveground, and roosting trees averaged 21.2 meters high.

Studies in Colorado (Kingery, 1998) indicate that there the birds breed most commonly in developed rural or residential areas, a situation comparable to the high populations of suburban eastern screech-owls studied by Gehlbach (1994a) in Texas, and in mature riverine cottonwood stands. There its breeding range extends out along river courses into the high plains well to the east of the mountains, such as along the Arkansas River, where local contact with the eastern screech-owl now evidently occurs in Bent County. Probably more than 1000 pairs breed in that state. A slow eastward expansion along the entire front range of the Rockies is apparently under way, from Alberta to New Mexico (Cannings and Angell, 2001).

In western Texas the birds tend to occur in somewhat drier and higher habitats than the eastern screech-owl. They usually are found in areas that are moderately to heavily wooded with cottonwoods and other hardwoods, mainly those occurring along riverine courses or in montane areas. Frederick Gehlbach (pers. comm.) estimated a mean (7-year average) density of 1.4 pairs per square kilometer in the owl-rich Cave Creek area of southeastern Arizona. In northern Mexico the birds use semiarid woodlands, riparian woods, areas with only scattered trees, and those with large cardon cacti (Howell and Webb, 1995).

Probably few generalities can be made from all this, except that the birds like open tree (or large cactus) growth having an abundance of insects and small mammals, available cavities for nesting, and a background of tree bark or other environmental factors with which their plumage blends well enough to keep them inconspicuous through the daytime hours.

Movements

Apparently few if any movements of significance occur among western screech-owls, other than the dispersal typical of birds during their first year of life. Hayward (1983) reported the home ranges of two radio-tagged birds to be 3–9 hectares and 29–58 hectares, based on mapped 75 percent and 95 percent contour interval estimates. He considered the birds to be nonmigratory in central Idaho.

Foods and Foraging Behavior

Early observations on the foods of the western screech-owl were summarized by Bent (1938). Observations on *kennicottii* from Washington indicate that there the birds feed mostly on mice, to which ants, beetles, and other insects are added, as well as crayfish, angleworms, and birds as large as northern flickers (*Colaptes auritus*) and Steller's jays (*Cyanocitta stelleri*). In the Puget Sound area the birds continue to feed on arthropods such as cutworms, crickets, beetles, and centipedes well into winter, but have also been found raiding farms and even attacking domestic ducks, bantams, and pheasants during cold winter weather. These large birds are clearly too big to be carried away but may nonetheless be seriously injured or even killed by these tiny owls. Similarly, in the Victoria area of British Columbia the birds are largely insectivorous for much of the year, favoring beetles, orthopterans, and the larvae of moths and butterflies, but in winter they may turn to small mammals and such birds as they are able to catch easily (Munro, 1925). In California, *bendirei* has been found to favor house sparrows (*Passer domesticus*) where they are numerous, but is also known to consume pocket gophers (*Thomomys*), meadow voles (*Microtus*), salamanders, and beetles (Bent, 1938).

A sample of winter foods from Utah indicated that of 80 individual prey identified, about 25 percent each were insects or of mammalian origin, and about 50 percent were avian (D. Smith and Wilson, 1971). Farther south, in the range of *aikeni*, these owls have been found to consume wood rats (*Neotoma*), kangaroo rats (*Dipodomys*), grasshopper mice (*Onychomys*), pocket mice (*Perognathus*), deer mice (*Peromyscus*), gophers, small birds, snakes, lizards, frogs, scorpions, grasshoppers, locusts, and beetles (Bent, 1938).

Hayward (1983) found that in Idaho the birds began foraging each evening within 45 minutes of sunset, and retired to daytime roosts within 30 minutes of sunrise, a pattern he also found to be typical of boreal and northern saw-whet owls. A limited pellet analysis suggested that small mammals (*Sorex* and *Peromyscus*) were primary prey. There the most mammalian prey types were in the 16–35-g weight range, and there was a high level of overlap in prey choice with the northern saw-whet owl (Hayward and Garton, 1988). Like the eastern screech-owl, the western species exhibits a tremendous degree of flexibility in its foods from place to place and season to season, so local samples are unlikely to provide a broad overview or basis for interspecies comparisons (Cannings and Angell, 2001).

Social Behavior

So far as is known, the pair-bonding behavior of western screech-owls does not differ in any significant way from that described for the eastern screech-owl. Courtship begins as early as January throughout the species' range, with the male calling from near his chosen nest site. Duetting and close encounters follow, with nest-showing, courtship-feeding, and allopreening important components of pair bonding. No cases of polygyny have been reported. Marshall (1967) described finding a mixed-species pair (female eastern, male western) tending young in the Rio Grande area of Texas, with the female responding vocally with her long trill to the male western's double-trill call, and indulging in billing and mutual head-preening behavior with him.

Copulatory behavior has been described by McQueen (1972). He attracted a pair to a nearby tree by a whistled imitation of the bouncing-ball call, whereupon both sexes responded with similar calls. Eventually the female approached her mate, and sat in contact beside him. The pair remained thus for more than 10 minutes, calling frequently and nibbling one another around the bill area. Suddenly the male changed his call to a rapid tremulato, consisting of a short phrase followed by a longer one of equal intensity. This call was repeated with more regular and shorter intervals than the earlier one, and during each interval the female uttered a short, unbroken tremolo at a higher pitch. This duetting was terminated when the male suddenly mounted the female, with copulation lasting about two seconds, the male flapping continuously to maintain his balance. After treading was completed the pair flew to separate trees.

Marshall (in Phillips, Marshall, and Monson, 1964) indicated that in Arizona the territories of western screech-owls average about 275 meters apart except in mesquite bosques, where they are about 90 meters apart. R. Johnson, Haight, and Simpson (1979) reported that pairs inhabiting cottonwood-mesquite riparian woodlands on the

Salt and Verde rivers of Maricopa County, Arizona, often are spaced only 45–50 meters apart, whereas those in the surrounding uplands, dominated by saguaros and paloverdes (*Cercium*), rarely support a pair per 275 meters of habitat. This general range is similar to that estimated by A. Miller and Miller (1951), who said that although territories may be less than 90 meters apart, they are often separated by 180–360 m. In a southern California study, 14 territories occurred along a 6.4-kilometer stretch of river, with nests averaging 420 meters apart. By comparison, 34 nests along the Snake River in southern Idaho averaged 3054 meters apart (Cannings and Angell, 2001).

Breeding Biology

Nest sites in the western screech-owl vary greatly according to habitat. Certainly natural tree cavities or those excavated by woodpeckers represent favorite choices. In the Rocky Mountain area, flicker holes in cottonwoods or large willows growing along streams are perhaps most commonly used, but junipers along dry arroyos are likewise exploited. In southern Arizona the giant saguaro is commonly used; indeed the desert form *"gilmani"* reportedly nests nowhere but in these cacti, at heights ranging from about 1.2 meters above the ground almost to the top of the plant, and mainly along the river bottoms and the bordering mesas. Nest sites from a variety of locations have ranged from about 1 to 12 meters above the ground, with a mean of one nationwide sample of 25 being 4.8 meters (Cannings and Angell, 2001).

Egg-laying periods in Arizona and southern Texas range from mid-March to April; in southern Idaho egg dates extend from late February to early April; and in British Columbia the range is from mid-March to late May. Seven of 10 Arizona nests were placed in cottonwoods, sycamores, or walnut trees, and of 24 British Columbia nests 16 were in natural cavities and 8 were in cavities made by woodpeckers. Preferred sites are those with entrance openings close to the bird's body size (about 7 centimeters in diameter), and depths of natural cavities generally range from 30 to 45 centimeters deep (Cannings and Angell, 2001).

Mean clutch sizes of western screech-owls mostly vary from about 3.4 to 4.0 eggs, with a slight trend toward larger clutches at higher latitudes except along the Pacific Coast, where a reverse trend is apparent. There is also an apparent trend toward reduced clutch sizes from the interior toward the Pacific Coast. An overall average of 435 clutches from western regions was 3.48 eggs. The eggs are deposited at intervals of 1–2 days. Observations by Sumner (1928) were provided on the development of the young and weight changes, which are not significantly different from those of eastern screech-owls. Incubation usually lasts 26–30 days. Eight measured periods in southern Idaho averaged 26 days, and a sample from Texas averaged 30.3 days. Clutches of 3–4 eggs in Washington hatched over 2–3 days (Cannings and Angell, 2001). The mean fledging period is about 28 days, but may extend to 35 days. Maximum known longevity under natural conditions is 12 years and 11 months (Patuxent Wildlife Research Center banding data).

Evolutionary Relationships and Conservation Status

The complex evolutionary and taxonomic relationships of the North American screech-owls have been commented on by many authors, who take such extreme positions as recommending that all the North American screech-owls be considered a single species, with no formal subspecies recognized (Owen, 1963b), or conversely finding more than 20 taxonomically distinctive forms discernible, including four "incipient" species in the single species *O. asio* (Marshall, 1967). Adding to the complexity of clinal variations in both size and plumage pigmentation is the presence of rufous and gray plumage morphs, the frequency of which is geographically highly variable. Early efforts by Hasbrouck (1893a, b) to explain such geographic variation in the screech-owls were met with swift and strong criticism by J. Allen (1893), and later efforts by Owen (1963a) concluded that "nothing is known of the adaptive significance" of this polymorphic variation. Marshall (1967) and others have pointed out that the general intensities of plumage pigmentation in North American screech-owls follow Gloger's Law (darker pigmentation in more humid regions), but rather than diverging in morphology and behavior during speciation the taxa have undergone parallel evolution since they became geographically separated. Marshall also noted that they have not exhibited any apparent divergence of traits associated with niche segregation in areas where they have come into secondary contact, but Gehlbach (in press) believes that divergence in at least one trait (culmen length) has occurred in such areas.

In many areas western screech-owls have suffered substantial habitat losses as riparian deciduous habitats have been destroyed during "development," while on the other hand relatively abundant foods and probable increased protection from enemies such as great horned owls has allowed screech-owls to become more abundant in city parks and suburban areas. There is a statistically nonsignificant annual population trend of +0.27 percent, based on annual Breeding Bird Surveys done between 1966 and 2000 (http://www.mbr-pwrc.usgs.gov/bbs/bbs.html). During the 1999 and 2000 Christmas Bird Counts the species was detected on an average of 133 counts, or about 7 percent of all U.S. and Canadian counts. In terms of all localities where the species was detected, the western screech-owl was the eighth most widely reported of the 18 North American owl species then encountered. The average total number counted among the 100 counts having the highest species totals was 586 birds (averaging 5.9 birds per count, with each count area equal to 175 square miles). This total represents about 6 percent of all owls reported in the tabulated summaries of counts for the United States and Canada during 1999–2000. The maximum number reported on a single CBC was 58, at Palo Alto, California, in 1987. The Canadian population has recently been estimated at 1000–2000 pairs (Kirk and Hyslop, 1998), and is considered as possibly stable.

MEXICAN SCREECH-OWLS (3 species)

Balsas Screech-Owl   *Otus seductus* Moore, 1941

Pacific Screech-Owl   *Otus cooperi* Ridgway, 1878

Vermiculated (Middle American) Screech-Owl   *Otus guatemalae* Sharpe, 1875

Other Vernacular Names (all three forms):
O. seductus: Balsas Basin screech-owl, Rio Balsas screech-owl Tecolote del Balsas (Spanish).
Also part of western screech-owl (when merged with *kennicottii*), *O. cooperi:* Cooper's
screech-owl, Tecolote de Cooper, Lechucita sabanera (Spanish). Also sometimes part of
western screech-owl (when merged with *kennicottii*), *O. guatemalae:* Guatemalan screech-
owl, Cassin's screech-owl (*cassini*), Mazatlan screech-owl (*hastatus*), Yucatan screech-owl
(*thompsoni*), Tecolote Guatemalteco (Spanish). Also often considered part of the
vermiculated screech-owl group (when merged with *O. vermiculatus*) as by the AOU (1998),
and sometimes part of variable screech-owl group (when merged with *O. atricapillus*), as by
Sibley and Monroe (1990). (The AOU [1998] calls this form the vermiculated screech-
owl, but inasmuch as the species limits of this group are still controversial, "Middle
American" or "Guatemalan" seems to be a better descriptive choice, as well as being
appropriate geographically, at least until conspecificity has been established.)

Range (See Figure 29.)

The form *seductus* ranges residentially from southern Jalisco and Colima southeast to Morelos and
central Guerrero. Specimens evidently have been
collected from only two Mexican states (Colima
and Michoacan) (Enriquez-Rocha, Rangel-
Sanchez, and Holt, 1993).

The form *cooperi* ranges residentially along
the Pacific slope of Mexico from southeastern
Oaxaca southeast to El Salvador and Honduras. If
lambi is included in this species, the range includes nearly all of Oaxaca.

The form *guatemalae* ranges residentially
along the Pacific slope of Mexico from southern
Sonora to Oaxaca, along the Atlantic slope from
southern Tamaulipas to Honduras and north-
central Nicaragua. In northwestern Costa Rica it
is geographically replaced by typical *vermiculatus*
Ridgway, 1887, which extends from there south
through Panama and an uncertain distance into
northwestern South America. Depending on accepted species limits, from northwestern Colombia it may extend south (as *napensis*) to Peru and
northwestern Bolivia, and may also continue east
to northern Venezuela and (as *roraimai*) southeast
to southern Venezuela and Brazil. These two apparently allopatric forms were classified as full
species by König, Weick, and Becking (1999), but
they are sometimes regarded as racial extensions
of *vermiculatus* (Voous, 1988; Hoyo, Elliott, and
Sargatal, 1999).

In contrast, Sibley and Monroe (1990) recognized *vermiculatus* as a distinct species ("vermiculated screech-owl") but excluded *guatemalae*. Instead, they regarded *guatemalae* as part of a large
and disjunctively allopatric South American assemblage that also included *hoyi, atricapillus,* and
sanctaecatarinae. They named this rather diverse
assemblage the "variable screech-owl," *O. atri-*

capillus (Temminck) 1822. These forms, although
quite variable in size, plumage, and habitat, were
thus grouped on the basis of having a common
song type, according to information provided Sibley and Monroe by J. T. Marshall Jr.

Considering these two different treatments
of *guatemalae*, as well as still other earlier ones, it
seemed prudent for the author to tentatively
limit the known range of this species to the region extending from Mexico to northwestern
Costa Rica, and the range of *vermiculatus* as an allopatric or parapatric semispecies, extending
from northern Costa Rica to northwestern South
America an uncertain distance. Stotz et al. (1995)
listed *guatemalae* (*sensu lato*) as occurring in at
least 13 Latin American countries, south to Brazil
and Bolivia.

North American Subspecies

The evolutionary relations of *lambi*, here tentatively judged a subspecies of *O. cooperi*, are controversial. König, Weick, and Becking (1999) considered *lambi* as a distinct species that may have
become isolated from ancestral *asio* during Pleistocene times. However, Binford (1989) followed
Marshall (1967) in considering the typical Oaxacan screech-owls to be represented by *O. c. cooperi*
in coastal areas of extreme southeastern Oaxaca,
and *O. c. lambi* over the rest of Oaxaca, with intergrades occurring between Laguna Inferior and
Mar Muerto, coastal estuaries located near the
Chiapas border.

If *lambi* is considered as specifically distinct,
it would be necessary to assume that two sibling
species of relatively arid scrub-adapted screech-
owls must exist parapatrically in Oaxaca, with an
asio-like *lambi* in the interior and a *kennicottii*-like
cooperi in the adjoining southeastern coastal lowlands, the two intergrading or possibly hybridizing

Figure 29. Distribution of Mexican screech-owls, including Balsas (horizontal hatching), Middle American (*guatemalae* racial group), and vermiculated (*vermiculatus* racial group) screech-owls (upper and lower cross-hatching). The presumptive South American range of the *vermiculatus* group is shown in the inset; its southern boundary is unsettled, and depends on variously perceived species limits. Mainly after Howell and Webb (1995). See Figure 28 for distribution of the Pacific screech-owl.

in their zone of contact. These would be in addition to the more forest-adapted whiskered screech-owl and the Middle American screech-owl, which, respectively, occur in interior highland pine or pine-oak forests, and in tropical evergreen forests of the Atlantic slope. At this stage it seems best to recognize the status of *lambi* as part of *cooperi*, until and unless more information is forthcoming.

No subspecies were recognized for *cooperi* by König, Weick, and Becking (1999), but Hoyo, Elliott, and Sargatal (1999) recognized the Chiapas population as a distinct race, namely *chiapensi* Moore, 1947. Both *seductus* and *cooperi* have often been considered as subspecies of *kennicottii*, and *seductus* has also been included within *O. vinaceus* Brewster 1888, which now likewise is usually considered part of *kennicottii*. The Colima population of *seductus* was separated as *colimensis* by Hekstra (1982), a form that is not now recognized by other recent authors.

The North American forms of *guatemalae* that were recognized by König, Weick, and Becking (1999) are as follows:

O. g. cassini (Ridgway). Tamaulipas south to northern Vera Cruz. Considered by Ridgway (1914) to be a distinct species. The coastal population has at times been separated as *pettingilli*.

O. g. guatemalae (Sharpe). Southern Vera Cruz to Guatemala and Honduras, including the Yucatan peninsula.

O. g. hastatus (Ridgway). Sonora south to Sinaloa. Considered by Ridgway (1914) to be a distinct species.

In his revision of *Otus,* Hekstra (1982) regarded *guatemalae* and *vermiculatus* as conspecific, and also considered the species to consist of several allopatrically distributed Latin American groups. These include the *cassini* group (including *pettingilli*) of the Caribbean slope of Mexico, the *hastatus* group of western Mexico, the *guatemalae* group (of lowland Mexico, the Yucatan region, and Central America), the *vermiculatus* group (with six mostly South American races extending to the Bolivian Andes), the *roraimae* group (one race in the highlands of Venezuela), and the *pacificus* group (with two Pacific-slope South American races). Hekstra considered *guatemalae* to be "the most primitive living" screech-owl, because of its apparent lack of a secondary song or of duetting behavior. He recognized no fewer than 19 races of *guatemalae.*

The AOU (1998) currently regards *guatemalae* as conspecific with the vermiculated screech-owl *O. vermiculatus* (Ridgway) 1887, which replaces it allopatrically (or perhaps parapatrically) in northern Costa Rica. As compared with *guatemalae, vermiculatus* has a less feathered or entirely unfeathered lower tarsus; it is less streaked vertically with black and more vermiculated or mottled with brown on the underparts. It also has a relatively short tail and has rich brown to rufous eyebrows that closely match its rufous brown face and dorsal color, rather than contrasting whitish ones.

Other authors have included *guatemalae* (and *vermiculatus*) within a larger species-level assemblage of allopatrically distributed South American forms (usually including *hoyi, sanctaecatarinae,* and *atricapillus*). These collective forms have been variously named the variable screech-owl, the long-tufted screech-owl, and the black-capped screech-owl, the last of these derived from the oldest available (1822) collective Latin epithet *O. atricapillus* (Temminck). However, these three strictly South American forms now all appear to be full species (König, Weick, and Becking, 1999). Additionally, the Puerto Rican screech-owl has been implicated as a probable close relative of *guatemalae* and possibly even conspecific with it (van der Weyden, 1975). Caution in establishing broad species limits on Central and South American *Otus* forms seems appropriate.

Other Mexican forms of screech-owls that have been described and at least tentatively associated with or racially included within *guatemalae* include *thompsoni* Cole 1906 from Yucatan, Campeche, and Cozumel Island; *tomlini* Moore 1937 from Sonora, Chihuahua, Sinaloa, and Durango; and *fuscus* Moore and Peters 1939 from central Veracruz. None of these was accepted by König, Weick, and Becking (1999), but all were adopted by Hoyo, Elliott, and Sargatal (1999).

Measurements

Wing (*cooperi*), one male 171 mm, females 173–178.5 mm (ave. of 2, 175.7); tail, male 86 mm, females 82–83.5 mm (ave. of 2, 82.7) (Ridgway, 1914).

Wing (*seductus*), both sexes (unspecified sample size) 165–183 mm; tail, both sexes 89.5–95.5 mm (König, Weick, and Becking, 1999).

Wing (typical *guatemalae*), males 157.5–160 mm (ave. of 3, 158.7), females 164.6–175 mm (ave. of 4, 170.2); tail, males 81.2–81.8 mm (ave. of 3, 81.5), females 86.4–88.5 mm (ave. of 4, 87.4) (Ridgway, 1914).

Birds here assigned to *guatemalae* from western Mexico (*hastatus* of Ridgway, 1914) have comparable measurements to those just given for typical *guatemalae,* but those from northeastern Mexico (*cassini* of Ridgway, 1914) have shorter wings and tails (wing 139–151.5 mm, tail 68–76.5 mm). However, wing-lengths of birds from Nicaragua average larger than Mexican birds (164–175 mm) (Hekstra, 1982), whereas those of the Panamanian population (thus representing typical *vermiculatus*) are also somewhat smaller than nominate *guatemalae,* especially in tail length (tail of 6 males 71.2–76.4 mm [ave. 74.1], 8 females 73.1–79.4 [ave. 76.8]) (Wetmore, 1968). No mensural information is available on the eggs for any of these forms, but they are likely to be similar in size to the eggs of the eastern and western screech-owls.

Weights

Three males of *cooperi* averaged 149 g (145–153 g), and two of *seductus* were 158 and 161 g (Dunning, 1993). No female weights are available; in the western screech-owl females average 10–20 percent heavier than males, so females of *cooperi* and *seductus* are likely to average about 170–190 g. An unsexed *seductus* was 165 g, and a *lambi* 130 g (Moore and Marshall, 1959). Nine unsexed birds identified as *"guatemalae"* from Peru (thus perhaps actually representing *vermiculatus roraimae,* using the taxonomy adopted in this book) averaged 107 g (range 91–123 g) (Weske and Terbaugh, 1981). Stiles and Skutch (1989) reported sex-unspecified weights of 170 g for *cooperi* and 150 g for *guatemalae.* These figures presumably refer to Costa Rican birds, and thus the latter one probably actually represents *vermiculatus.* A reported weight of 100 g for *guatemalae* (= *vermiculatus*) from Panama is considerably smaller (W. Robinson, Brown, and Robinson, 2000).

Identification

In the field. These forms can best be identified in the field by their vocalizations. *Otus seductus* utters a bouncing-ball series of gruff notes that accelerate to a trill, presumably the primary or alpha advertising song of territorial males. In typical *O. cooperi* there is a trilled song that slows toward the middle and ends with a series of evenly spaced notes. Single gruff *woof* notes have also been heard. The (primary advertising?) song of the confusing taxon *lambi* may be different in that it is said to be a guttural, grunting trill, ending in a staccato series of *gok* notes, whereas the male's presumptive secondary song is a whinny similar to that of the eastern screech-owl. The female's vocalizations are similar but are higher pitched.

In the hand. Of the three types listed here, *guatemalae* is the only one with bare, unfeathered toes that are relatively short and weak (middle toe to 20 mm in males, 24 mm in females). It exists in both rufous and wood brown to grayish brown plumage morphs, whereas the other two seem to lack rufous morphs. It is also small enough to be separated from the others by a combination of wing and tail measurements alone (see above). The even smaller whiskered screech-owl (maximum wing-length 153 mm) has feathered toes.

The other two species *O. cooperi* and *O. seductus* have overlapping wing and tail measurements, but *seductus* has unique brown eyes, whereas *cooperi* has yellow ones like the other more widespread screech-owls occurring farther north. Both forms are known to occur only as grayish brown plumage morphs, with a few rufescent tones present, and both are largely adapted to relatively dry climates, where rufous colors would not be expected. The whiskered screech-owl of the same region in central Mexico is usually even grayer overall in plumage tone, but more than 20 percent of the birds are rufous morphs (Gehlbach and Gehlbach, 2000). The Middle American screech-owl is usually much more rufous, even in its most grayish plumage morph. *O. cooperi* is slightly paler on average than *seductus* and has a more definitely yellowish bill. The bill of the taxonomically puzzling Oaxacan form *lambi* is also mostly yellowish green, with a yellower tip. A grayish to blackish bill occurs in the widespread Mexican race (*suttoni*) of the western screech-owl, whereas *mccallii* has the typical yellow or olive bill of the eastern screech-owl.

Vocalizations

Vocalizations of these three species rather closely correspond to those of the western screech-owl, at least in the case of the first two; *guatemalae* may be more distinctive.

In *cooperi* the male's primary song is of the bouncing-ball type, with 5–15 rather gruff notes, these notes typically increasing in loudness and becoming slower in the middle, then falling away and becoming more rapid. A typical phrase has 12 notes and lasts about 1.5 seconds, with interphrase intervals of similar length. Each note is slightly inflected, producing a rougher effect than in *kennicottii*. There is also a two-motif phrase, with the first part so rapidly uttered as to sound like a roll, and the second part distinctly slower. Duetting by the two sexes regularly occurs, and young birds utter whining notes. A tape recording of the song and trill from Costa Rican birds was provided by J. Hardy, Coffey, and Reynard (1999),

as well as comparable vocalizations of *lambi* from Oaxaca. The *lambi* vocalization consists of four bouncing-ball phrases in 25 seconds, the phrases sounding distinctly gruff and having 10–12 notes. According to van der Weyden (1975) *cooperi* songs have 6.7–13.2 notes per second, and 8–31 notes per phrase, each phrase ("motif") thus lasting up to about two seconds. Their highest pitches range from 510 to 840 Hz. These compare with 3.5–11.4 notes per second, 4–20 notes per phrase, and highest pitches of 500–650 Hz in *kennicottii*.

In *seductus* the male's primary vocalization (or so-called alpha-song) resembles the bouncing-ball song of *kennicottii* but may be gruffer in tone and thus more like *cooperi*. It similarly tends to accelerate to a trill. The female's comparable phrase is similar, but higher in pitch. Both sexes utter screaming whinny-like notes that may be used in intersexual encounters (the secondary, or "beta-song," type). A tape recording from Colima of the bouncing-ball and trill calls was provided by J. Hardy, Coffey, and Reynard (1999). The bouncing-ball sequence had three phrases in 20 seconds, each having 10–12 notes. According to van der Weyden (1975) *seductus* songs have 5.1–7.7 notes per second, and 6–14 notes per phrase, thus lasting about two seconds. Their highest pitches range from 700 to 750 Hz.

Vocalizations of *guatemalae* include the primary song, which begins extremely softly but builds to a uniform volume and lasts 5–10 (extremes 2.5–19) seconds on a single pitch, ending abruptly. Each song phrase is a pulsed sequence of notes uttered at a rate averaging about 14 notes per second and with fairly long pauses between individual song sequences. Asynchronous duetting by the two sexes occurs. Van der Weyden (1975) reported that in *guatemalae* the trilled song lasts 3–9 seconds, with the female's vocalizations about three tones higher than the male's. In birds from Costa Rica south, these phrases are said to be shorter and their overall pitch higher. Tape recordings of both Mexican (*hastatus*) and northern Costa Rican (*vermiculatus*) populations were provided by J. Hardy, Coffey, and Reynard (1999). The Mexican sequence has three monotone trills in 25 seconds, each trill lasting 4–6 seconds. The Costa Rican sequence has four monotone trills in 70 seconds, the trills lasting longer (up to 12 seconds) and having longer intervals between, contrary to the statement made earlier about possibly shorter phrase lengths. According to van der Weyden (1975) *guatemalae* songs have 12.6–15.8 notes per second, and 40–144 notes per overall phrase. Their highest pitches range from 570 to 1060 Hz. The song of the Puerto Rican screech-owl is also similar to that of *guatemalae*, but is shorter and more rapidly uttered. On this basis alone the Puerto Rican birds could be considered conspecific with *guatemalae* (Van der Weyden, 1974), although the two populations are widely allopatric and conspecificity seems unlikely.

Duetting (synchronized in *cooperi* and *seductus*) is common in all three of these species, and secondary songs have been reported for all but *guatemalae,* which exhibits both synchronized and nonsynchronized duetting. Duetting and secondary songs also occur in *asio* and *kennicottii.* In contrast, duetting is rare in the flammulated owl, and no secondary song is known to occur (van der Weyden, 1975).

Habitats and Ecology

The endemic Mexican form *seductus* is mainly associated with thorn forest. Schaldach (1963) stated that in Colima it is strictly limited to thorn forests; none was seen or heard in thorn scrub or on the coastal plain. It has also been reported in second-growth woods, edges of cropland, and in desertlike habitats with scattered large cacti or mesquite. Its altitudinal range is from sea level to at least 1200 meters (Enriquez-Rocha, Rangel-Salazar, and Holt, 1993).

The lowland form *cooperi* occurs in Mexico in arid to semiarid woodlands, mangrove edges, swamps, tropical lowlands, and in various semi-desert habitats with scattered trees or treelike cacti. Its altitudinal range is from sea level to about 1000 m, or possibly only to about 100 meters (Enriquez- Rocha, Rangel-Salazar, and Holt, 1993). In Costa Rica at the southern end of its range it has a similar ecological and altitudinal distribution, favoring lowland and foothills forests (especially deciduous forests), savannas, second-growth woods, and mangroves, from sea level to at least 1000 meters (Stiles and Skutch, 1989).

The form *guatemalae* is mostly associated in Mexico with dense, scrubby woodland and evergreen to semideciduous forests, including riparian scrub, but Schaldach (1963) found it in Colima only in dense thorn forest. Additionally *guatemalae* has been seen in dense second-growth woods, plantations, and dense scrub. Its Mexican altitudinal range is from sea level to about 1500 m. Somewhat farther south, as in Guatemala, it has been reported from open forests and plantations, at elevations between sea level and 750 m. In Honduras it has been described (perhaps incorrectly) as a cloud forest species, with a minimum altitudinal limit of about 600 meters (Monroe, 1968). Other observers suggest it does not occur in cloud forest but rather is confined to warmer habitats and is a "thermophilous" species that does not extend beyond the frost line (Burton, 1973; Voous, 1988). In Costa Rica its geographic

replacement form (*vermiculatus*) is mostly a wet lowland to middle-altitude forest species, ranging from sea level to about 1000 meters (Stiles and Skutch, 1989). Near the southern end of the range of typical *vermiculatus* in Panama it is associated with humid forests and with second-growth woodlands in lowlands and foothills up to about 900 meters (Ridgely and Gwynne, 1989). In Panama its territory size has been estimated at 15 ha, and its breeding density at four pairs per 100 ha (W. Robinson, Brown, and Robinson, 2000).

Movements

Apparently residential and sedentary.

Foods and Foraging Behavior

All screech-owl forms seem to feed to a large degree on insects, at least when they are readily available, as well as on some small vertebrates, especially small rodents. Few specific prey items have been documented, but the birds are likely to be very much like both eastern and western screech-owls in their prey choices, with somewhat smaller average prey sizes. The eyes of these screech-owls do not "glow" in reflected light, suggesting that they are as much crepuscular as nocturnal.

These Mexican species average slightly smaller than the U.S. screech-owls, and their toes are somewhat shorter, so their prey are also likely to be rather smaller. The long downward-pointing bristles that are present at the base of the mouth in the whiskered screech-owl and that probably aid in insect catching are not apparent in the Middle American screech-owl, so it would be interesting to know how important insects might be in the latter's diet. At least in Oaxaca both species may be regionally sympatric but they tend to be separated attitudinally, with the Guatemalan screech-owl occurring up to about 1100 meters of elevation and the whiskered screech-owl extending from about 1250 meters upward (Binford, 1989).

Social Behavior

No detailed studies on the social behavior of any these forms have been done, but they must certainly closely resemble the better studied North American species in this regard.

Breeding Biology

The Pacific screech-owl probably breeds during the spring dry season, since a nest with both eggs and young present was reported in Oaxaca during late March (Binford, 1989). Old woodpecker holes are the likely usual site, as in other screech-owls. The Balsas screech-owl's breeding biology is still undescribed.

The Middle American screech-owl breeds in Mexico mainly during March and April, with incubating birds recorded as late as June in Yucatan. Its typical nest site is an old woodpecker hole. Frederick Gehlbach (pers. comm.) found males advertising nesting cavities in Belize during mid-February. There they occurred at an estimated density of 1–2 per square kilometer. In Costa Rica (these birds probably representing nominate *vermiculatus*) March and April seem to be the usual breeding time (Stiles and Skutch, 1989). Three sets of eggs in the Western Foundation of Vertebrate Zoology are of 2, 3, and 4 eggs, and are from April 3 (Costa Rica) to May 17 (Sonora). The Costa Rican nest was in an old trogon nesting cavity about 5 meters aboveground.

Evolutionary Relationships and Conservation Status

Cytochrome b gene sequence analysis (maximum likelihood) evidence now places *guatemalae* (evidently sampled from the *vermiculatus* genotype) close to *Otus asio* (sampled from the southwestern form *hasbroucki*), and progressively less close to the allopatric South American forms *sanctaecatarinae, atricapillus,* and *hoyi* (cladogram on p. 45 by Michael Wink and Petra Heidrich, in König, Weick, and Becking, 1999). These associations might suggest a North American rather than

South American origin for *guatemalae,* and Voous (1988) proposed that the Middle American screech-owl is a geographic replacement form for these two more northerly species. However, a separate haplotypic cladogram by Wink and Heidrich (p. 49) shows *guatemalae* as quite distinct from *asio,* as well as from the South American *hoyi* and *atricapillus.* They have suggested that *guatemalae* and *asio* may both belong to the general South American *atricapillus* assemblage.

Until the geographical, biochemical, and behavioral relations between *guatemalae* and *vermiculatus* are better known, and the further relationships of *vermiculatus* to these additional South American forms (as well as the West Indies populations) have been resolved, it seems prudent to regard *guatemalae* and *vermiculatus* as separate allospecies within the superspecies *guatemalae.* Their closest affinities may be with the North American screech-owls (as has been inferred from cytochrome b gene sequence data), or with the Neotropical owls of the *atricapillus* assemblage (as is suggested by geography and song types).

All three forms are believed still to be common to fairly common in Mexico, and their scrub forest habitats are still widespread, if not increasing. Unlike larger owls, screech-owls have proven to be able to live in close proximity to humans without even being noticed, which favors their long-term survival opportunities. In Mexico *O. seductus* has been reported from 3 states, *O. cooperi* from 6 states, and *O. guatemalae* from 18 states (Enriquez-Rocha, Rangel-Salazar, and Holt, 1993).

Whiskered Screech-Owl   *Otus trichopsis* (Wagler) 1832

Other Vernacular Names:
Arizona whiskered owl; spotted screech-owl.

Range (Adapted from AOU, 1983, and J. Peters, 1940.)

Resident from southeastern Arizona, northeastern Sonora, Chihuahua, Durango, San Luis Potosí, and Nuevo León south through the mountains of Mexico, El Salvador, and Honduras to northern Nicaragua. (See Figure 30.)

Subspecies (As recognized by Marshall, 1967.)

O. t. aspersus (Brewster). Southeastern Arizona, Sonora, and Chihuahua.

O. t. trichopsis (Wagler). Southern Mexican Plateau from Michoacan to Veracruz, Chiapas, and Oaxaca. Hekstra (1982) also recognized *ridgwayi* (Michoacan to Jalisco) and *guerrerensis* (Guerrero).

O. t. mesamericanus (van Rossem). Central El Salvador to Nicaragua. Hekstra (1982) also recognized *pumilis* (Nicaragua and Honduras) and *inexpectatus,* a new form seemingly linking *trichopsis* with *O. guatemalae.*

Measurements

Wing, males 139.5–151.5 mm (ave. of 14, 143.2), females 141–151 mm (ave. of 9, 145.7); tail, males 64–75.5 mm (ave. of 14, 69.7), females 68.5–75.5 mm (ave. of 9, 71.5) (Ridgway, 1914). Gehlbach and Gehlbach (2000) have provided wing and culmen measurements for all three subspecies. The eggs average 33 × 27.6 mm (Bent, 1938).

Weights

Earhart and Johnson (1970) reported that 23 males averaged 84.5 g (range 70–104), while 8 females averaged 92.2 g (range 79–121). Gehlbach and Gehlbach (2000) give respective male:female mean weights of 83.6:96.3 g for *asperus,* 83.7:96.6 g for *trichopsis,* and 92.9:102.7 g for *mesamericanus.* The estimated male:female mass ratio is 1:1.09. The estimated egg weight is 13.1 g, representing an egg-to-female proportional ratio of 14 percent.

Identification

In the field. Overlapping in the United States and Canada only with the very similar western screech-owl and flammulated owl (in Arizona), from which it can be separated by the whiskered screech-owl's territorial call of about eight notes of fairly uniform pitch and timing, lasting up to almost 3 seconds, rather than a speeded up bouncing-ball type of call with a greater number of notes. The duetting secondary song between mated birds is a syncopated series of long and short notes on the same pitch, typically consisting of two "dots" and two to five "dashes," the series often repeated three times and terminating with an extra "dash." Females have higher-pitched and more melodious voices than males, and additionally utter a descending *kew* note. Typically found in dense oak groves of the pine-oak zone in southeastern Arizona, where flammulated owls and western screech-owls might also occur. A comparison of typical whiskered screech-owl and flammulated owl postures is shown in Figure 25.

In the hand. In-hand separation from the western screech-owl is possible by the whiskered screech-owl's shorter toes and claws (middle toe no more than 14 mm to base of claw) and smaller bill (culmen from cere 10.5–13 mm vs. at least 13.5 mm in *O. kennicottii*). Compared with the western screech-owl the ear tufts are shorter; there is coarser and darker patterning on the dorsal and underpart feathers; a more definite pattern of white spots is present on the lower hind neck, scapulars, and wing coverts; and the facial "whiskers" are more conspicuous. There are no sexual differences in plumage, and adult measurements greatly overlap, so the presence of a brood patch may provide the only useful clue for sexing breeding adults. Juveniles of the gray-morph birds have dull grayish brown upperparts, indistinctly barred or transversely mottled with dusky and dull grayish white, the latter on tips of the feathers; underparts dull white, broadly barred with grayish brown. Upperparts of juvenile rufous-morph birds (which are rare or absent in the United States) resemble adults, but have indistinct black streaks, and the underparts are pale cinnamon buff to light cinnamon rufous, with narrow and indistinct dusky bars on the sides and flanks. In all types the flight feathers and rectrices of first-year birds are more uniform in color and wear than in older birds, and the outer primaries are initially broader near their tips than in adults, these tips gradually becoming more pointed through wear (Pyle, 1997b).

Vocalizations

According to Marshall (1967), the syncopated duetting song and the uniformly pitched territor-

104

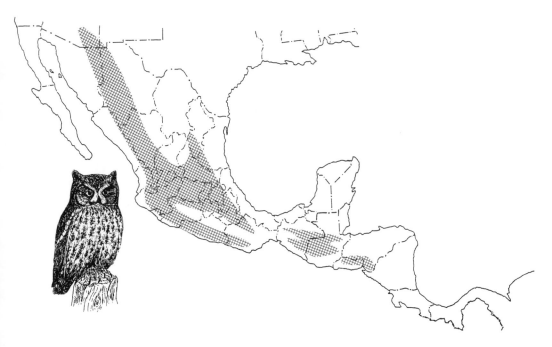

Figure 30. Distribution of the whiskered screech-owl. Mexican distribution after Howell and Webb (1995).

ial song of about 8 notes provide a certain separation of this species from *asio* (including *kennicottii*). Weyden (1975) reported that the highest pitch of the advertisement call is at 610–830 Hz, that 4.6–8.8 notes are uttered per second, and that the total number of notes per motif ranges from 4 to 16. There are a variety of other male calls, including barks, and the female utters a similar array of calls and barks but at a higher pitch, and additionally has a descending *kew* note. Like other owls, the birds snap their bills when disturbed, and they utter alarm *chang* notes or mournful *choo-you-coo-coo* when greatly disturbed (Jacot, 1931). The *chang* note is similar to a cat's meow and is uttered by the female during copulation as well as when she is waiting for the male to return with food (D. Martin, 1974). This note lasts nearly half a second, and has an average fundamental frequency of about 1200 Hz. A trilled note was described by Marshall (1957), and a high-pitched scream was once uttered by the male during copulation (D. Martin, 1974).

The best descriptions of vocalizations in this species come from D. Martin (1974), who provided sonograms of several calls. He stated that the typical song consists of 4–8 notes, and is the most prevalent vocalization during territory establishment and pair formation, but declines during the period of copulation and egg laying. The syncopated song was heard when the birds were excited, usually sexually, as when a male pre-

sented food to a female prior to copulation, and during copulation. The song did not appear to be important in territory defense and more probably serves a role in individual recognition.

Habitats and Ecology

In general, this species prefers higher elevations and denser groves of trees than does *O. asio* (*sensu lato*), typically occurring in dense groves of oaks within the pine-oak zone of montane vegetation. It ranges down into dense oaks below the pine zone in Arizona and some parts of Mexico, and upward through the pine forest to the lower portions of the cloud forest in southern Mexico and El Salvador. In the latter areas it also occurs within coffee plantations, where it breeds in tall shade trees above coffee bushes and occupies groves and thickets remaining after incomplete logging has opened up the denser forests. In some areas of southern Arizona, Sonora, and Chihuahua it occurs in contact with western screech-owls and flammulated owls at about 1700 meters elevation, but normally the two latter forms are more associated with open woodlands and tall pine forests, respectively.

Movements

So far as is known, this species is entirely sedentary, and it probably occurs too far south to be significantly affected by winter reductions in insect populations. The northern limit of the species' range occurs at the terminus of the dense, continuous oak woodlands in southern Arizona. It is possible that some vertical migration to lower elevations occurs during winter, at least at the northern end of the range.

Foods and Foraging Behavior

According to Marshall (1967), the smaller and weaker feet of this species as compared with *O. asio* are a probable reflection of its being adapted to eating smaller invertebrates, or perhaps the smaller feet provide better perching on small twigs. Early information on the foods of the species, as summarized by Bent (1938), indicates a high level of insect consumption, especially of crickets and caterpillars, but foods also include moths, grasshoppers, mantids, beetle larvae, and centipedes. Centipedes are evidently an important winter food.

A similar high level of insect dependency was indicated for this species by Ross (1969), who examined stomachs of 23 birds collected in various parts of the species' range. All the identified prey were arthropods, mostly orthopterans, beetles, and moths. A few other insects, spiders, and a centipede were also found. Most of the arthropods were insects about 15 millimeters in length, but the range was 6–75 millimeters. Evidently the birds forage primarily by aerial captures of winged insects, with some additional captures of vegetation-dependent forms and occasional descents to capture nonflying forms on the ground. There is only one record of a vertebrate (a mouse) among all the foods so far reported for this species.

Social Behavior

Marshall (1957) reported that this was the most pugnacious of all the territorial birds he encountered in the pine-oak woodland community, and strong antagonism could be generated by imitating territorial songs. In some areas it overlaps territorially with western screech-owls, although it favors more densely vegetated areas having extensive screening masses of foliage. A sketch map of whiskered screech-owl territories made by Marshall suggests that they are similar in size to those of flammulated owls and perhaps average about 300 meters in diameter.

D. Smith, Devine, and Gendron (1982) reported that a pair of wild whiskered owls were heard singing a syncopated duet for approximately two minutes while perched on a juniper branch about three meters above the ground. Shortly after the duetting stopped the birds were observed to be copulating, which lasted only a few seconds. After dismounting, the male approached the female and began bill rubbing. Mutual bill rubbing occurred for 7–10 seconds and was followed by mutual preening for about 35 seconds. This was followed again by a brief period of bill rubbing and then a nibbling of the shoulder and breast feathers. Finally the male flew off, followed by the female.

D. Martin (1974) observed three copulation sequences, which exhibited little variation. The male fed the female throughout the night but did not present any food when he approached for copulation, which occurred on the lower branches of the nesting tree or a nearby one. Prior to two of the copulations the female uttered her catlike call, and before the third she had just left the nest cavity. In every case the male flew toward the female, uttering the syncopated song; the female answered with the same song. Both birds continued this duetting for some time after the male had landed beside her. The female began her catlike call at about the time the male mounted her and continued to utter this call during copulation, while the male uttered his syncopated song. After dismounting the male flew off, in two cases while uttering the syncopated song and in the third case silently, while the female usually continued her "meow" call.

Breeding Biology

Fewer nests of this species have been found than of any other species of North American owl; only some seven sets of eggs exist in the major museum collections of North America. Bent (1938) was able to find information on four nests, two of which were taken in the Chiricahua Mountains and two in the Huachuca Mountains of Arizona. All were found in trees (oak, walnut, juniper, sycamore) between 1650 and 1950 m, either in flicker holes or natural cavities, and mostly located between 5.5 and 6.8 meters aboveground. All were found between May 1 and 9 and had either 3 (3 cases) or 4 (1 case) eggs present. The eggs averaged almost exactly the same size (34.2 × 29.3 mm, maximum 35 × 31 mm) as those of western screech-owls, which in that area (*aikeni*) average 34.3 × 28.8 mm. The sets varied from virtually fresh to highly incubated suggesting that egg laying probably occurred in April. A nest mentioned by D. Martin (1974) was in a sycamore tree at an elevation of 1675 meters in the Chiricahua Mountains and had a fresh egg on April 18.

There is no information on incubation or

fledging periods in this species, but they are likely to be essentially the same as those of eastern and western screech-owls.

Evolutionary Relationships and Conservation Status

Marshall (1967) postulated that the whiskered screech-owl was derived from *asio* (*sensu lato*) stock and, since separation, has evolved in parallel with the *kennicottii* group, apparently because of the same ecological forces selecting for cryptic coloration. However, it has evolved smaller feet, a different timing pattern of its advertisement song, and a preference for denser woods at higher altitudes. Marshall supposed that its somewhat smaller size than typical *asio* might be an advantage in maneuvering through heavy foliage, and that its smaller feet might be related to taking smaller invertebrates or for perching on smaller twigs. The three currently recognized races seem to conform to three different climatic conditions (arid and winter cold in Arizona, where gray-morph birds occur; high altitude and cold in the southern Mexican Plateau; and high humidity in Honduras and Nicaragua, where a red morph is fairly common). Hekstra (1982) has postulated that *trichopsis* originated from the *pacificus* group of *O.* (*vermiculatus*) *guatemalae*.

Little is known of the population or its trends in this species, but clearly these are dependent upon the future of fairly dense montane forests within its range. During the 1997–2000 Christmas Bird Counts the species was detected on an average of only 2.0 U.S. and Mexican counts. The maximum number reported on a single count was 8, at Yacora, Sonora, in 2000. In terms of all localities where the species was detected, the whiskered screech-owl was the 16th most widely reported of the 18 North American owl species then encountered. The average total number counted per year was 7.0 birds.

Crested Owl *Lophostrix cristata* Daudin, 1800

Other Vernacular Names:
Strickland's owl (*stricklandi*), Búho corniblanco, Búho penachudo, Tecolote crestadado
(Spanish)

Range

Residentially distributed from southern Mexico
from southern Veracruz to Honduras and along
the Pacific slope from eastern Oaxaca south
through Guatemala, El Salvador, and the rest of
Central America from Colombia and Venezuela
south and east to eastern Peru, eastern Bolivia,
and the Amazonian drainage of Brazil. (See Figure 31.)

North American Subspecies

None recognized by König, Weick, and Becking
(1999). Hoyo, Elliott, and Sargatal (1999) recognized *stricklandi* Sclater and Salvin as the type occurring from Mexico south to western Colombia.

Measurements

Wing, males 303–310 mm (ave. of 6, 307.5 mm),
females 298–320 mm (ave. of 8, 306.6); tail,
males 175–186.5 mm (ave. of 6, 179.2), females
169.5–182.5 mm (ave. of 8, 176.5) (Ridgway,
1914). Six eggs of the nominate race ranged from
43.7 to 48.5 × 37.2 to 38.6 mm (Schönwetter
1967).

Weights

One female weighed 620 g, and two males were
425 and 510 g (Dunning, 1993), suggesting a
male:female mass ratio of about 1.2–1.45. Stiles
and Skutch (1989) reported a sex-unspecified
weight of 400 g; W. Robinson, Brown, and Robinson (2000) similarly reported 510 g. The estimated egg weight is 35.7 g, (Schönwetter, 1967),
representing about 8 percent of the adult female's mass.

Identification

In the field. Easily identified by its distinctive
white ear tufts that extend down to form white
"eyebrows," set off against an otherwise all-dark
brown face. The eyes are typically reddish brown,
although Haverschmidt (1968) described them as
yellow, and Guatemalan specimens were also described as having yellow eyes (Wetmore, 1968), so
regional or age variation might exist. The (usually?) brownish eyes are thus similar in color to
the rufous facial disk, unlike the bright yellow
eyes of the spectacled owl, which also has white
eyebrows but no ear tufts. The only other large
white markings on these generally dark-colored

Figure 31. Distribution of the crested owl. Mexican distribution after Howell and Webb (1995).

upperparts are rounded spots on the scapulars and wing coverts. The underparts are slightly lighter than the upperparts, but lack distinctive markings. It is usually found in dense forests. See also "Vocalizations."

In the hand. This is the only New World owl with long, mostly white, and gradually tapering ear tufts that join at the base of the upper mandible, forming a conspicuous white V that is set against an otherwise uniformly dark brown head. It is a fairly large owl, with wing-lengths usually in excess of 300 mm and adult weights of about 400 grams (0.9 lb) or more. Its toes are naked and the external ear openings are small, symmetrical, and nonoperculated. Linear measurements and weights should serve to identify sex among adults. Natal and juvenile plumages remain undescribed in detail.

Vocalizations

The likely advertising call, or "primary song," is a deep, emphatic growl repeated about every 5 seconds or so. It has also been described as a frog-like croaking call that begins as a stuttering rattle and increases to a croaking rattle, the last and loudest part sounding like *ohrrrrr, gurrrrr,* or *kwarrrr.* The early stuttering notes of the sequence are quite soft and may not be audible at any great distance. A tape recording of the species' song from Venezuela was provided by J. Hardy, Coffey, and Reynard (1999). The song phrase was uttered five times within 20 seconds, and each was a low-pitched croaking growl with about four almost inaudible preliminary syllables, *rr,rr,rr,rr,KWARR.*

Habitats and Ecology

This is a bird of humid evergreen forests and secondarily from riparian forests, semideciduous forests, and partially cleared forests in Mexico, ranging from sea level to about 1800 meters, or possibly to only about 1000. In Guatemala it has a similar vertical range and a preference for heavy forest (Land, 1970). In Honduras it occurs from near sea level in lowland rain forest to 1980 meters in cloud forest (Monroe, 1968). In Panama it occurs both in mature forest and second-growth woodland, to altitudes of about 900 meters (Ridgely and Gwynne, 1989). In Costa Rica the crested owl favors cacao orchards, calling from

the canopy layer, roosting in the midcanopy, and foraging mostly on invertebrates (Enriquez-Rocha and Rangel-Salazar, 1997).

Movements

Apparently residential and sedentary.

Foods and Foraging Behavior

Little is known of the foods of this elusive owl. Wetmore (1968) stated that in one stomach from Panama he found fragments of roaches, locustids, and three large cerembycid beetles. In another specimen from the same area were a caterpillar, orthopteran remains, and fragments of beetles. It seems unlikely that such a large owls would forage mostly on insects, but too little information is available to prove otherwise.

Social Behavior

Almost nothing can be said on this subject. This is a nocturnal species that seems to be able to exist in an area without generating much evidence of its presence.

Breeding Biology

There are no descriptions of breeding in Mexico, but in South America this owl reportedly nests in tree hollows or may even (in Guyana) nest within the loft of a house. It is thought to nest from late in the dry season to the early wet season, breeding in Colombia from February to May (Hilty and Brown, 1986).

Evolutionary Relationships and Conservation Status

This monotypic genus has been thought by some to be most closely related to *Bubo* and by others to *Otus* (Ford, 1967), but it may actually be closest to *Pulsatrix* (Storer, 1972; Hoyo, Elliott, and Sargatal, 1999).

Currently the species is believed to be common to fairly common in Mexico, although the humid evergreen montane forests in which it lives are under concerted attack by loggers, and no real information as to its actual abundance is available. Stotz et al. (1995) listed it as occurring in at least 17 Latin American countries, and as uncommon to fairly common.

Spectacled Owl   *Pulsatrix perspicillata* Latham, 1790

Other Vernacular Names:
sooty spectacled owl (*saturata*), Búho de Anteojos, Lechuza de Anteojos (Spanish)

Range

Residentially distributed along the Atlantic slope of Mexico from southern Veracruz to Honduras and on the Pacific slope from eastern Oaxaca southeast through Guatemala, El Salvador (rare), and the rest of Central America to Colombia, western Ecuador, and tropical and subtropical forests of northern and eastern South America south to northern Argentina. (See Figure 32.)

North American Subspecies

P. p. saturata Ridgway. Southern Mexico to northern Costa Rica.

Measurements

Wing (of *saturata*), males 314–347 mm (ave. of 11, 333.4), females 317–360 mm (ave. of 10, 339); tail, males 177.5–215 mm (ave. of 11, 195.2), females 164–204 mm (ave. of 10, 192.9) (Ridgway, 1914). Two eggs from the endemic Trinidad race measured 50–50.4 × 42.5–42.7 mm (Belcher and Smooker, 1936). Four eggs from Trinidad averaged 50.6 × 42.6 mm (Ffrench, 1973). Haverschmidt (1968) gave somewhat smaller measurements (45.2–45.5 × 37.5–38.1 mm) for eggs (unspecified sample size) of the nominate race from Surinam. Schönwetter (1967) reported average measurements of 45.8 × 37.8 mm, and an associated estimated weight of 35.6 g for two eggs.

Weights

Weight of males 591–761 g, females 765–980 g (unspecified sample sizes) (Haverschmidt 1968; König, Weick, and Becking, 1999). This would suggest an average male:female mass ratio of about 1:1.3. Thirteen birds of both sexes weighed 591–1250 g (average, 873 g) (Dunning, 1993). Two females 816 and 982 g (S. Russell, 1964). One male from Guatemala, 655 g; one of unstated sex from Argentina, 800 g (Field Museum of Natural History). Voous (1988) gave an overall range of 600–1000 g. The estimated egg weight is 51 g, representing about 5–6 percent of the adult female's body mass.

Identification

In the field. One of the most visually distinctive and largest of all Mexican owls, this species is

Figure 32. Distribution of the spectacled owl. Mexican distribution after Howell and Webb (1995).

unique in being dark chocolate brown from the breast up to the rounded and earless crown, except for a broad white throat-band and white eyebrows that coalesce to form a conspicuous white moustache. The eyes are bright yellow, and the underparts below the brown breast-band are pale yellow. It is found in relatively dense forests. Juvenile birds are mostly white except for their wings and tail, with blackish facial disks and contrasting yellow eyes. Several years are evidently required to attain the definitive adult plumage, judging from observations of captive-raised birds. At night the spectacled owl's eyes reflect a dull red color, and it can often be stimulated to return an imitated call, which usually consist of three or four deep, abrupt *hu* notes. See also "Vocalizations."

In the hand. This large tropical owl (600–1000 g) is readily identified by the plumage features just mentioned. Additionally, the tarsi and toes are densely feathered and the toes very large; the bill is wide and strong, The external ears are small, symmetrical, and lack opercular flaps. A combination of linear measurements and weights should readily separate the sexes of adults; brood patches are presumably present in breeding females. Juvenile birds have a distinctive bicolored plumage, with dark sooty brown to blackish orbital regions, similar sooty brown on the chin and throat, and whitish down persistently present elsewhere on the body. The iris color probably changes gradually over time to yellow or yellow orange, but details remain undescribed.

Vocalizations

The commonest vocalization and presumed primary song of this species is an accelerating and fading series of deep, resonant whooping or knocking notes, consisting of up to about 10–20 *wump* or *pup* or *gog* elements. A tape recording of the species' song from Peru was provided by J. Hardy, Coffey, and Reynard (1999). It has three song phrases occurring within 30 seconds, each a rapid series of 10–15 *wump* notes that remind me of a steam engine letting off steam and losing pressure, the last notes progressively faster and softer. The female's corresponding call is reportedly similar but higher in pitch. The notes are sometimes uttered as doublets, and birds may engage in duets. The song has also been described as sounding like a woodpecker knocking on wood or a piece of metal such as a saw blade that is rapidly being flexed back and forth, but these descriptions don't capture the accelerating and fading aspects of the recorded soundtrack just mentioned. Adults may also at times utter a single low-pitched, dovelike *whooo*. Short whistled notes may be uttered while the bird is in flight; these

each last about a half-second, and are repeated every 5–10 seconds. Young birds also produce whistled notes.

Habitats and Ecology

Like the crested owl, this is a lowland evergreen forest species, ranging in Mexico up to about 700 meters, in Guatemala to 450 meters (Land, 1970), in Honduras to 750 meters (Monroe, 1968), and in Costa Rica to 1500 meters (Stiles and Skutch, 1989). Semideciduous forests, forest edges, and plantations are sometimes also used, as are wooded river courses, but forest destruction can be very damaging to the species.

Movements

Apparently residential and sedentary.

Foods and Foraging Behavior

The foods of the spectacled owl appear to be catholic, with a probable emphasis on small mammals, birds, and lizards. Wetmore (1968) noted that they were attracted at night to the squeaking sounds he made to attract smaller owls and that a specimen he shot contained a small arboreal rat (*Oryzomys*), the remains of a larger rat (*Tylomys*), a lizard, and a large orthopteran. These owls have also been found raiding the nests of chestnut-headed oropendulas (*Psarocolius wagleri*) on Barro Colorado Island in Panama. Gomez de Silva, Perez-Villafana, and Santos-Moreno (1997) examined 19 pellets from northern Oaxaca and found that most of the remains were from the naked-tailed climbing rat (*Tylomys nudicaudus*), a large rat weighing up to 326 grams, or about half the mass of the adult owl. There were also remains of other rodents, a bat, a few birds (*Momotus* and *Leptotila*), some insects, and small crustaceans. They have also been reported to take crabs, caterpillars, skunks, opossums, and birds up to the size of jays (Voous, 1988). Others have reported that agoutis (*Dasyprocta*), bats, frogs, beetles, and other insects may be taken (Hoyo, Elliott, and Sargatal, 1999). It is often found near water, and the fact that a local name is "crab owl" suggests that these invertebrates may at times also comprise part of the prey spectrum.

Social Behavior

Although this species is thought to be largely nocturnal, the contrasting plumage colors and bright yellow eyes suggest that some social interactions must occur under somewhat lighted conditions. Wetmore (1968) noted that it is common to encounter males and females close to one another outside the breeding season, suggesting that a prolonged pair bond exists.

Breeding Biology

Like other tropical American owls, this species seems to breed late in the dry season and into the early part of the wet season, its breeding probably thus timed so that its young are raised during the period of peak prey availability. This period extends mostly from April to June in Central America but has been reported during August in Suriname and from September to October in Panama. In Trinidad a nest with two eggs was found during April in a tree cavity about 9 meters aboveground (Belcher and Smooker, 1936), and other breeding records from Trinidad extend from January to May. Incubation requires about 5 weeks, and brooding another 5–6 weeks. Typically only a single chick survives to fledging. The youngster may remain with its parents for at least a year after hatching, and may require up to five years to attain full adult plumage (Bohmke and Macek, 1994; Grossman and Hamlet, 1964). This is one of the few tropical American owls regularly maintained and bred in captivity in North America and elsewhere.

Evolutionary Relationships and Conservation Status

Michael Wink and Petra Heidrich (in König, Weick, and Becking, 1999) reported that cytochrome b gene sequence studies indicate (using the maximum likelihood method) a possible position between *Strix* and *Bubo,* but in an alternative and perhaps less reliable analysis (neighbor-joining method) this genus clustered as a sister group with the South American *Otus* species. Neither association had high statistical reliability (in the form of bootstrap support), so these affinities must remain questionable. *Pulsatrix* has also been allied with *Strix* on the basis of its osteology (Ford, 1967). A possible relationship with *Lophostrix* has also been suggested (Hoyo, Elliott, and Sargatal, 1999).

The spectacled owl is believed to still be common to fairly common albeit inconspicuous in Mexico, and besides occurring in heavy forest it can also survive in tall second growth and forest edges. Heavy logging with no forest replacement could be a serious threat. Stotz et al. (1996) listed it as occurring in at least 20 Latin American countries, and fairly common.

Great Horned Owl   *Bubo virginianus* (Gmelin) 1788

Other Vernacular Names:
dusky great horned owl (*saturatus*); Labrador great horned owl (*heterocnemis*); Montana great horned owl (*occidentalis*); northwestern great horned owl (*lagophonus*); Pacific great horned owl (*pacificus*); St. Michael great horned owl (*algistus*); tundra great horned-owl (*subarcticus*); western great horned owl (*pallescens*).

Range (Adapted from AOU, 1983.)

Breeds from western and central Alaska, central Yukon, Nunavut, northern Manitoba, northern Ontario, northern Quebec, Labrador, and Newfoundland south throughout the Americas to Tierra del Fuego. (See Figure 33.)

North and Central American Subspecies
(Adapted from AOU, 1957, and J. Peters, 1940.)

B. v. subarcticus (Hoy). British Columbia and Mackenzie Valley east to northern Ontario. Identical to *wapacuthu* (Gmelin), sometimes improperly applied to this race. See also *occidentalis* and *scalariventris* below.

B. v. occidentalis (Stone). Southern Alberta and Montana east to Isle Royale, and south to northeastern California and central Kansas. The birds of the central Great Plains were considered synonymous with *subarcticus* by Dickerman (1993),

but a new name may nonetheless be required for part of this large and still-inadequately defined population (Houston, Smith, and Rohner, 1998).

B. v. scalariventris (Snyder). Northern and western Ontario; probably should be included in *subarcticus*.

B. v. heterocnemis (Oberholser). Northern Quebec, Labrador, and Newfoundland.

B. v. virginianus (Gmelin). Minnesota east to Nova Scotia and south to eastern Oklahoma and Florida.

B. v algistus (Oberholser). Western Alaska. Possibly an invalid race (Houston, Smith, and Rohner, 1998).

B. v. lagophonus (Oberholser). Interior Alaska and Yukon south to Oregon and northwestern Montana.

B. v. saturatus (Ridgway). Coastal southeastern Alaska south to coastal California.

Figure 33. North American distribution of the great horned owl, showing residential ranges of races *virginianus* (va), *algistus* (al), *elachistus* (el), *heterocnemis* (he), *lagophonus* (la), *mayensis* (ma), *occidentalis* (oc), *pallescens* (p), *pacificus* (pa), *saturatus* (sa), *scalariventris* (sc), and *subarcticus* (su). Racial boundaries are approximations only; crosshatching indicates regions of intergrades or of uncertain racial status. Extralimital distribution shown in inset.

B. v. pacificus (Cassin). Central California.

B. v. pallescens (Stone). Interior southeastern California east to north-central Texas, and south to northern Tamaulipas.

B. v. elachitus (Brewster). Baja California.

B. v. mayensis (Nelson). Mexico from Jalisco, San Luis Potosí, and southern Tamaulipas south to western Panama. Here includes *mesembrinus* (Oberholser), which if separately recognized occurs from the Isthmus of Tehuantepec south.

Measurements

Wing (all subspecies), males 305–370 mm, females 330–400 mm; tail, males 175–235 mm, females 200–252 mm (Ridgway, 1914). Range of wings of five subspecies, males 316–372, females 315–390. Snyder and Wiley (1976) indicated an average wing-length (chord) for 125 males (all subspecies) as 338.7 mm, and 118 females as 356.5 mm. The eggs of various races average from 53.3 × 43.7 mm in *elachistus* to 56.1 × 47 mm in *virginianus* (Bent, 1938).

Weights

Earhart and Johnson (1970) reported that 18 males and 18 females of *occidentalis* averaged 1154 and 1555 g, respectively, while 22 males and 29 females of *virginianus* averaged 1318 and 1769 g, and 18 males and 12 females of *pallescens* averaged 914 and 1142 g (male:female mass ratio = 1:1.24–1.35). A sample of 895 males and 772 females not separated by subspecies averaged 1304 and 1706 g, respectively (male:female mass ratio = 1:1.31) (J. Craighead and Craighead, 1956). A sample of 21 eggs from southern California averaged 51.3 g, representing an egg-to-female proportional mass ratio of 3.3 percent (Houston, Smith, and Rohner, 1998).

Identification

In the field. This very large owl is difficult to misidentify, at least if its ear tufts can be seen, as no other very large owl has prominent "ears." It is usually inconspicuous during the day, roosting silently in dense vegetation and taking off silently if disturbed. Perched birds often show a white throat mark, and their underparts are barred rather than striped. The most common call is a five-syllable, low-pitched, soft dovelike hooting with a cadence something like "Don't kill owls; save owls." However, shorter three-note sequences ("Don't kill owls") are common, and long series, by the addition of single- or double-note trailing phrases, are also frequent.

In the hand. No other brownish North American owl is as heavy as this species (more than 1200 g), and no other very large North American owl has long ear tufts. Linear measurements and weights are probably adequate for judging sex of most adults, and brood patches would confirm the sex of breeding females. Females tend to be darker and more heavily barred on the underwing coverts than males but have narrower barring on the secondaries and inner primaries than males (Pyle, 1997b). Nestlings are initially covered with white down. The downy head, neck, and body plumage of older nestlings is ochraceous or buff, with detached and rather distant black bars. Remiges and rectrices of first-year birds are more uniform in color and wear than in adults; the bars on these feathers usually are more numerous, and the distances between them shorter, than in adults. The rectrices are relatively narrow and tapered, with irregular barring (see Figure 8A). The iris color is initially brownish yellow to yellow orange; this color becomes clear yellow by the third year (Pyle, 1997b).

Vocalizations

The vocalizations of the great horned owl are rather difficult to characterize, as they are quite variable in their number of syllables and lack the strong accenting typical, for example, of barred owls. Additionally, male calls are more prolonged and elaborate than female calls, as well as being relatively rich, deep, and mellow (Austing and Holt, 1966). When the birds are excited the hooting may be preceded by a short barking note, although this is not so strong as in the barred owl, and the cadence or accenting may then also be more apparent. When attacking an enemy, the owl utters angry, growling *krrooo-ooo* notes, and screaming sounds have also been described under these circumstances. It is likely that these or similar screams are used as the food calls by young and dependent birds to attract the attention of adults (Bent, 1938).

During nest defense, females have uttered short, laughing *wha-whaart* notes. Similar chuckling noises that sound like *whar, whah, wha-a-a-ah,* with the accent on the last syllable, have been heard during disturbance. Adult birds utter whistling notes, with a rising inflection, when the young are first flying, apparently as a means of keeping in contact with them (Bent, 1938).

Emlen (1973) described the patterned calling typical of paired owls, observing synchronized calling of this type from late December through March. The female typically initiated these sequences, which sometimes lasted more than an hour. Her call lasted about 3 seconds and consisted of about 6 notes (dash, dot, dot, dot, dash, dash), with call intervals of 15–20 seconds initially, gradually increasing to 30–50 seconds. The male's response consisted of 5 notes (dash, short dash, short dash, dash, dash). His call also lasted about 3 seconds and often began before the female had finished calling or followed the female's call by only a few seconds. Occasionally the female would follow the male in calling, but at quite variable intervals, and sometimes the male would respond to other males' calls. Emlen believed that such mutual calling between mates may serve to coordinate and strengthen the pair bond.

Habitats and Ecology

Probably no other North American owl lives in so many habitats and under so many climatic variations as the great horned owl; thus it is difficult to characterize habitat requirements for the species. At minimum the birds need a nesting site, a roosting site, and a hunting area. Roosting sites are chosen that allow maximum concealment during daylight hours and often are trees that are more or less segregated, by size, type, or location, from other trees in the area. Conifers are favored over deciduous trees, but in their absence trees that tend to hold clusters of dead leaves through winter, such as oaks and beeches, are favored. A wide variety of nesting sites are suitable, including old stick nests of other birds, snags, large tree hollows, deep and broad crotches in giant cacti or holes in them, cliff ledges, and caves (Austing and Holt, 1966). Hunting areas are typically relatively open, but also include some woodlands or groves, or at least scattered trees for perching.

Baumgartner (1939) believed that for breeding territories the birds prefer mature timbered areas that border water and are surrounded by more open habitats suitable for hunting. Hagar (1957) reported that favored nesting habitats consist of wooded areas larger than 7.7 hectares and typically comprised of mature deciduous forests, with scattered conifers for roosting. Petersen (1979) noted that for nesting the birds preferred the interiors of woodlots to open woods, gallery forests, or forest edges. Based on telemetry data, Fuller (1979) observed that fields and forest edges were preferred hunting habitats, but upland oak forests were preferred for all other activities. McGarigal and Fraser (1985) judged that old forest stands near farmlands were preferred habitats. Apparently older stands not only provide more potential nest sites for this species and the barred owl but also offer more subcanopy flying room because of fewer low branches that might impede their flights.

In an Alberta study of nesting distribution of great horned owls and red-tailed hawks, McInvaille and Keith (1974) found that the habitat diversity in areas representing 1.94 square kilometers around each active owl nest (their presumed hunting ranges) mirrored almost exactly the general availability of various habitat types over the entire area, suggesting that the birds were not having to choose, for example, between agricultural areas and such natural habitats as forest, brush, bog meadows, or aquatic sites, which all occurred in varying degrees of abundance.

Petersen (1979) estimated from telemetry data that annual great horned owl home ranges averaged 329 hectares and tended to be larger among successful breeders than unsuccessful. The average home ranges decreased for both of these categories of birds in spring, apparently as prey availability increased. During summer the home ranges of adults gradually expanded, and those of fledglings also increased from 22 hectares in July to 31 hectares in September. Throughout the year, radio-tagged owls showed a marked preference for using upland and lowland hardwoods, with winter roosting sites typically situated in upland stands of white oaks (*Quercus alba*) or black willows (*Salix nigra*) on lowland sites. In the spring and summer, marsh and related shrub cover became an important habitat for hunting, especially where scattered trees provided perching sites, and woodland edges were also primary hunting habitats. On average, 27 percent of the entire annual home range was of woodlands or marsh–wet shrub habitats, and these were apparently the most essential cover components. In Michigan the distribution of large, mature woodlots seems to determine the distribution and density of great horned owls (J. Craighead and Craighead, 1956).

It seems quite clear that breeding great horned owls disperse and strongly defend recognizable territories, as suggested some time ago by A. Miller (1930). He estimated the radius of each of two males' calling territories as about 40 kilometers, for an area of 60 hectares, about the same as the nesting territory size estimated by Baumgartner (1939). Fitch (1958) estimated that the breeding territory consisted of about 65 hectares, and D. Smith (1969) estimated that the nesting territories of three pairs averaged 105 hectares. Similarly, McInvaille and Keith (1974) suggested that the regular and dispersed distribution of owl nests reflected their territoriality, although this behavior apparently did not limit owl densities, which instead varied with prey abundance. On their Alberta study area of

162 square kilometers from 5 to 16 resident pairs were present in various years, representing a maximum owl density of a pair per 10.1 square kilometers. This is considerably below the estimated breeding densities reported by Gates (1972; approximately one pair per sq km), Baumgartner (1939; a pair per 2.6 sq km of Kansas creek-bottom woods, and about one-third that density in central New York), and by D. Smith (1969; a pair per 6.5 sq km). However, it is similar to densities reported by Hagar (1957; a pair per 11.2 sq km), by J. Craighead and Craighead (1956; a pair per 13.7 sq km in Michigan), and by Orians and Kuhlman (1956; a pair per 12–20 sq km). Houston (1975) reported a maximum nesting density of a breeding pair per 5.7 square kilometers on a 50-square-kilometer study area. Another reported estimate of fairly high breeding densities is that of Errington, Hamerstrom, and Hamerstrom (1940) of a pair per 5.2 square kilometers. In a summary of estimated breeding densities from 15 study areas throughout North America, the minimum recorded density was 0.002 pairs per square kilometer in a large (3000-sq-km) Alberta study area, and the maximum was 0.34 pairs per square kilometer in a small (13.6-sq-km) area in Saskatchewan. Commonly the density ranges from 0.10 to 0.20 pairs per square kilometer (5–10 sq km per pair) (Houston, Smith, and Rohner, 1998). Aspen parklands and grasslands in the Canadian prairie provinces, the central plains of Nebraska, Kansas and Colorado, and the savanna-like habitats of southern Texas all have relatively dense winter (and presumably all-season) populations (Root, 1988).

D. Smith (1969) reported that great horned owl nests were evenly distributed throughout his study area, with sites averaging about 1.6 kilometers apart and situated no closer together than 1.2 kilometers. In the latter case the nests were on opposite sides of the same mountain range, allowing for nonoverlapping hunting ranges. The owls sometimes nested in fairly close proximity to red-tailed hawks, ferruginous hawks (*Buteo regalis*), golden eagles (*Aquila chrysaetos*), and peregrines (*Falco peregrinus*). However, other large owls such as long-eared and short-eared owls were absent and possibly excluded from the great horned owls' nesting territories. As might be expected in sedentary birds, the territories were also occupied during fall and winter months. Smith was of the opinion that in his area great horned owl nesting density may have been influenced by the availability of suitable nesting sites, but might also have been influenced by relative food availability, a possibility that was not specifically investigated.

Movements

Although generally regarded as fairly sedentary birds, great horned owls do exhibit some significant movements, especially toward the northern parts of their breeding range. The banding study by Houston (1978) in Saskatchewan documented this trend. Of 34 recoveries of immature birds obtained prior to August, only 11 had moved more than 11 kilometers from their points of banding. During September and October, 35 of 37 recoveries were within 40 kilometers. However, during November and December there was a very substantial movement of birds (including some of older age categories) to the southeast, with some individuals moving as far as Iowa and Nebraska, and with 17 of 35 recoveries occurring beyond 250 kilometers from the point of banding. Evidently the owls tend to move farther south during years of decreased reproductive success and presumably reduced food supplies than in years of population buildup. Of the total 209 recoveries, 36 involved birds that had moved more than 250 kilometers.

A study by Stewart (1969) of 434 band recoveries from great horned owls banded in various parts of their breeding range indicated that 93 percent were obtained within 80 kilometers of the point of banding. Fewer southern-banded owls had traveled long distances at the time of band recovery than had northern birds, and young birds were more prone to travel than adults. Of the total banding sample, a minimum of 52 percent (possibly as high as 86 percent) of the birds had been shot and another 10 percent trapped. Possibly as many as 96 percent of the birds had thus been intentionally killed by humans in one way or another. Only a very few died of "natural" causes. Similarly, of the 301 banding recoveries reported by Houston (1978), 69 birds were found dead of unknown causes; 62 had been shot; 44 had been trapped; 59 had been hit by cars or otherwise found dead on highways; and 20 had been electrocuted. In a larger sample of 481 Saskatchewan-banded birds, 16.2 percent were killed by vehicles and 9.2 percent died from electrocution. Starvation kills many first-year birds, perhaps almost half in some years of low food supplies. The annual survival rate of adult birds after their second year of life was estimated at 68–72 percent in two studies (Stewart, 1969; Houston, 1978). More recent studies by Houston and Francis (1995) in Saskatchewan indicate variations in annual survival relative to snowshoe hare abundance. In good hare years the survival rates for immatures, yearlings, and adults are 58, 74, and 88 percent, respectively, but in poor hare years 37, 59, and 81 percent. In a noncycling Ohio population the corresponding annual survival rates were

68, 76, and 85 percent (Houston, Smith, and Rohner, 1998).

Foods and Foraging Behavior

One of the early major studies of great horned owl foods was that of Errington, Hamerstrom, and Hamerstrom (1940), who reported that rabbits and hares comprised the major part of this species' prey in the north-central part of the United States. However, J. Craighead and Craighead (1956) found a substantially lower percentage of rabbits (estimated by Petersen, 1979, as representing 41 percent of the total prey biomass intake) in their winter pellet samples from Michigan and a larger consumption (estimated at 27 percent of total biomass by Petersen) of such small rodents as *Peromyscus* mice and meadow voles (*Microtus*). The Craigheads attributed this difference to the local flexibility of the owl in adapting to available food sources.

Bent (1938) has provided an extended list of great horned owl foods, which because of its great diversity is scarcely worth repeating. The foods span a size spectrum from insects and scorpions to domestic cats and woodchucks (*Marmota monax*) among mammals and to geese and herons among birds. It is perhaps more interesting to compare the foraging ecology of the great horned owl with other coexisting owls and hawks, as was done by the Craigheads (1956) and by Marti (1974). Marti found that great horned owls in Colorado took prey averaging substantially larger (mean 177 g) than the prey of three other sympatric owls, and ranging in weight from less than 1 to nearly 3000 grams. The great horned owl's prey spectrum exhibited the widest range of the four species, both in size and variety. It has the strongest talon grip of the four, requiring 13,000 grams of force to open the talons, and the largest talon spread (ca. 200×100 mm). It was apparently the least successful of the four species in capturing prey by hearing alone in complete darkness but was able to find prey visually at low light levels (13×10^{-6} foot-candles) comparable to those levels observed for common barn owls. Marti judged that the great horned owl hunts primarily by perching on vantage posts and making short (up to 100 m) flights out to capture prey only after it has been detected. In a similar comparison between the great horned owl and common barn owl, Rudolph (1978) found that these two species differed in their relative nocturnality, in the proportions of prey types taken, and in their hunting behavior and related habitat preferences. Some of these differences are discussed in Chapter 2.

In a broad survey of 10 North American food studies (Houston, Smith, and Rohner, 1998), lago-morphs regionally constituted from 54.2 to 70.5 percent of estimated prey biomass; large rodents, 9.1–39.4 percent; mice and voles, 7.3–22.7 percent; ducks and galliform birds, 3.8–7.7 percent; passerine birds, 1.5–3.9 percent; and other prey types, generally less than 1 percent. Mammals collectively averaged about 90 percent of prey biomass. Regionally or locally, mean prey biomass varies considerably in North America, from as little as 28 grams to as much as 98 grams. Large rabbits and hares are among the largest of regularly taken prey, but mammals as large as raccoons (*Procyon lotor*), skunks (*Mephitis* and *Spilogale*), marmots and woodchucks (*Marmota* spp.) are sometimes also killed. Many of these larger prey species are well beyond the lifting capacity of the owl. Such prey are dismembered and partly eaten, with the bird possibly returning later to finish consumption. Prey as large as snowshoe hares may be temporarily stored in the nest. There is no information on possible differential prey sizes associated with the substantial difference in adult owl body masses (Houston, Smith, and Rohner, 1998).

Social Behavior

Although some details of courtship in this species are still inadequately understood, it is well known that during courtship display both sexes hoot while bowing and while simultaneously drooping the wings somewhat and cocking the tail upward, sometimes almost at a right angle to the body (Spiers, 1961; Austing and Holt, 1966). During hooting the white feathers below the chin are fluffed or expanded in synchrony with the calls and are highly conspicuous (Figure 34, and photo in Austing and Holt, 1966, p. 37), which probably serves as an effective visual signal under low-light conditions. The female's hooting posture is similar to that of the male's, but her voice is higher in pitch and faster in rhythm. She typically calls less frequently than the male and often responds a few seconds after his calls. Throughout a long session of mutual calling the birds may rub their bills together, and bill snapping may also occur. Some representative "expressions" of adult great horned owls are shown in Figure 35.

Copulation has only recently been well described; it occurred after a period of intense hooting, especially by the male. Both sexes called in the forward-leaning posture described above, and this was followed by the male's preening the female. Both birds then adopted an arched-back and tail-raised posture, followed by prolonged duetting by the pair. The female then flew to a high perch, followed shortly thereafter by the male, who landed on her back. Copulation lasted

Figure 34. Defensive posture of
nestling (*A*), and hooting posture (*B*)
of male great horned owl (after photos
in Austing and Holt, 1966). Also shown
(*C*) is copulatory posture of a male
Eurasian eagle owl (after a drawing in
Glutz von Blotzheim and Bauer, 1980).

4–7 seconds, with both birds hooting rapidly, the male nuzzling the female's nape and occasionally flapping his wings to maintain balance (Houston, Smith, and Rohner, 1998). A similar precopulatory courtship sequence occurs in the closely related Eurasian eagle owl (*Bubo bubo*). While treading, the male eagle owl maintains a distinctly "puffed" appearance, especially around the neck feathers (see Figure 34, adapted from a drawing of an imprinted male eagle owl).

Paired great horned owls are fairly sedentary, with unpaired birds serving as a more mobile population available to form pair bonds with birds that have lost their mates, sometimes in a fairly short time (Petersen, 1979). During much of the year great horned owls are territorial, according to Baumgartner (1939) and later observers.

Paired birds thus tend to be dispersed, their nests forming a regular spaced pattern that is not only related to the presence of other great horned owls but also, judging from some studies (McInvaille and Keith, 1974; Petersen, 1979), affected by the distribution of breeding red-tailed hawks.

From early July until early December, adult owls in Petersen's Wisconsin study area led a nearly solitary existence, apart from some juvenile-female contacts. During the fall young fledged owls hunted and roosted almost entirely on their own but remained within their parents' home ranges before dispersing. By early December apparent pairs began to roost together, and courtship was actively under way by early January. During the four-week period before laying, females began to examine potential nest sites. Up

Figure 35. Great horned owl, behavior, showing facial expressions of a hand-raised bird, including (*A*) curious, (*B*) angry, (*C*) at ease, and (*D*) sleepy. After sketches by Heinrich (1987).

to about two weeks prior to laying, females began to restrict their nocturnal activity to the immediate vicinity of the nest site, but rarely roosted near the nest prior to this time (Petersen, 1979).

With the onset of laying in late February, their mates began roosting within 75 meters of the nest. While their mates supplied them with food, the females performed all the incubation and brooding. Newly hatched and even fledged birds remained dependent upon their parents for food until early June, when paired birds began roosting and hunting separately. Territorial boundaries began to break down in July. By August 1 pair bonds seemed to be nonexistent, although the two birds continued to exhibit home range overlap. Fall habitat preferences of males continued almost unchanged from those of summer, casting doubt on the commonly held view that it is the male that initiates territorial defense and courtship. However, higher use of woodland habitats by presumably unpaired males may reflect their greater interest in territorial defense and courtship (Petersen, 1979). Baumgartner (1939) believed that males assume a dominant territorial role by vigorous hooting during a six-week period in early winter (by early January in Wisconsin and by December in Utah), which probably serves to disperse potential competitors and may also attract the attention of unmated females.

Breeding Biology

Nest sites used by this species are highly variable, apparently depending upon environmental variations in site availability, relative prey distribution, and perhaps relative freedom from human disturbance. Of 13 nests found by Bent (1938) in eastern North America, all were in the heaviest available timber and as far as possible from human habitation. In this general region, large stick tree nests of various hawks seem to be the preferred sites, with live trees probably preferred over dead ones, and some types of trees such as oaks (*Quercus*) and elms (*Ulmus*) possibly preferred above others (Petersen, 1979). Coniferous trees with witches'-broom fungal growths are also favored. Among 1236 breedings in Ohio, 533 were in nests of red-tailed hawks; 527 were in hollows, snags, or rotted cavities of old beeches (*Fagus*) and maples (*Acer*); and artificial platforms were used 125 times. Various other sites such as old squirrel nests were used at lower frequencies. Most nest sites were used only a single year, but one red-tailed hawk nest was used for eight years, and a dead snag nest site was used for nine years (Holt, 1996). Of 10 sites described by D. Smith (1969) in unforested and arid habitats of Utah, cliff faces

and rock outcrops were apparently favored sites. Probably no generalities can be made about favored nest sites that apply throughout broad portions of North America, but perhaps old hawk nests are most commonly chosen when they are available. Great horned owls not only usurp hawk nests, they are also deadly predators on nestlings of many hawk species as large as red-tailed hawks and goshawks, and tend to reduce nesting success of such species when they are nesting within owl territories (Houston, Smith, and Rohner, 1998).

Clutch sizes of great horned owls tend to be rather small, averaging regionally from 1.86 to 2.59, with only slight latitudinal or longitudinal trends evident (Murray, 1976). Some small samples from Alberta suggest that average clutch sizes may vary some from year to year; additionally the yearly percentage of birds attempting to nest varies with relative prey availability (McInvaille and Keith, 1974; Adamcik, Todd, and Keith, 1978). In the latter study there were also marked yearly differences in the mortality rates of unfledged young and in the average number of eggs hatched per successful nest. The average clutch size increased significantly and the mean hatching date was earlier during a year (1969) when hares and mice were abundant, but the clutch size declined the next year, when rodent (but not hare) populations declined. Houston (1971, 1975) found that brood sizes were largest and eggs were laid earlier during years of prey abundance, and the incidence of nonbreeding likewise varied reciprocally with prey abundance. He also observed that in years of prey abundance the nests were located in more diverse (less remote and sheltered) locations than in years of prey scarcity. Some nesting sites have been in use for at least 25 years (Houston, Smith, and Rohner, 1998), although it is probable that more than one pair was involved during such a long duration.

D. Smith (1969) suggested that there may be yearly variations in clutch size and average egg-deposition data that are related to variations in winter temperature and severity, based on 2 years of data. Variations in clutch size as related to female age have not been studied, but it is believed that only about 20 percent of yearling females attempt to nest at all (Henny, 1972), and this percentage may be much lower or even nil in years of low prey availability. Field study records indicate that at least 4 females are known to have bred at 1 year, and 1 male at 2 years of age (Houston, Smith, and Rohner, 1998). Petersen (1979) reported that during 4 years of study, an average of 8.8 out of 11.3 occupied territories supported active nests, suggesting that an overall average of about 78 percent of the sexually mature females in the area attempted to nest during this period, with the percentage ranging annually from 64 percent to 91 percent. In an Ohio study, 62.3 percent of pairs attempted to breed annually over a 28-year study involving 1777 territorial pairs (Holt, 1996). Nonhooting "floaters" that seem to be common in populations may be sexually mature birds unable to compete successfully for territories. In one study the home ranges of 7 floaters averaged about 725 hectares, overlapping with territorial holders, whose home ranges averaged only about one-fifth as large (Rohner, 1997).

If the first clutch of eggs laid by a female is removed or destroyed, she may lay a replacement set, the second clutch usually being of fewer eggs and the eggs themselves often smaller. In at least one case, successive clutches of 4, 3, and 2 eggs were taken from a single female of *occidentalis*, and the nesting site was changed for each nesting attempt (Bent, 1938). The eggs were laid at variable intervals; such intervals may vary from 1 to 7 days. Incubation begins immediately and lasts about 26–35 days, with the longer estimates perhaps reflecting interrupted incubation during cold weather. In one carefully timed case, 2 eggs were laid 3 days apart, and the incubation of each lasted 30 days (Gilkey et al., 1943). In another nest, 2 eggs required 34 and 35 days, and a third at least 33 days (Hoffmeister and Setzer, 1947). Relatively few eggs in successful nests remain unhatched; Holt (1996) reported that 5.8 percent of 1667 eggs didn't hatch, but most of these were in nests that produced no young.

During the nestling period the young are provided with prey that seem to differ little if at all from that consumed by adults during the same interval. The relative weight or biomass of food brought to the young each day does seem to differ according to the number of young present, the hunting skills of the individual parents, and relative prey abundance. In general, about 300 grams of food per day are provided to nests having single young, and nearly 900 grams to a brood of three young (McInvaille and Keith, 1974; Petersen, 1979).The mean number of young present in the nest at the time of banding in three studies involving a total of 4543 nests was 2.13 young; predation, starvation, and siblicide all contribute to chick losses (Houston, Smith, and Rohner, 1998).

During the first week after hatching the young are brought only quite small mammals and birds, but later larger prey is brought, with its identity apparently largely determined by its relative local abundance and its appropriate size. Hoffmeister and Setzer (1947) noted that during a 45-day period at one Kansas nest site, 91 individual prey animals representing 16 species of birds and mammals were brought to the young, with cottontails and rock doves (*Columba livia*) the most frequent prey species. During the first 25–28 days there was a rapid increase in the weight of the young, but thereafter the weight changes were quite varied, apparently as a result of varied feeding rates. By the time the young are 3–4 weeks old they are starting to lose their down, and at 5–6 weeks they are well feathered on the wings. At this time they regularly leave their nest, climbing into nearby trees where they spend their time hiding among the branches while waiting for food. The fledging period is quite indefinite or variable in this species, with the young birds sometimes initially flying for short distances at about 45 days after hatching when about three-quarters grown (Hoffmeister and Setzer, 1947), but they are not proficient at flight until they are 9 or 10 weeks old.

The young birds remain dependent on their parents for a considerable period as they slowly acquire hunting skills. They probably initially capture such easy prey as large insects, but gradually develop hunting skills and associated independence from their parents and eventually move out of their parents' home range to disperse during fall. This period of dispersal may begin in Wisconsin as early as October, or it may be delayed until late January, the latter extreme probably as a result of renewed courtship and territorial activity by resident pairs (Petersen, 1979). Maximum known longevity under natural conditions is 27 years and 7 months (Patuxent Wildlife Research Center banding data).

Evolutionary Relationships and Conservation Status

There is no doubt that this species and the Eurasian eagle owl are very closely related and represent a superspecies if not an allospecies. Other eagle owls also occur in the world, but in distribution and appearance *bubo* and *virginianus* are clearly derived from common ancestral stock. Based on molecular data, Wink and Heidrich (1999) regarded *B. bubo* and *B. virginianus* as distinct species, and also separated the South American *magellanicus* from *virginianus*.

In spite of persecution over most of its range, the great horned owl continues to survive almost throughout the continent, perhaps by virtue of its highly secretive nature and its capacity for ecological adaptability. D. Smith (1969) stated that habitat disruption, road kills, and indiscriminate shooting are major causes of its population decline in Utah, and it is likely that these decimating factors are generally applicable elsewhere in its range. There is a statistically nonsignificant annual population trend of +16.8 percent, based on annual Breeding Bird Surveys done between 1966 and 1993 (Price, Droege, and Price, 1995). A more recent Internet summary of these data suggest a 0.9 percent population decline from 1966 to 2000, which is also not statistically significant (http://www.mbr-pwrc.usgs.gov/bbs/bbs.html). During the 1999 and 2000 Christmas Bird Counts the species was detected on an average of 1240 counts, or about 69 percent of all U.S. and Canadian counts. In terms of all localities where the species was detected, the great horned owl was the most widely reported of the 18 North American owl species then encountered. The average total number counted among the 100 counts having the highest species totals was 2207 birds, the highest of all such averages (averaging 22 birds per count, with each count area equal to 175 sq mi, or 453 sq km). This total represents 23 percent of all owls reported in the tabulated summaries of counts for the United States and Canada during 1999–2000. The maximum number ever reported on a single CBC was 153, in Cumberland County, New Jersey, in 1998. This number suggests a mean density close to a bird per square mile, still well below the species' maximum reported density of a pair per square kilometer. The Canadian population has recently been estimated at 100,000–150,000 pairs (Kirk and Hyslop, 1998), and is considered to be fluctuating or increasing.

Snowy Owl *Nyctea scandiaca* (Linnaeus) 1758

Other Vernacular Names:
American snowy owl; arctic owl; great white owl; white owl.

North American Range (Adapted from AOU, 1983.)

Breeds in North America in the western Aleutians, on Hall Island, and from northern Alaska, northern Yukon, and Prince Patrick and northern Ellesmere islands south to coastal western Alaska, northern Northwest Territories, Nunavut, northeastern Manitoba, Southampton and Belcher islands, northern Quebec, and northern Labrador. Winters irregularly from the breeding range in North America south to southern Canada, Minnesota, and New York, casually or sporadically to central California, southern Nevada, Utah, Colorado, Oklahoma, central and southeastern Texas, the Gulf states, and Georgia. Also occurs widely in northern Eurasia. No subspecies recognized. (See Figure 36.)

Measurements

Wing, males 394–422 mm (ave. of 14, 408.1), females 425–465 mm (ave. of 11, 446.4); tail, males 220–244 mm (ave. of 14, 230.2), females 235–275 mm (ave. of 11, 254) (Ridgway, 1914). The eggs average 56.4 × 44.8 mm (Bent, 1938).

Weights

Earhart and Johnson (1970) reported that 27 males averaged 1642 g (range 1320–2013), while 30 females averaged 1963 g (range 1550–2690). The sexual adult mass ratio is 1:1.2. Winter weights of 23 males averaged 1806 g, and of 21 females 2279 g (Kerlinger and Lein, 1988a), representing a mass ratio of 1:1.26. Mikkola (1983) reported the average weights of 13 males and 27 females of the Eurasian population as 1726 and 2239 g, respectively, producing a male:female mass ratio of 1:1.3. Field weights of eggs have ranged from 60 to 65 g, averaging 58 g (Watson, 1957). The egg-to-female proportional mass ratio is thus 2–2.6 percent. This is the smallest such ratio of all North American owls, helping to account for the species' sometimes large clutch size. An 11-egg clutch would represent about 30 percent of adult body weight, roughly comparable to a 2-egg clutch of the elf owl.

Identification

In the field. The large size and mostly white plumage of this species make it unmistakable. It is often seen perched on large rocks or haystacks, and less often on posts, stumps, or in trees. It is almost always completely silent except during breeding. Then a low-pitched hooting is conspicuous, and can be heard at great distances (for more then 10 km under favorable conditions).

In the hand. The predominantly white plumage and large size (wing at least 375 mm, weight at least 1300 g) sets this species apart from all other New World owls. The ear tufts are absent or rudimentary, the ear openings are relatively small and symmetrical, and the five outer primaries are distinctly emarginated. Adults can be fairly reliably sexed by the darker plumage and substantially larger measurements of females (but young males also resemble females in relative plumage darkness). Breeding females are also always heavily soiled on their underparts from nesting. Nestlings are covered with white down for the first 10 days after hatching; this soon becomes sooty gray except on the throat, legs, toes, and facial disk. Juveniles are uniformly brown with scattered white tips of down; the scapulars are brown with whitish bars; and the rectrices and remiges are white with brown vermiculations and crossbars (Mikkola, 1983). The remiges and rectrices are uniform in color and relative amounts of wear; the central rectrices are tapered and have 4–6 (rather than up to 3) dark bars, including any indistinct bars. The iris is initially brownish to brownish yellow, gradually becoming a clear yellow or orange yellow by the third year (Pyle, 1997b).

Vocalizations

Outside the breeding season snowy owls are very quiet, but breeding birds are more vocal, especially near the nest. Disturbed owls, particularly males, produce a repeated *kre* call, uttered in flight. During various situations, such as before and after being fed by the male, during distraction display, and during displacement coition, the female produces a loud, intense whistling or mewing note. Another common call is a low, rapid, and repeated cackling *ka,* uttered by the male under various conditions of excitement. The female produces a higher-pitched but similar *ke* note under similar kinds of stimulation, such as before copulation. Various other minor calls or call variants have also been heard, and bill snapping together with hissing are commonly associated with threat (Watson, 1957).

Figure 36. North American breeding distribution of the snowy owl. The long dashes indicate usual southern wintering limits; the short dashes indicate limits sometimes reached by wintering vagrants. Extralimital distribution shown in inset.

A call that is probably entirely confined to the breeding season is male hooting, a loud, hollow booming sound that may be uttered from the ground, from a perch, or in flight, and that may be heard for several kilometers. The call is usually a double *hoo, hoo,* but it is sometimes a single note or may be uttered in a series of 6 or more, with the last the loudest. The notes may be uttered at 1- or 2-second intervals, and may be very difficult to localize. When hooting the throat is greatly swollen, and the tail is variably cocked, sometimes almost vertically. Hooting by females has also been reported but is seemingly rather rare. It clearly serves the function of song in that males often hoot in response to one another's calls from adjacent territories, or toward human intruders, but hooting may also be performed toward a female as her mate approaches her or leaves her. Nonbreeding owls evidently do not hoot (Watson, 1957).

Habitats and Ecology

Typical breeding habitats consist of lowland tundra, which on Baffin Island are usually valley floors below 200 meters elevation. However, in southern Norway the birds breed on high mountains about 1000 meters elevation or higher. In both cases the breeding habitat corresponds to the zone where lemmings are most abundant. In a Swedish study area nests similarly were broadly spread between about 600 and 1000 meters elevation, on high mountain plateaus (Wiklund and Stigh, 1986). In general the breeding distribution of the species corresponds to the distribution of arctic or subarctic rodents, especially lemmings (Watson, 1957). Snowy owls are among the most cold-adapted of all birds owing to the heavy coat of highly insulating feathers, resulting in an extraordinarily low level of thermal conductance, indeed among the lowest level reported for all birds (Gessaman, 1972).

Breeding densities of owls vary by location and also during different years in the same location, depending on prey abundance. Population densities in the area studied by Watson (1957) varied in different valleys and in different parts of the same valley, reflecting local variations in lemming density. Some breeding territories were limited by river flats or glacial moraines, while others extended to include both sides of a river. The birds rarely flew up slopes and apparently did not defend them. Eleven breeding territories were found in a river valley floor of about 32 linear kilometers. In the central part of the valley, where lemmings were most abundant, the average territory was about 2.6 square kilometers. The smallest territory in that area was about 1.3

square kilometers, and the largest in the entire valley was 5–6.5 square kilometers. One bigamous male had a territory of about 1.6 square kilometers in the area of densest lemming abundance.

Similar estimates of population densities and territory sizes have been made by others. In Alaska, breeding territories averaging about 5.2 square kilometers have been found, with nests up to about 1.6 kilometers apart (Pitelka, Tomich, and Trichel, 1955b), and elsewhere on Baffin Island territories of 1.6 kilometers in diameter have been estimated (Sutton and Parmelee, 1956). On Banks Island, during good lemming years densities of an owl per 2.6 square kilometers have been found, as compared with one per 26 square kilometers in poor years (A. Manning, cited in Watson, 1957).

Wiklund and Stigh (1986) reported that the breeding territories of snowy owls in Sweden in different areas and years averaged from 2.86 to 6.53 square kilometers (total range 0.3–9.6 sq km), but 38 percent of the territories in one area were estimated to be less than 1.0 square kilometer during a good breeding year. Lower overall food availability during one year was apparently compensated for by larger territory sizes and greater distances between nests. The average minimum internest distances varied from 1390 to 2875 meters in different study areas and years. During a good breeding year on Southampton Island, Parker (1974) observed a general nesting density of a nest per 22 square kilometers over a large land area, and a mean distance between nests of 4.5 kilometers. Following a lemming crash, no breeding owls were found in the same area.

Winter territoriality has also been reported in snowy owls (Keith, 1964). Boxall and Lein (1982c) reported that near Calgary, Alberta, male snowy owls appeared to be nomadic, but females defended territories of from 150 to 450 hectares for periods of up to 80 days. Territory size was inversely related to the proportion of preferred hunting habitats in the form of stubble and edge present within the territory. Juvenile females were found to defend territories that averaged larger than those of adults but had lower proportions of preferred hunting habitats within them.

Movements

The movements of this species, like other arctic-breeding raptors, are irregular and irruptive in nature, rather than being regular migrations. Thus in some years there are major incursions of the birds into southern Canada and the northern United States, such as during 1945–1946, when a widespread invasion occurred from coast to coast (Gross, 1948), and again during 1966–1967, when there was a major influx into the Pacific Northwest (Hanson, 1971). Over a 64-year period (1882–1946), irruptions occurred at a mean frequency of 3.9 years. These incursions have varied greatly in intensity, the geographical area concerned, and the total amount of territory affected. However, they generally occur at intervals of about 3–5 years, often coincidentally with cyclic or periodic declines in lemming populations. Such declines often occur during a year in which there has been relatively little snow, which may reduce breeding activity and also probably subjects the lemmings to increased levels of predation (Gross, 1948). Additionally, during years of sparse snow cover and depleted available forage, the lemmings apparently frequently die from "cold weather starvation," or increased winter mortality associated with malnutrition and exposure, especially in genetically weakened populations resulting from a rapid population buildup (Parker, 1974).

Studies by Kerlinger and Lein (1986, 1988b) have shown that there are major significant differences in winter movements and distributions among age classes and the two sexes of snowy owls in North America. On average, immature males (which are the lightest age-sex class) winter farthest south, and adult females (the heaviest age-sex class) the farthest north, with the other two age-sex classes occupying intermediate positions. Apparently it is typical for males to arrive first in wintering areas of southern Canada or northern United States, only to be evicted later as the larger and socially dominant females arrive, forcing the males to move farther south. The age-class distribution of wintering owls suggests that adults regularly winter south to the northern Great Plains, while a much larger proportion of the birds wintering to the east or west of this region are immatures. Thus the snowy owl should probably be considered a regular overwintering migrant to the northern plains; winter influxes elsewhere (farther south and on both coasts) are more likely to be the result of periodic and irregular irruptions (Kerlinger, Lein, and Sevick, 1985). States reporting snowy owls every year on Christmas counts held between 1952 and 1981 include Montana, the Dakotas, and the New England states from New York north. Owl invasions into the northern United States during winter are erratic, and movements such as these are not always directly correlated with lemming population cycles (Parmelee, 1992). Recent radio-tagging of snowy owls in western Alaska by the Center for Conservation Research and Technology, Baltimore, Maryland, has also documented transcontinental movements of adult birds across the Bering Strait into northeastern Siberia.

Foods and Foraging Behavior

While on the breeding grounds, there is no doubt that over much of the species' range lemmings (*Lemmus* and *Dicrostonyx*) or voles (*Microtus*) are almost the exclusive food of snowy owls. Watson (1957) noted that all but one of 56 items of fresh food seen at owl nests and perches were lemmings, as were all of the many dozens of feed items observed being caught and fed to young. Evidently the owls do not select for any particular size of lemming while their young are in the nest, and the two species of lemmings are apparently caught in ratios reflecting their general abundance locally. However, on the Shetland Islands (where lemmings and voles are absent), M. Robinson and Becker (1986) observed that the birds fed on rabbits (*Oryctolagus cuniculus*) in years when they were available, but during years when they were rare the chicks of various wading birds served as the primary prey.

In Finland and Scandinavia the role of lemmings is apparently secondary to that of *Microtus* voles as a food source during the breeding season; only about a third of more than 2700 prey items identified in various studies there were lemmings, and most (50.6 percent) were *Microtus*. Birds comprised less than 2 percent of the total sample in those areas (Mikkola, 1983).

In an early review of the winter foods of North American snowy owls, Gross (1944) stated that birds and mammals are the primary prey species. Among mammals, rats (*Rattus*) are the most commonly reported prey, with mice, voles, moles, rabbits, hares, and large rodents all being identified in varying quantities. Among birds, ring-necked pheasants (*Phasianus colchicus*) are the most common prey species in southern Canada and the Midwest; but grouse, quails, waterfowl, domestic poultry, rock doves (*Columba livia*), and various alcids such as dovekies (*Alle alle*) have been reported in some quantities. In 127 stomachs from owls killed in New England, 71 mammal remains (mostly of *Rattus* and *Microtus*) were present, as well as 77 bird remains representing at least 18 species. Winter stomach samples from 87 Maine owls had rats and mice present in 35 percent, snowshoe hares (*Lepus americanus*) in 20 percent, and passerine birds in 10 percent (Mendall, 1944).

In an analysis of prey remains obtained from five snowy owl winter territories in British Columbia, grebes and ducks comprised an estimated 90 percent of the prey intake when analyzed by prey weight, which usually consisted of birds in the 400–800-gram range of adult body size (Campbell and MacColl, 1978). In a Michigan study, the winter hunting range of one owl was estimated at about 1 square kilometer, and its hunting activities tended to be concentrated in the earliest and latest hours of the day. Still-hunting by watching for prey from an elevated perch was used most often; ground-hunting by walking or hopping over the snow surface and apparently listening for prey immediately below was done occasionally; and hunting by extended coursing was observed only rarely. The primary prey were voles (*Microtus*), followed in diminishing order by mice (*Peromyscus*) and shrews (*Blarina*). Several unsuccessful attacks on birds were also witnessed (Chamberlin, 1980). *Microtus* and *Peromyscus* also comprised the majority of winter food items in an Alberta study, although year-to-year variations in weather affected prey choice considerably. Additionally the smaller males were prone to feed almost entirely on such rodents, whereas females took a wider array of prey, including some larger mammal and bird species (Boxall and Lein, 1982a).

Social Behavior

In apparent contrast to most owls, the snowy owl has a seemingly weak pair bond. Beyond the normal condition of monogamy, several cases of bigamous matings of one male and two females are known (Watson, 1957; Hagen, 1960; M. Robinson and Becker, 1986), and in Norway there are cases of females having either two mates simultaneously or mating successively with three males in an 18-day period (Mikkola, 1983). On the other hand, it is believed (from plumage traits) that the same pair of owls nested on Fetlar, in the Shetland Islands, during nine successive years from 1967 to 1975, after which the male disappeared.

P. Taylor (1973) has provided a detailed description of courtship in snowy owls, based on observations on Bathurst Island. There the birds arrive in late April or early May, when the tundra is still snow covered. The birds begin courting in early May, and soon scatter out over the tundra. The most conspicuous aspect of courtship is aerial display by the male, which was seen from early May until July. In this display (Figure 37) the male flies with a marked delay at the top of an exaggerated wing stroke, forming a high V position and causing the bird to sink slightly. With rapid wing strokes the bird would regain its lost height, then again assume the V position, thus proceeding in an undulating flight path. Toward the end of the flight the male would climb a few meters, then set its wings in the V position and glide to earth, or flap in almost vertical descent. No calls were heard by Taylor during these moth-like flights. They were initiated when the female was visible, but were variably oriented toward, away from, or past her, and varied in length from a few dozen meters to more than a kilometer. At times a lemming would be carried in the male's bill.

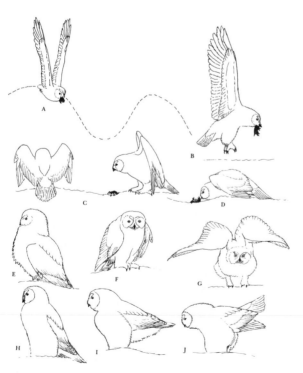

Figure 37. Snowy owl behavior, including male display flight (*A–B*), mantling display and closely following posture (*C–D*), defensive postures (*E–G*), resting posture between hooting (*H*), pause in hooting (*I*), and hooting posture (*J*). After P. Taylor (1973).

Upon landing after such a flight, the male usually began a ground display in which he often dropped the lemming he was carrying and remained fairly erect, keeping his wings partly spread with the wrists held high, often walking and turning around within a few feet of the starting point, and usually orienting the back toward the courted female, producing a highly conspicuous white surface with his back and horizontally stretched wings (an "angel-like" conformation). As the display progressed the male would lean farther forward, partially fanning the tail and lowering the head. At this time the female would usually approach the male, sometimes landing immediately behind him, and possibly begin begging for food. The male invariably leaned farther forward, as if trying to hide the lemming from her view, and often held the wing nearest the female higher than the farther one. Then the male might fly to a new perch to repeat the same courtship sequence. It is possible but not certain that the lemming is eventually fed to the female by the male, probably just prior to copulation. In any case, the female eventually assumes a receptive posture, with tail partly raised, wings held loosely to the side, and body tilted forward. This posture is maintained for a brief period after treading, and some males may continue ground display at that time.

Watson (1957) observed more than 100 cases of what he described as "displacement coition," which occurred after the female had been disturbed on the nest and flew away, followed closely by the male. On landing the male would immediately copulate with her. In these cases the treading was seemingly nonfunctional in that cloacal contact was apparently lacking. Tulloch (1968) also observed copulation behavior following disturbance, as did P. Taylor (1973). However, the copulations observed by Taylor in this context did not appear abnormal to him, and he believed them functional rather than "displacement." Females would usually sway the head laterally before and sometimes after copulation, as well as during normal begging for food, either for their own consumption or prior to passing it on to the young. Watson also observed that males would often rub the face and breast of a female with food prior to passing it on to her, especially if she initially refused to accept it. Although it is known that the female will preen her chicks, and older owlets will allopreen one another, mutual allopreening between pair members has apparently not been specifically noted.

When uttering territorial hoots from the ground, males assume a nearly horizontal posture, with head held forward and tail raised (Figure 37), similar to the posture of great horned owls, but when hooting is used as a threat toward human intruders the associated posture is more upright and lacks the lifted-tail component (P. Taylor, 1973).

During defensive threat, snowy owls lower the body, partly extend the wings, and stretch the head low and forward, with most body feathers fluffed. The wings are also variably raised, thus producing a larger surface area and an almost circular outline when viewed from in front. Although males most often exhibit this posture, females sometimes do as well, as when they are disturbed on the nest during incubation. Both sexes perform distraction displays, leaving the nest and walking toward the intruder while beating the wings, trailing and dragging them, or raising and waving them. This behavior is probably most typical of birds with young in the nest, but may also occur during the incubation period (Watson, 1957).

Judging from observations by Boxall and Lein (1982b), courtship activities (hooting and ground display) may begin as early as midwinter, in areas well removed from the breeding grounds, thus some birds may arrive on nesting areas already mated and ready to begin nesting as soon as conditions permit. It is generally believed, however, that pair bonds are weaker and less permanent in this species than in many other large owls, possibly as a reflection of its semimigratory behavior and seemingly rather nomadic breeding.

Breeding Biology

Watson's (1957) observations on Baffin Island provide an excellent overview of breeding biology in this species. All the nests he observed were in sites having a good view, unsheltered from wind. Six were on large boulders; another was on a small crag; and yet another on a dry hillock in an area having no boulders. There, the greater safety from foxes provided by nesting on boulders as compared with hillocks might be significant. However, Wiklund and Stigh (1986) noted that the nests they found in Sweden were more often on hillocks then boulders, both of which serve as lookout perches and provide early snow-free areas. Most of their nests were also within 100 meters of a watercourse, the dense associated vegetation of which provided excellent lemming habitat. Similarly, Pitelka, Tomich, and Trichel (1955a) found that 8 of 10 nest sites were on high earth polygons or the raised portions of well-drained sites, which apparently not only offer early snow-free ground but might also allow the bird to arrive on and leave the nest in the opposite direction from possible enemies without ready detection. Snowy owl nests are sometimes found close to snow goose colonies; probably the geese choose to nest near the owls to gain protection from other predators; owls and other raptors rarely attack prey close to their own nest sites, and it is quite common for vulnerable bird species to select nest sites remarkably close to those of large owls, hawks, and eagles. Judging from Watson's observations, new nest sites are typically chosen each year, although these may be close to the previous year's site. He found three old sites and one active one within a radius of about 360 meters. The two nest sites of a bigamous male were less than 1600 meters apart, with one of the females initiating breeding about three weeks later than the other.

The breeding season of snowy owls is surprisingly constant over about 30 degrees of latitude, with the first eggs typically laid between May 10 and 22, and laying ending by early June. This seems to be related to the average peak abundance of prey rodents in July, when the owlets are being fed, and the usual amelioration of winter weather in May, allowing for nesting and improved prey access. Replacement clutches following egg loss have not been reported but have been suspected. The clutch size of snowy owls varies considerably according to local or temporal variations in food supply, with means during peak lemming years ranging in various studies from 7.5 to 9.2 eggs (Watson, 1957) but often only 3–5 eggs in years of scarce prey. It is also possible that young females may lay smaller clutches than older ones nesting in the same area (M. Robinson and Becker, 1986). Watson observed a maximum clutch of 9 eggs in his sample, but believed that earlier reports of up to 15 eggs in a single nest were reliable. Thus in Finnish Lapland, clutches ranging from 5 to 14 eggs have been reported, the mean of 66 nests being 7.74 eggs (Mikkola, 1983). The eggs are laid at variable intervals of 2–4 days, usually averaging nearer the former, but are sometimes delayed during periods of severe weather (Tulloch, 1968; M. Robinson and Becker, 1986).

Incubation and brooding are performed entirely by the female, who is provided food by the male. Typically prey caught by the male is delivered to the female at the nest; she may either take it to a feeding station to eat it or store it in a food depot. Males apparently never tear up prey to feed to the nestling young directly, but the female does this soon after the male delivers it. However, after the young have left the nest and scattered the male may deliver whole prey items to them (M. Robinson and Becker, 1986).

The incubation period lasts on average 32–33 days, with some unconfirmed estimates of as long as 38 days and others of as little as 27 days

(Parmelee, 1992). Watson (1957) reported an average hatching interval of about 42 hours, or almost exactly the same as the average egg-laying interval reported by M. Robinson and Becker (1986), indicating that intense incubation must begin with the laying of the first egg. Assuming an average hatching interval of about 2 days, and a clutch of at least 7 eggs, it is apparent that a 2-week or even 3-week span may separate the ages of the oldest and youngest owlets in a fully hatched nest. However, it is unlikely if not impossible that the first should hatch by the time the last is laid, and unlikely that the first young should be almost fledged (perhaps having already left the nest) before the last is hatched, as had been thought by some early observers (Watson, 1957).

The owls are cloaked in white down for up to 10 days, after which grayish down appears. The wing quills begin to appear at about a week of age, and there is a rapid weight increase for the first 4 weeks, by which time the young weigh in the range of 1300–1600 grams. This extremely fast rate of growth may reflect the potential 24-hour period of available hunting at very high latitudes, but on Fetlar, hunting was most intense during the period of least light and lowest during midday. On Baffin Island the birds were less active around midday and midnight than at other times (Watson, 1957), but the timing of hunting may largely reflect relative abundance and activity patterns of local prey. On Fetlar a 4-year average of about 340 grams of prey were brought to the nest per day by the male during the incubation period, compared to about 710 grams per day following hatching. On Baffin Island, Watson estimated that the young nestling birds averaged about 2 lemmings each per day, or about 160 grams of food apiece daily. By comparison, adults evidently consumed about 150–350 grams per day. Following nest departure the young probably consumed about 200 grams a day, so that each probably consumed about 1500 lemmings between hatching and the time of complete independence from their parents. Even with this substantial lemming harvest, it is likely that the daily intake of 3 adults and 15 young on a bigamous male's territory did not harvest as much as 0.2 percent of the available lemming population (Watson, 1957).

Watson (1957) observed that the young began to leave their nests at 18–28 days of age, usually at about 25 days, well before they could fly. Similar estimates have been made in Alaska (Pitelka, Tomich, and Trichel, 1955a), compared with only 14 days elsewhere on Baffin Island (Sutton and Parmelee, 1956) and 16–26 days on Fetlar (M. Robinson and Becker, 1986). The young soon scattered over an area of at least a square kilometer, but stayed within the boundaries of their parents' territory and away from its edges. Actual fledging occurred as early as 30 days, but the young were unable to fly well until more than 50 days old, and could fly almost as well as the adults when more than 60 days old (Watson, 1957). On Fetlar the average owlet's age at the time of its first flight was 45 days, with a range of 43–50 days (M. Robinson and Becker, 1986).

Watson found that of 32 eggs laid in 4 nests during a good lemming year, 31 owlets hatched and all of these fledged, apparently an unusually high breeding success rate. Only 2 of 19 pairs he studied certainly did not breed at all. Higher mortality rates of eggs and young have generally been reported elsewhere (Pitelka, Tomich, and Trichel, 1955b; Sutton and Parmelee, 1956). Out of 12 nesting efforts during 9 years on Fetlar Island, a total of 56 eggs were laid by 2 pairs, and 44 young hatched (79 percent) in 9 successful nesting efforts. Of these, 23 young fledged (fledging success 52 percent; overall breeding success 41 percent), with most of the deaths occurring during the first 10 days after hatching. Most (18) juveniles survived to their first winter. More young fledged during years when rabbits were abundant than during years when their food was primarily provided by shorebird chicks (M. Robinson and Becker, 1986). Following fledging, there seems to be a substantial degree of dispersal by young birds. Among 7 young banded on Victoria Island, one was recovered a year later in eastern Ontario, and another 4 months afterward on Sakhalin Island in eastern Siberia. Such dispersal tendencies of young birds help explain why the species is both nomadic and monotypic (Parmelee, 1992).

Sexual maturity is reached by the time the birds are a year old in captive birds, but it is possible or even probable that wild owls usually do not breed until they are at least 2 years old (Parmelee, 1992). Maximum known longevity under natural conditions is 10 years and 9 months (Patuxent Wildlife Research Center banding data). In the wild, most documented mortality has been attributed to collisions, typically with automobiles or unknown objects. Gunshot wounds and electrocution also account for a significant number of deaths. Captive birds have lived up to at least 28 years (Parmelee, 1992).

Evolutionary Relationships and Conservation Status

The genus *Nyctea* is a monotypic one, with no obvious near affinities, but now is increasingly considered a probable close relative of *Bubo* (Parmelee, 1992). On the basis of osteological traits, Ford (1967) included *Nyctea* with *Bubo* in his proposed tribe Bubonini. Based on molecular studies, Wink and Heidrich (1999) have suggested that these two genera shared a common ancestor about 4 million years ago. Similarities in male calling postures and courtship behavior might tend to support this association.

Most of the snowy owl's breeding range places it well out of contact with disruptive human influences, and thus its overall status is presumably little changed in North America in recent years. However, the unpredictability in its prey-dependent breeding densities and wintering distributions make such statements of doubtful value and essentially impossible to test. In Europe it is believed to have generally decreased for uncertain reasons, but perhaps as a reflection of long-term climatic trends (Mikkola, 1983; Voous, 1988). During the 1997–2000 Christmas Bird Counts the species was detected on an average of 73 counts, or about 4 percent of all U.S. and Canadian counts. In terms of all localities where the species was detected, the snowy owl was the 11th most widely reported of the 18 North American owl species then encountered. The average total number counted during these 4 years was 146 birds. The 2000 total count was 238 birds, or substantially above the previous 3-year average of 126. The maximum number reported on a single CBC was 79, at Padilla Bay, Washington, in 1974. The Canadian population has recently been estimated at 10,000–30,000 pairs (Kirk and Hyslop, 1998) and is believed to be fluctuating.

Northern Hawk-Owl *Surnia ulula* (Linnaeus) 1758

Other Vernacular Names:
American hawk-owl; day owl; hawk owl; Hudsonian owl

North American Range (Adapted from AOU, 1983.)

Breeds in North America from the limit of trees in western and central Alaska, central Yukon, northwestern and central Mackenzie, southern Keewatin (now part of Nunavut), northern Manitoba, northern Ontario, northern Quebec, central Labrador, and Newfoundland south to south-coastal Alaska, southern British Columbia, south-central Alberta, central Saskatchewan, southern Manitoba, northern Minnesota (and perhaps rarely northern Wisconsin), northern Michigan, south-central Ontario, southern Quebec, and New Brunswick. Winters from the breeding range southward, in North America irregularly to southern Canada and northern Minnesota, and casually to western Oregon, Idaho, Montana, South Dakota, Iowa, Wisconsin, southern Michigan, northern Ohio, Pennsylvania, and New Jersey. Also distributed widely in northern Eurasia. (See Figure 38.)

North American Subspecies (Adapted from AOU, 1957.)

S. u. caparoch (Müller). Distributed in North America as indicated above.

S. u. ulula (Linnaeus). Resident in Eurasia; accidental in western Alaska.

Measurements

Wing (of *caparoch*), males 218–251 mm (ave. of 14, 239.6), females 223–240 mm (ave. of 8, 232.9); tail, males 160–191.5 mm (ave. of 14, 176.2), females 172.5–182 mm (ave. of 8, 178.8). The eggs average 40.1 × 31.9 mm (Bent, 1938).

Weights

Earhart and Johnson (1970) reported that 16 males averaged 299 g (range 273–326) and that 14 females averaged 345 g (range 306–392). J. Duncan and Duncan (1998) reported 42 males averaging 301.2 (range 242–375) and 54 females as 339.8 g (range 250–454). Mikkola (1983) reported the average weights of 22 males and 20 females of the Eurasian population as 282 and 324 g, respectively. There is no indication of clinal weight variation and only slight geographic plumage variation. These weight data suggest a male:female mass ratio of 1:1.13–1.15. The estimated egg weight is 21.2 g. The estimated egg-to-female proportional mass ratio is thus 6–6.5 percent.

Identification

In the field. This distinctly hawklike owl is usually seen perched in the dead branches of trees, often perched at an angle as if about to take flight. The relatively long tail is sometimes cocked, and the small whitish facial disk is outlined in black. The territorial call is a prolonged series of whistled *ululu . . .* notes that may last for more than 10 seconds, and a hawklike series of repeated *kee* or *kip* alarm notes is often uttered in flight, which is rapid and rather falconlike. Most hunting is done during daylight hours, which adds to the bird's hawklike impression.

In the hand. The relatively long (at least 160 mm) tail, which is banded with white and somewhat tapering in shape, is distinctive, as are the straight, narrowly pointed primaries (the eighth the longest and the outer four variably emarginated on one or both webs) and the densely feathered toes. The soft filaments at the tips of the wing feathers are poorly developed. Adults cannot be reliably sexed by plumage differences, although females might average somewhat more rufous brown. Body measurements also overlap too greatly to permit sexing by this means, but brood patches might allow for recognition of breeding females. The young are initially covered with white down having a yellowish buff tint. Upperparts of juveniles are dark sooty brown or sepia, the feathers of crown and hind neck tipped with dull grayish buff, which forms the predominating color. The underparts are dull whitish, deeply shaded across the chest with dark sooty brown, the rest being broadly but rather indistinctly barred with brown. The juvenal remiges and rectrices are uniform in color and degree of wear, with the outer primaries gradually becoming more tapered through wear. The central rectrices are narrow and more tapered than in older birds and have bolder, more complete white cross-bars but a narrower whitish terminal fringe (Pyle, 1997b).

Vocalizations

The most typical vocalization is the male's advertising song, a long, bubbling or trilled whistle that may last up to 14 seconds, with intervals of up to 5 minutes between song phrases or given at

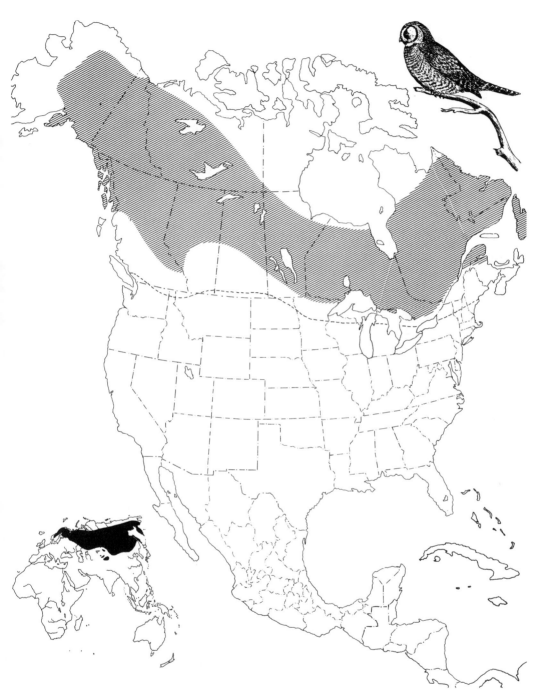

Figure 38. North American breeding distribution of the northern hawk-owl. The dashes indicate usual southern limits of wintering vagrants. Extralimital distribution shown in inset.

shorter intervals of only a few seconds. The song is usually uttered during display flights over the territory, as well as when showing nest sites to a female or when summoning an incubating female. It may be heard from as far as 500 meters away. The corresponding female call is shriller, shorter, and less sonorous, often having a wheezing or melancholic aspect. Both sexes utter a series of short, sharp, and trilled call notes when a human approaches the nest; this call varies from soft and purring in males to harsher, more bleating, and metallic in females, somewhat resembling a squeaky ungreased machine shaft. A softer, more kittenlike version of this call is uttered during pair formation as a contact call, as a greeting call between mates, during copulation, and as a duet when prey is being passed between the mated birds (Cramp, 1985; J. Duncan and Duncan, 1998).

Other adult calls include a repeated, drawn-out screech, uttered as an alarm call toward humans, when being mobbed, or toward rival birds; a descending whinny that perhaps serves as a long-distance contact call; and a strident and repeated yelping call that is typically uttered in flight, such as during aerial attack or when defending a nest. The same or a similar call has been described as a "hunting call" uttered by perched and alert adults while scanning their surroundings. Some minor calls include a "lure call," uttered by either sex as it enters a potential nest site, a copulation solicitation call by the female, an injury-feigning call by the female, and various calls associated with bringing food to the nest or with passing food to the young. In addition to these, several calls are produced by young birds, including hissing sounds, suggesting that this species is a highly accomplished vocalist (Mikkola, 1983; Cramp, 1985; J. Duncan and Duncan, 1998).

Habitats and Ecology

Typical breeding habitat of this species consists of open to moderately dense coniferous or mixed coniferous-deciduous forests bordering marshes or other open areas such as those cleared by lumbering. The northern limits of breeding closely approximate the southern limits of the snowy owl's breeding range, in timberline fringes of forest taiga or forest tundra, while in mountainous areas the birds extend upward to timberline, sometimes as high as about 2000 meters elevation. The species' southern breeding limits extend to the edges of forest steppe and to cultivated regions. Access to open hunting areas, in the form of muskegs, dry ridges, stunted krummholz trees at timberline, and burned areas are fa-

vored, particularly where there are broken-off stumps, snags, or bare tree branches available to serve as convenient lookout perches (Mikkola, 1983; Cramp, 1985). Open-canopy forests are preferred over closed-canopy woods, and storm-damaged trees provide nest sites, as do burned trees whose decaying wood favors woodpecker excavations and rotted cavities. Open woods such as burned-over areas also offer greater opportunities for visual small-mammal hunting; relative prey abundance may override other factors in habitat selection (J. Duncan and Duncan, 1998).

Outside of the breeding season the birds are found more widely, such as around wooded farmlands and sometimes even reaching prairie areas, where they may perch on haystacks, on posts, or in such trees or bushes as may be available. Territoriality is evidently lacking in winter. Territories are established by the males a few weeks prior to the breeding season and are largely advertised by display flights and singing.

Breeding densities are evidently low; an area of 200 square kilometers surveyed by Hagen (1956) in Norway supported only four pairs. Good habitat in Sweden may support only about a pair per 500 square kilometers (Cramp, 1985). Apparently, territories are correspondingly large, with nest sites well isolated from one another. Six internest distances in North America ranged from 1.8 to 8 kilometers apart, averaging 5.2 kilometers (J. Duncan and Duncan, 1998). Shorter internest distances have been reported from Scandinavia, these distances probably varying with population densities. Robiller (1982) described the territorial behavior of a breeding pair in Lapmark. Indirect evidence suggests that at times the birds may sometimes obtain prey from at least a kilometer away from the nest (Mikkola, 1983). Home range estimates from Baekken, Nybo, and Sonerud (1987) for three males and two females varied from 140 to 848 hectares, as determined by the convex polygon estimation method.

Movements

Like other boreal owls, this species is relatively dispersive and irruptive, the birds apparently moving freely through the coniferous forest zone according to the local abundance of prey, especially small rodents. Thus there are periodic changes in breeding densities and distributions, as local or regional rodent populations rise and crash, often at intervals of 3–5 years. During years of normal rodent density the birds typically winter well to the north, but during winters following peak rodent years many birds disperse southward (Cramp, 1985; Mikkola, 1983). The majority of those involved in such southward invasions are juveniles; 85 percent of those studied during an irruption year in Sweden were young (Edberg, 1955), and nearly 90 percent of birds collected in Finland from areas south of the breeding range were also young (Forsman, 1980). Byrkjedal and Langhelle (1986) reported that seasonal movements away from the breeding range were most pronounced in adult females, least in adult males and juvenile females, and intermediate in juvenile males. This was believed to result from the effects of male competition for breeding territories and social dominance. Although movements of birds in North America are still largely unstudied, most fall southward dispersals here evidently occur from mid-October through mid-November, with a northward return from February to early April. Depending on the region, the directions of these postbreeding movements seem to be either north–south, or northwest–southeast (J. Duncan and Duncan, 1998). Other information is available for the Eurasian population, which is presumably comparably mobile. Thus nestlings banded in Sweden have been recovered in Finland and the former USSR in directions mainly to the northeast and southeast of banding and at distances of up to 1860 kilometers displacement (Osterlof, 1969). No such long-distance movements have been documented in North America, where only eight banded birds had been recaptured or recovered through 1992 (J. Duncan and Duncan, 1998).

Foods and Foraging Behavior

As with movements, the best data on foods derive from studies in Europe, especially Scandinavia. Mikkola (1983) and Cramp (1985) have summarized these data very well. Based on studies at nest sites in Norway, Finland, and Russia, in all cases voles (Microtidae) made up at least 93 percent of the identified prey animals (average 95.7 percent for all 3 areas), with the largest percentages comprised of *Microtus* and *Clethrionomys,* and a few other voles (*Arvicola*) or lemmings (*Lemmus, Myopus*) also represented. Shrews (Soricidae), birds, and other prey comprised the remaining total. When considered as percentage of live weight, microtine rodents represented 75.6–96.3 percent of the biomass intake.

Outside the breeding season the food of hawk-owls evidently changes drastically; like snowy owls the percentage of birds taken tends to increase sharply as rodents presumably become less easily captured. Thus birds as large as the willow ptarmigan (*Lagopus lagopus*) and hazel grouse (*Bonasa bonasia*) are apparently regularly taken, as are smaller species (including the Teng-

malm's owl). Data on 37 hawk-owls from Finland and Russia indicate that birds may comprise about 30 percent of the total food items, voles about 57 percent, and shrews about 11 percent (Mikkola, 1983). If relative biomass is also considered, the role of birds as an important food source is emphasized even more; data from Finland suggest that at least 90 percent of the food intake by weight may be avian (Mikkola, 1972; reanalyzed by Cramp, 1985).

By comparison, North American data are relatively scarce, but information summarized by Bent (1938) indicates that a similar array of rodents such as mice, lemmings, and ground squirrels are consumed during summer, plus some insects, and a comparable shift in winter to readily available birds such as ptarmigans (*Lagopus* spp.) occurs. There are a few reports of hawk-owls' killing or at least being associated with the remains of other fairly large mammals, such as weasels (*Mustela*) and snowshoe hares (*Lepus americanus*), and birds as large as sharp-tailed grouse (*Tympanuchus phasianellus*) and ruffed grouse (*Bonasa umbellus*). Axelrod (1980) observed in Minnesota that ruffed grouse remains were present, along with those of *Microtus* and *Blarina,* in some hawk-owl pellets, but was uncertain whether the grouse remnants were a result of predation or scavenging. In a summary of 8 North American studies involving a total of 1600 identified prey items, *Microtus* comprised 62.3 percent, *Clethrionomys* 22.4 percent, *Synaptomys* 2.9 percent, *Phenacomys* 2.3 percent, and *Lepus* hares 2.3 percent. Birds contributed 1.4 percent, *Peromyscus* and *Lemmus* 0.2 percent each, and other prey types comprised 6.1 percent (J. Duncan and Duncan, 1998). This summary might underestimate the importance of birds during the winter season, as many of the analyzed pellets came from summer nest studies, and only 3 of the 8 studies were done during winter. Surprisingly, however, the highest incidence of bird remains (18 percent) came from a summer nesting study involving 651 prey items from central Alaska (Kertell, 1986). By comparison, in a Yukon nest study, birds comprised only 3 percent of 449 prey items found in pellets (Rohner et al., 1995). Evidently the birds are highly opportunistic in their prey selection.

The hawk-owl's typical mode of hunting involves visual searching from a convenient and elevated lookout, followed by a rapid pursuit flight. Swift swoops down, followed by a return to a favored perching site, are apparently typical. It has also been observed hovering, presumably over prey, as in some other owl species. Rarely prey may be taken in flight, or even stalked or chased on the ground. Scavenging carrion is sometimes also performed. It is one of the most diurnal of owls, sometimes hunting during bright daylight, but it will also hunt during early morning and early evening hours, though rarely if ever in total darkness (Mikkola, 1983; Cramp, 1985). Prey can be detected visually from as far away as 800 meters, but also can be captured when it is concealed by snow, suggesting a keen binaural hearing ability. However, unlike in great gray owls, snow-plunging is relatively rare in this species (J. Duncan and Duncan, 1998). Perch-hunting is often done from a site about 5 meters aboveground, and the average strike distance for 18 attempts was around 8 meters. The capture success rate for 28 male attempts was 68 percent, but was much lower for females undergoing tail molt (Kertell, 1986).

Social Behavior

The general appearance of this species (Figure 39) is characterized by a streamlined, rather hawklike aspect, resulting in part from the long tail, which is often quickly raised and more slowly lowered. The long tail is reminiscent of pygmy-owls; the similarity is increased by the black nape patterning, which somewhat resembles the "false eyes" pattern of pygmy-owls. At times, as when

warm, the birds hold their folded wings slightly out from the sides of the body. When alarmed by a potential predator, a sleeked upright posture may be assumed by adults as they stare at the source of danger. When being mobbed, a similar posture is adopted but with the eyes reduced to slits. A similar upright posture has also been observed in young birds (James and Nash, 1983). Head cocking is also commonly performed, perhaps as a means of helping to estimate distance visually. As with other owls, mantling of captured prey is typical.

Hawk-owls probably become sexually mature at one year, and the pair bonding of this species is presumably one of seasonal monogamy. No data are available on the possible incidence of monogamy persisting more than one breeding season, or on nest site fidelity, although the nomadic tendencies of the species would tend to make these unlikely attributes (Cramp, 1985). In Alaska, however, one nest site was used for two consecutive years (J. Duncan and Duncan, 1998). McKeever (1997) noted that captive birds form only seasonal pair bonds, and by October the birds fail to recognize their partners of the prior spring, however, females appear to form an attachment to particular nest sites. The first known

Figure 39. Northern hawk-owl behavior, including (*A*) sleeked upright posture, (*B*) normal resting posture, (*C*) mantling of prey, (*D*) head cocking (for prey location?), and (*E*) clinging to nesting tree. After various sources including Mikkola (1983) and Cramp (1985).

case of polygamy (bigyny) was documented by Sonerud et al. (1987a), in which two females had separate breeding territories within a single male's territory. The second female abandoned her nesting effort after the nest of the male's initial mate hatched, requiring all the male's food-getting abilities to tend its first brood successfully. Similar bigyny has been reported in captive birds. Mutual billing or nuzzling behavior occurs in paired birds, as does antiphonal duetting. (Glutz von Blotzheim and Bauer, 1980).

In central Finland male hawk-owls have been heard calling from mid-February until mid-April, and both singing and copulation in North America have been observed as early as February. Males perform display flights above their territories while calling, and associated wing clapping has also been reported. Advertising calls are often uttered from perches as the head is slightly raised, drawing visual attention to the black and white throat. Females respond to male advertising calls by uttering a similar call, and later the paired birds perform antiphonal duetting of trilled calls. The pair may also make wheeling and turning flights close to one another, and in one case the female landed and displaced the perched male, who then approached the female from behind and above, as if to attempt copulation (J. Duncan and Duncan, 1998). Nest site selection is achieved by the male sitting below or near a prospective site and attracting his mate with the advertising call. Tree holes, the tops of snags or stumps, nest boxes, and occasionally old stick nests of raptors or crows (*Corvus* spp.) may all be chosen as potential nest sites. The pair may then inspect the site together, or the female may inspect the site as the male continues to utter advertisement calls while staring fixedly at the site. However, the female may also utter her "lure call" and thus take the initiative in attracting the male to a potential site, and presumably the final choice of nest site is the female's (Glutz von Blotzheim and Bauer, 1980).

Copulation has been observed around midday and also at dusk. It is typically preceded by a loud duetting and may be initiated by either sex. The female utters a series of soliciting calls while in a receptive horizontal posture, with her wings drooped and her tail cocked. The male then nudges her flank and flutters his wings while uttering his trilling call. Trilling and other calls occur during treading, and after dismounting the male assumes a rather stiff posture prior to flying off. The female may also remain in a somewhat stiff posture for a period, making slow tail-pumping movements. At about the time that copulations begin, the male also begins courtship feeding and food caching, both at the nest and away from it. Males announce their arrival with food by uttering their trilled calls, and females may also summon their mates by making wide circling movements of the head and uttering the soliciting call (Glutz von Blotzheim and Bauer, 1980).

Breeding Biology

Rather few egg dates are available for North America. There are 10 Alaskan and arctic Canadian records from April 28 to June 14, with half the records between May 4 and 17, and 38 Alberta records from April 1 to June 4, with half between April 13 and 28 (Bent, 1938). More recent records generally fall within these parameters. Six Labrador and Newfoundland records are from May 3 to June 30. Collectively, North American egg dates range from April 1 to June 30 (J. Duncan and Duncan, 1998). It would seem that most laying in subarctic Canada occurs during April, whereas in arctic Canada and northern Alaska laying during May is more typical. There is no information on possible renesting following clutch loss in North America, but egg replacements have been reported in Europe.

More than half of 16 nest sites found by Sonerud (1985) in Scandinavia were in open forests with only scattered trees (clear-cut areas or bogs); about 30 percent were in open spruce forests; and the rest were in closed spruce forests or in ecotone areas between open and closed forests. Nest sites in Alberta, where the birds are usually found in muskeg areas, are most often in natural cavities or enlarged northern flicker (*Colaptes auritus*) or pileated woodpecker (*Dryocopus pileatus*) holes in dead stubs, but birds have been found using old crows' nests as well. Other North American sites mentioned by Bent include the tops or hollows of stubs or snags, natural tree cavities, and even cliffs, although the last-named site would be distinctly unusual if not unlikely. In height the nests mentioned by Bent ranged from as low as 1.6 meters aboveground to at least 13 meters high. Clutch sizes in North America usually range from 3 to 9 eggs, with clutches of 7 commonly found (Bent, 1938). The mean of 51 North American clutches summarized by J. Duncan and Duncan (1998) was 5.8 eggs, not very different from Scandinavian figures. In Lapland the average of 101 clutches was 6.56 eggs, versus 5.94 for 18 clutches in central Finland and 5.13 for 13 clutches in southern Finland, indicating an increased average clutch size at higher latitudes. Additionally, clutches may average somewhat higher during good vole years than in poor ones.

There was an overall countrywide average of 6.31 eggs for 135 clutches (Mikkola, 1983).

The eggs are laid at intervals of 1–2 days, averaging about 1.6 days, and incubation begins immediately with the laying of the first egg. Incubation probably requires from a minimum of 25 to no more than 30 days. As with other owls, all the incubation is apparently done by the female. The male does bring food to the nest but rarely enters it after incubation is under way. Huhtala, Korpimäki, and Pullianien (1987) reported that the young of hawk-owls gain weight very rapidly (averaging 9 g per day), about three times the rate of growth of Tengmalm's owls, as a result of feeding the young at a rate of three to four times more often than other northern owls. This is achieved by temporary caching of excess foods caught by the male and by the daylight hunting behavior of the species that is adaptive in these long-day latitudes. Thus a very early fledging of the young is possible during the short available breeding season. The young require 25–35 days to fledge, but typically leave the nest and move out onto nearby branches when about three weeks old, where they are fed by both parents. The mean number of fledged young from 49 successful North American clutches summarized by J. Duncan and Duncan (1998) was 4.2, or 72 percent of mean clutch-size estimates.

Until they are about two months old the fledglings remain in the vicinity of the nesting tree, and probably they are not fully independent until nearly three months of age, when dispersal of the juveniles may begin (Mikkola, 1983; Cramp, 1985). By October or November many young birds may be appearing well outside their breeding range during years of hawk-owl invasions. Typically, young birds and females are more prone to disperse, and to travel greater distances, than adult males, which are more likely to remain near their nest sites. Maximum reported lifespan in the wild is at least 8 years; birds have lived up to 10 years in captivity There is little information on causes of mortality in the wild (J. Duncan and Duncan, 1998).

Evolutionary Relationships and Conservation Status

The genus *Surnia* is traditionally placed near *Glaucidium* in taxonomic sequence, and certainly in some conformational and behavioral aspects (such as tail length, wing shape, and mode of hunting) these two genera appear to be quite similar. An osteological study by Ford (1967) concluded that these two genera are not only closely related but also part of a larger assemblage (tribe Surniini in Ford's taxonomy) that also includes *Athene*. All of these are visual hunters, with relatively simple ear structure. Molecular studies by Wink and Heidrich (1999) also place *Surnia* near *Glaucidium,* and somewhat less close to *Athene*.

As with the snowy owl, it is difficult if not impossible to assess the population status or trends of this species, as their breeding ranges are so far removed from most human population centers, and at least in the case of the hawk-owl the birds would be almost impossible to census visually on the breeding areas. It is thought that North American populations have remained fairly steady during the past century (J. Duncan and Duncan, 1998). However, in Europe it is believed that the species has declined markedly during that same period, since the more recent periodic invasions southward have been of fairly small scale as compared with earlier ones (Mikkola, 1983). During the 1997–2000 Christmas Bird Counts the species was detected on an average of 25 counts. In terms of all localities where the species was detected, the northern hawk-owl was the 13th most widely reported of the 18 North American owl species then encountered. The average total number counted over the 4-year period 1997–2000 was 48 birds, but the total 2000 count was 143 birds, or substantially higher than the average of 17 for the previous 3 years. The maximum number ever reported on a single CBC was nine, at Sax-Zim, Manitoba, in 1992. The Canadian population has recently been estimated at 10,000–50,000 pairs (Kirk and Hyslop, 1998) and is believed to be fluctuating but stable.

Northern Pygmy-Owl *Glaucidium gnoma* (Wagler) 1832

Other Vernacular Names:
mountain pygmy-owl (*californicum* group); California pygmy-owl (*californicum, sensu stricto*); coast pygmy-owl (*grinnelli*); cape pygmy-owl (*hoskinsii*); Mexican pygmy-owl (*gnoma* group) Rocky Mountain pygmy-owl (*pinicola*); Vancouver pygmy-owl (*swarthi*).

Range (Adapted from AOU, 1983.)

Resident from southeastern Alaska, British Columbia, southwestern Alberta, and western Montana south, mostly in mountainous regions, to southern California, and east to central Colorado, central New Mexico, and extreme western Texas. Also (*gnoma* group) from southern Arizona south through the interior of Mexico to Guatemala. A disjunctive population (possibly a distinct species) also occurs in southern Baja California. Species limits in this pygmy-owl are controversial and unsettled. (See Figure 40.)

Subspecies (Adapted from AOU, 1957, and J. Peters, 1940.)

G. g. grinnelli (Ridgway). Southeastern Alaska south to coastal southern California.

G. g. swarthi (Grinnell). Vancouver Island.

G. g. californicum (Sclater). Northern interior British Columbia south to southern California and east to Colorado and northern Coahuila. Includes *pinicola*. Considered specifically distinct from *gnoma* by Sibley and Monroe (1990) and König, Weick, and Becking (1999). This putative species, including *swarthi* and *grinnelli,* has been called the "mountain pygmy-owl" to distinguish it from the *gnoma* and *cobanense* groups.

G. g. gnoma (Wagler). Southern Arizona south to Guerrero and Chiapas. Sometimes (with *cobanense*) called "Mexican pygmy-owl" to distinguish it from previous forms.

G. g. cobanense (Sharpe). Highlands of Guatemala. Has been synonymized with *gnoma* (Ridgway, 1914).

G. g. hoskinsii (Brewster). Baja California (Cape District). Considered specifically distinct ("Cape Pygmy-owl") by Howell and Webb (1995) and König, Weick, and Becking (1999), but possibly best considered part of *gnoma* group.

Measurements

Wing (of *gnoma*), males 82–92 mm (ave. of 9, 87.4), females 89.5–98 mm (ave. of 5, 93.7); tail, males 57–63 mm (ave. of 9, 59.2), females 58–63.5 mm (ave. of 5, 59.4) (Ridgway, 1914). The eggs of *californicum* average 29.6 × 24.3 mm (Bent, 1938).

Weights

Earhart and Johnson (1970) reported that 42 males of *californicum* averaged 61.9 g (range 54–74), and 10 females averaged 73.0 g (range 64–87). Mean of 3 males 57.8 g (range 41.5–69); 1 female 70.0 g (Holt and Petersen, 2000) The estimated male:female mass ratio is 1:1.18–1.21. N. Johnson and Russell (1962) indicated that 8 breeding males of *californicum* averaged 62.8 g (range 57.3–80). The estimated egg weight of *californicum* is 9.1 g. Holt and Petersen (2000) reported 6 eggs of 1 clutch as 6.2–9.1 g. The estimated egg-to-female proportional mass ratio is thus about 8–13 percent.

Identification

In the field. Except in the very limited area of possible overlap with the ferruginous pygmy-owl, this tiny species is unlikely to be mistaken for any other owl; it has a long, narrow tail, dark brown streaks on its white underparts, and two black patches on its nape that vaguely resemble a pair of extra eyes. Its usual calls include a distinctive series of three (sometimes two or four) single-note hoots that are uttered periodically (usually terminally) among a series of other more rapid notes. In central and southern Arizona these are usually double-noted hoots; there confusion with the sympatric but ecologically separated ferruginous pygmy-owl (which has a more "popping" call) is possible. Where both species might possibly occur these call differences and the less rufous coloration of the northern pygmy-owl should help to identify it.

In the hand. Distinction from all other owls but the ferruginous pygmy-owl is possible by the combination of very small size (wing under 130 mm), a crown that is plain brown or lightly spotted with white, a mostly grayish brown upperpart coloration, and a relatively long tail with five to eight incomplete white bars. Adults cannot be reliably sexed by plumage or by measurements, although females tend to be more rufous brown, especially in Mexico, and breeding females would have brood patches. The nestlings are initially covered with whitish down. Juveniles are similar to adults, but the crown is unspotted brownish gray, in marked contrast with browner color of back, brown on sides of breast unspotted, and texture of plumage softer. The general plumage

Figure 40. Distribution of the northern pygmy-owl, showing residential ranges of the races *californicum* (ca), *cobanense* (co), *gnoma* (gn), *grinnelli* (gr), *hoskinsii* (ho), and *swarthi* (sw); indicated racial limits are only approximations. Taxonomy unsettled; some populations (*gnoma, hoskinsii* and *swarthi*) may eventually be separated specifically from remaining forms.

markings of young birds may be somewhat weaker (less contrasting) than in older age-groups, and the central rectrices might be somewhat narrower and have more closely spaced cross-bars than in adults, but these possible age differences require confirmation (Pyle, 1997b).

Vocalizations

The primary advertising call of this species of pygmy-owl over most of its U.S. range is a monotonous, repetitive series of *hoot* or *hoot-hoot* notes, usually uttered at 1- to 2-second intervals. At least in the case of *californicum*, two or three (rarely four) final and well-spaced hoots are uttered after a preliminary more rapid series of soft whistlelike *too* or *kew* notes uttered at an even pitch, the sequence something like *too-too-too-too-too-too-too-too; toot; toot; toot.* At times the group of spaced notes may be uttered independently, or a single toot note may be produced, and the series of soft staccato whistles may also be uttered separately. Judging from limited information, the races *pinicola, grinnelli,* and *hoskinsii* are similar to those of *californicum* in uttering a series of single-noted and evenly spaced *toot* notes, often following a staccato sequence, but in southern Arizona the local form *gnoma* tends to deliver its notes in groups of twos, interspersed with single notes (Phillips, Marshall, and Monson, 1964). In Colorado the birds have been heard producing an extended series of single-noted hoots at a rate of about 2 per second for periods of up to 20 seconds (Bailey and Niedrach, 1965). Both single-noted and double-noted sequences have been heard in Colorado, the single-note sequences with longer internote intervals. These songs, like those from Montana and Oregon, closely resemble the song of the European species (Holt and Petersen, 2000).

Judging from observations on the closely related Eurasian pygmy-owl (*G. passerina*), this repeated single- or double-noted tooting is a territorial song and may be uttered throughout the year but is most commonly heard in spring and again in autumn. Singing is also most frequent near sunrise and again just after sunset. It sometimes occurs during the day as well, but rarely after dark. In this species calling bouts may last for several minutes, and softer versions of the advertising call are uttered by the male when bringing food to his mate while she is incubating or brooding. The female has a "stuttering" version of the male's call, which is more cackling in quality. This call may be uttered in response to the male's song, when defending its territory, or when the nest tree is threatened by humans (König, 1968; Schönn, 1978). Observations on the American species are more limited, but duetting is also

known to occur, the male uttering the primary "tooting" song and the female responding with lower-pitched notes, the notes of the two sexes mostly alternating. When excited, the male's single toots may shift to couplets, such as when a female approaches or enters a nesting cavity for the first time (Rashid, 1999). Tooting and trilling notes that are higher pitched in females have also been reported. As in the European species, most singing occurs around sunrise and sunset (Holt and Petersen, 2000).

The second major call of the Eurasian pygmy-owl is the "scale song." This series of 5–11 notes of ascending pitch and increasing cadence is uttered by both sexes. It is used throughout the year by both sexes but is most common in autumn, when the birds are defending territories and driving out their most recent brood of young. This call might correspond to the northern pygmy-owl's staccato whistled notes, which also may begin slowly at first and gradually accelerate in tempo (Taverner, cited by Bent, 1938). The American species has a similar "trill call" consisting of a rapid series of notes that might reflect agitation or even precede copulation (Holt and Petersen, 2000). This call may only be a variant of the staccato call described by Taverner.

The female's food-begging call in the Eurasian species is a high-pitched *siiih* or *tseeh,* the pitch initially rising and then falling and the call lasting almost a second. It may be uttered in response to a male's song, when begging food from him, or when distributing food to young. Males infrequently produce a lower-pitched and weaker version (Cramp, 1985). In the American species, the female's food-begging call usually consists of 5–7 rapid *kewing* notes, followed by 3–4 slower ones (Rashid, 1999).

Both sexes of the Eurasian pygmy-owl produce a trilled series of *tu* notes that are rapidly repeated (about 8 per sec) and usually of higher pitch than the advertising song. These notes are uttered when the birds are carrying food, or by the male prior to copulation or when he is showing a nest to the female. Females utter a high-pitched twittering during copulation. Both sexes also utter a high-pitched squeaking call, usually during food transfer, and a contact call like that of a soft human whistle is also produced. An accelerating series of *kiu* notes is used by the female to solicit copulation and as a danger signal to young. Single alarm notes are also used under the latter circumstance (Cramp, 1985).

Habitats and Ecology

A broad spectrum of woodland and forest habitats are used by this species in western North America; in the western Sierra Nevadas of California, for example, they extend from blue oak (*Quercus douglasii*) savanna habitats in the foothills to mixed montane conifer forests, becoming scarce above 1800 meters and preferring sites with low to intermediate canopy coverage. There they are most common near the edges of meadows, lakes, and other similar clearings (Verner and Boss, 1980). Similarly, in the Rocky Mountains they are usually to be found in the vicinity of meadows or other sizable openings in the forests, probably never occurring in unbroken dense forests (Bent, 1938; Webb, 1982a). Nests in Colorado have been found at 1645–3000 meters and in British Columbia from 490 to 1220 meters (Holt and Petersen, 2000; Kingery, 1998). The elevational range of 29 of 31 singing males near Boulder, Colorado, was 1700–2600 meters, with the highest densities occurring in foothills canyons. Breeding-season habitats in Colorado are most commonly ponderosa pine forests, especially open forests, followed closely by upland aspen forests (Kingery, 1998). Probably the availability of suitable nest sites, most often woodpecker holes, is more important than specific tree composition.

In the central and northern Rocky Mountains the birds may range up as high as 3600 meters, but during winter months they may be forced down to lower elevations, including prairie foothills sometimes well away from forested areas (Bailey and Niedrach, 1965). In Arizona the population that has been separated as *pinicola* (and here considered part of *californicum*) is primarily associated with montane coniferous forests in northern Arizona, but *gnoma* is more typically found on more arid south-facing mountain slopes dominated by oaks in southern Arizona (Phillips, Marshall, and Monson, 1964). There, open ponderosa pine forests with thin understories of Gambel's or evergreen oaks are favored (Monson, 1998a). The forms *gnoma* and *californicum* also have some vocalization differences that have led to the suggestion that they might perhaps represent separate sibling species (AOU, 1998; Holt and Petersen, 2000). Farther south in Mexico *gnoma* occupies similar montane pine–oak forests, from 1800 to 37,000 meters elevation (Holt and Petersen, 2000).

Favored-habitats of the Eurasian pygmy-owl in central Europe include a wide range of woodlands, varying from virgin coniferous forests to forest clearings, forest remnants, and even completely deforested areas. Territories are evidently defended throughout the year by the male, with assistance from the female, and their boundaries tend to coincide with natural landscape features. The estimated sizes of 50 territories in different forest types of Austria averaged 1.4 square kilo-

meters (Scherzinger, 1974). Other European estimates of territory sizes have ranged from 0.45 to 4.0 square kilometers (Cramp, 1985). With such remarkably large territories, average breeding densities are correspondingly low, with various estimates of from 2.2 to 4.2 pairs per 10 square kilometers in Germany, diminishing northward to only 0.17 territories per 100 square kilometers in southern Finland (Glutz von Blotzheim and Bauer, 1980). There are few comparable estimates available for the northern pygmy-owl, but Rashid (1999) estimated one Colorado territory as encompassing 75 hectares, and other estimates have judged average interpair distances as about 1.6 kilometers apart, but with singing males spaced as closely as 600 meters and nests as close as 1.25 kilometers apart (Holt and Petersen, 2000).

In theory, the northern pygmy-owl should be a serious competitor with the ferruginous pygmy-owl, but the species are apparently ecologically isolated in southern Arizona, the only place north of Mexico where they could possibly coexist and compete (Phillips, Marshall, and Monson, 1964).

Movements

Apart from the limited wintertime altitudinal movements down mountainsides to adjacent foothills and plains areas, no specific information is available on movements in northern pygmy-owls, which are believed to be essentially sedentary. No banding recovery data are available to test this assumption, but both seasonal elevational movements and suspected dispersals have been reported for Wyoming and British Columbia. Home ranges of three breeding males in Washington were 1.7–2.3 square kilometers; those of three nonbreeding females were larger (Holt and Petersen, 2000).

In Europe the Eurasian pygmy-owl is also primarily sedentary, with some winter movements to lower elevations in mountainous areas, although at the northern edge of their range the birds tend to be irregularly irruptive. These irruptions are apparently triggered by a combination of cold weather and low rodent populations. After such irruptions banded nestlings have been recovered as far as 230 kilometers south of the nesting site within three months of banding, and some that were banded as autumn migrants were recovered more than 100 (maximum 300) kilometers away during the following winter (Cramp, 1985).

Foods and Foraging Behavior

The foods of northern pygmy-owls have been summarized by Bent (1938) and include a rather long list of mammals and birds, plus other minor items. A complete list would probably include all the smaller mammals, birds, reptiles, amphibians, and larger insects within their range, but with a concentration on mice, large insects, and small to medium-sized birds. Bent listed 12 species or groups of birds that he believed to be taken mainly during the nesting season, with small mammals, lizards, and small snakes more common the rest of the year. Birds as large as Gambel's and California quails (*Callipepla gambelii* and *C. californica*) have been reported killed by these tiny owls, which are less than half the weight of this prey (Kimball, 1925; Balgooyen, 1969). Holt and Petersen (2000) listed 5 lizard genera, 34 bird genera, and 11 mammalian taxa as known vertebrate prey. Insect prey included beetles, butterflies, cicadas, dragonflies, grasshoppers and moths.

Extensive data are available on the foods of the Eurasian pygmy-owl, which are probably relatively comparable to the American species. Samples taken during the breeding season in Finland indicate that voles (Microtidae) comprise about half of the total prey items, other small rodents, bats, and weasels about 5 percent, and birds about 44 percent, most of the latter being of species weighing no more than 35 grams. Outside the breeding season the incidence of bird consumption is slightly lower (32 percent) and that of nonvole mammals (mainly shrews) higher, but microtid voles still comprise nearly half of the total prey items. Breeding season studies from Europe indicate that, on a relative biomass basis, microtine or murine rodents make up nearly 60 percent of prey biomass and birds about 28–40 percent of the remainder, with shrews a minor prey item. Winter studies in the former USSR indicated that birds then comprised nearly half of the total biomass; small voles, about 40 percent; and shrews, most of the remaining 10 percent (Cramp, 1985).

Pygmy-owls use the technique of surprise attack on their prey, gliding and diving down from an elevated perch after first locating the prey visually. If unsuccessful in its first effort, the owl terminates the chase. When flying between perches this owl reminds one of a shrike (*Lanius*), often coursing just above the ground and then rising quickly to a new perch. Unlike more nocturnal owls the wings of pygmy-owls produce audible sounds in flight. These wing sounds prevent a completely silent approach, and by hunting primarily during the day it is also unable to use the cover of darkness to approach its prey unseen. Pygmy-owls are essentially daytime or crepuscular hunters, and good information on their activity patterns is available for the Eurasian species.

Studies in both Austria and Finland indicate peak hunting periods around sunrise and sunset, with some hunting continuing through daytime hours and with a minor breeding-season peak near midday. This midday activity may be a result of the young becoming hungry at that time. Outside the breeding season the birds are also mainly daytime hunters, with activity (in Austria) starting about 40 minutes before sunrise and ending about 35 minutes after sunset. Hunting may continue after sunset on moonlit nights. However, in Finland and Sweden the activity patterns of pygmy-owls correspond closely with those of their major vole prey (*Microtus* and *Clethrionomys*), which are mainly nocturnal in summer and diurnal in winter. Activity changeovers occur during spring and fall for both the prey species and the owl. At these high latitudes enough light persists through the summer nights to allow the owl to hunt, even though it has perhaps the poorest night vision of all European owls (Mikkola, 1983).

Probably as a reflection of its small size and corresponding low energy reserves relative to metabolic needs, food-caching behavior is well developed in pygmy-owls and in the case of the Eurasian pygmy-owl probably better developed than in any other owls of the western Palearctic. Indeed, where prey caching occurs in other species, as in northern hawk-owls and snowy owls, it seems to be limited to surplus items that have been brought to the nest. In summer the pygmy-owl's caches tend to be small and may be stored in various open sites. In winter they are relatively large and are usually placed in hidden cavities, especially nest boxes with fairly small holes that effectively exclude larger owls (Solheim, 1984b; Cramp, 1985). Comparable food caching has been described for the northern pygmy-owl (Bent, 1938).

Social Behavior

Pygmy-owls, more than other owls, are seemingly rather nonsocial, tending to remain solitary or in highly dispersed pairs (or family groups) throughout the year. This is perhaps a reflection of their small body size and consequent high metabolic rate, requiring almost constant hunting. Being among the smallest owls, they are often preyed upon by larger species, and a common concealment response to danger is to assume a sleeked upright posture (Figures 41 and 42), with the wing nearer the danger source drawn in front of the flanks, the head turned toward the danger, the small ear tufts raised, and the eyes variably closed, sometimes to mere slits. A different apparent concealment posture—with ear tufts erect, the eyes widely open, and the body feathers

Figure 41. Eurasian pygmy-owl behavior, including (*A*) male singing posture, (*B*) tail cocking, (*C*) attentive posture, (*D*) concealing posture, and (*E*) defensive posture. After drawings in Glutz von Blotzheim and Bauer (1980).

slimmed—may be used in response to the sight of a cat or tethered falcon (Holt and Petersen, 2000). A flock of crows may bring about the same response (Scott Rashid, pers. comm.). Or the bird may close its eyes and ruffle its feathers in a pseudosleeping posture. If unable to escape, it will ruffle all its body and head feathers and spread its tail in an upright position, but will not spread its wings in the usual manner of cornered owls. However, when excited, males may sometimes land on a perch with their wings extended and slightly trembling. The tail is also used as a sign of excitement; it may be cocked, or waved horizontally, vertically, or in a more circular manner. When singing, the male Eurasian pygmy-owl stands in a rather diagonal posture, with the tail held in line with the body and the neck feathers distinctly expanded or fluffed (Glutz von Blotzheim and Bauer, 1980; Cramp, 1985). A comparable posture is used by the northern pygmy-owl.

All the species of pygmy-owls have false "eye spots" on their napes (Figure 41), which are generally believed to have social significance in possibly warding off surprise attacks from the rear by larger predators. Although this explanation is suppositional only, the spots do not have any other apparent signal functions during intraspecific social encounters. Similar nape spots occur on the closely related and similar-sized Old World owlets of the same genus, and it is interesting to note that the northern hawk-owl has a pair of di-

agonal black stripes extending down the sides of the nape that might easily be imagined as the evolutionary precursor of this interesting plumage feature.

Little information is available on the social behavior of the northern pygmy-owl, but probably most of what is known of the Eurasian pygmy-owl's general behavior patterns, calls, and postures is applicable to this species. The Eurasian pygmy-owl is monogamous, with pair bonds lasting through a single breeding season but with cases known in which the same pair of birds remained paired for four years, using the same nesting site throughout. Not surprisingly in these small owls, sexual maturity and breeding occur at one year of age. Courtship in the Eurasian pygmy-owl begins when the males begin singing (tooting) from perching sites to proclaim territorial ownership and advertise potential nest sites. In the northern pygmy-owl, comparable nest site advertisement and even copulation have been observed as early as February (Holt and Petersen, 2000). The two sexes appear to be shy of one another and initially show alternating fear and aggression toward their potential mate, including pursuit flights and attacks. Reciprocal singing activity by paired birds (called "tooting" and "trilling" by Holt and Petersen, 2000) diminishes gradually, as males begin nest-showing behavior. Nest-showing, which starts up to 8 weeks prior to laying, may continue right up to the onset of laying. The male will fly into a potential nesting hole, utter his trilling call, and peer out. He may also take food into the hole. The female also enters the hole, inspects it, and calls as she leaves it. Eventually she selects a nesting hole, cleans it, and remains near it through the day (Scherzinger, 1970; Glutz von Blotzheim and Bauer, 1980). Copulation occurs on a branch near the nest site and is preceded by the female's solicitation call, uttered in a tail-raised and horizontal body posture. The male flies to her, hovers briefly over her back, and may grip her nape during treading. One or both sexes may call during copulation; afterward the male flies off calling (Jannson, 1964; Cramp, 1985). Allopreening has been observed among broodmates and between pairs of the northern pygmy-owl.

Breeding Biology

Eggs of the Eurasian pygmy-owl are laid at approximately 2-day intervals, and copulation typically ends after the laying of the last egg (Glutz von Blotzheim and Bauer, 1980). Limited information on the northern pygmy-owl suggests daily egg laying (Holt and Petersen, 2000). Egg dates for the northern pygmy-owl in California are from April 11 to June 28 (20 records), in Arizona from May 19 to June 14 (15 records), in Colorado from May 17 to June 22 (5 records), and in Montana and Oregon from April 18 to May 20 (4 records) (Bent, 1938; Bailey and Niedrach, 1965; Holt and Petersen, 2000; and records from Western Foundation of Vertebrate Zoology).

Based on published information and that available from the Cornell Nest-Record Card Program and the Western Foundation of Vertebrate Zoology, the average size of 18 northern pygmy-owl clutches was 3.2 eggs (range 3–5 eggs, 12 of 18 having 3 eggs). Holt and Petersen (2000) reported means of 3.7 (10 clutches) and 5.0 (6 clutches). The Eurasian species has a somewhat higher average clutch size than does the northern, averaging about 5.5 eggs (Mikkola, 1983). Its clutch size is also influenced by the relative availability of small mammal prey prior to laying, tending to be higher in peak rodent years than in low ones. It is also generally higher in the coniferous forest zone of southern Norway than in the

Figure 42. Northern pygmy-owl behavior, including adults in alert-apprehensive or "concealment" posture (*A* and *C*) (after photos by Art Wolfe and Scott Rashid), and fluffed or relaxed posture (*B* and *D*) (after photos by Art Wolfe).

mixed forest zone, the former of which has greater fluctuations in small rodent densities (Solheim, 1984a).

Nineteen northern pygmy-owl nests averaged 6.3 meters aboveground (range 2.3–20 m). Of a total of 18 nests, 12 were in woodpecker holes, 5 in dead tree cavities, and 1 in the old nest of a cactus wren (*Campylorhynchus brunneicapillus*). Among 19 tree nests, 6 were in sycamores, 4 in pines, 3 in oaks, 2 in firs, and 1 each in locust, cedar, alder, and cottonwood trees. Holes used are typically about 6 centimeters in diameter and 15–50 centimeters deep. This predominance of broad-leafed nest trees is in contrast to the Eurasian pygmy-owl, which usually nests in mature spruce forests. However, the choice of nest tree is more likely to reflect the relative abundance of woodpecker holes or other cavities than specific tree type. One California nest site was occupied every year for three years, and a Montana site for two consecutive years, but the identities of the birds were not established (Holt and Petersen, 2000). Scott Rashid (pers. comm.) observed the same nest site being used for three consecutive years.

Pygmy-owls are among the few owls that might not begin incubation until the clutch is complete, thus the owlets would hatch over an interval of only a day or two. Some North American observations suggest a nearly simultaneous hatching, while others indicate asynchronous hatching (Holt and Petersen, 2000). The duration of the northern pygmy-owl's incubation is thought to be 28 days, similar to the 29-day duration reported for the Eurasian species (Holt and Petersen, 2000). Based on observations of the Eurasian pygmy-owl, it is likely that the male responds to his discovery of hatched young in the nest by increasing his food supply to the female, who then feeds the young. When they are 14 days old the owlets have already reached about 60 percent of adult weight, and by 25 days they are fully feathered. They leave the nest and are able to fly at 30 days, but parental protection does not end until about 20–30 days later. They begin to show signs of sexual maturity when only about 5 months old (Bergmann and Ganso, 1965; Mikkola, 1983). In the northern pygmy-owl the young leave the nest within a day or two of one another, at about 23 days after hatching (Holt and Petersen, 2000). In Colorado two members of one brood remained within 270 meters of the nest for at least 3 weeks (Rashid, 1999). Another brood was lured by the female away from the nest site for a distance of about 300 meters only a few days after fledging; this nest was under constant harassment by songbirds (Scott Rashid, pers. comm.).

Evolutionary Relationships and Conservation Status

There are no significant morphological differences between *gnoma* and *passerina,* and ecologically and behaviorally they appear to be virtually identical, based on the rather limited available information on these latter topics. However, Wink and Heidrich (1999) reported that genetic differences would suggest they were isolated nearly 8 million years ago and that the northern pygmy-owl's nearest relatives are instead with the tropical New World forms rather than Old World pygmy-owls. It has also been proposed that perhaps in the southwestern United States (Arizona) there are two sibling species (*californicum* and *gnoma*) present, the two groups having ecological and vocalization differences possibly warranting specific separation. Preliminary genetic data do not yet support this premise (Holt and Petersen, 2000), although Wink and Heidrich (1999) found that cytochrome b samples of *gnoma* (apparently from Central America) were clearly separable genetically from those of North American *californicum.*

The overall status of the northern pygmy-owl is difficult to assess. Breeding Bird Surveys done between 1966 and 2000 indicate a significant annual population increase of 3.1 percent, the largest rate reported for any North American owl (http://www.mbr-pwrc.usgs.gov/bbs/bbs.html). The species is tolerant of a broad range of habitats in western North America and is not harassed by humans or seemingly seriously affected by their activities, even when the birds are being rather closely studied. If anything, partial forest clearing may improve hunting opportunities for it. Certainly it is largely dependent upon woodpecker populations to maintain a supply of suitable nesting cavities, which may be a limiting factor for the birds in some areas where other types of nesting sites are unavailable. During the 1999–2000 Christmas Bird Counts the species was detected on an average of 125 counts, or about 7 percent of all U.S. and Canadian counts. In terms of all localities where the species was detected, the northern pygmy-owl was the ninth most widely reported of the 18 North American owl species then encountered. The maximum number ever reported on a single CBC was 27, at Ano Nuevo, California, in 1980. The average total number counted among the 100 counts having the highest totals was 239 birds, or about 2.5 percent of all owls reported in the tabulated summaries of counts for the United States and Canada during 1999–2000. The Canadian population has recently been estimated at 2000–10,000 pairs (Kirk and Hyslop, 1998) and is considered possibly stable.

MEXICAN PYGMY-OWLS (3 species)

Central American Pygmy-Owl   *Glaucidium griseiceps* Sharpe, 1875

Tamaulipas Pygmy-Owl   *Glaucidium sanchezi* Lowery and Newman, 1949

Colima Pygmy-Owl   *Glaucidium palmarum* Nelson, 1901

Other Vernacular Names:
G. griseiceps: gray-headed pygmy-owl; sometimes part of least pygmy-owl, Mochuelo
Enano, Tecolotito Centroamericano (Spanish)
G. sanchezi: sometimes considered part of least pygmy-owl, Tecolotito Tamaulipeco
(Spanish)
G. palmarum: palm pygmy-owl; sometimes considered part of least pygmy-owl, Tecolotito
Colimense (Spanish)

Range (See Figure 43.)

The form *griseiceps* ranges residentially along the
Atlantic slope of Mexico from northern Oaxaca
and southeastern Veracruz south to Honduras,
and locally along the Pacific slope in Guatemala
and possibly also Chiapas. It also (if considered
broadly, and as part of the *minutissimum* superspecies) occurs in Panama and continues south at
least to northwestern Colombia. Depending on
the broadness of accepted species limits, other
South American populations may also exist; the
type specimen of *minutissimum* came from southeastern Brazil.

The form *sanchezi* ranges residentially along

the Atlantic slope of Mexico from southern
Tamaulipas to eastern San Luis Potosí.

The form *palmarum* ranges residentially along
the Pacific slope of Mexico from central Sonora to
Oaxaca, and inland to Morelos along the Balsas
River drainage.

North American Subspecies

None recognized by König, Weick, and Becking
(1999). *G. palmarum* has been subdivided into
three subspecies, including *oberholseri* Moore 1937
and *griscomi* Moore 1947. Howell and Robbins
(1995) synonymized *oberholseri* with nominate *palmarum* but recognized *griscomi* (from the interior

Figure 43. Distribution of the Mexican pygmy-owls, including Tamaulipas (vertical hatching), Colima
(diagonal hatching) and Central American (cross-hatching) pygmy-owls. The inset map shows the South
American distribution. Mexican distribution partly after Howell and Webb (1995).

145

Rio Balsas basin) as distinct. All these taxa have often been merged into the collective species *minutissimum.*

Measurements

Wing of *palmarum,* mean of 79 males, 82.9 mm (SD 1.9), mean of 17 females, 85.9 mm (SD 1.7); tail, mean of 74 males, 52.5 mm (SD 1.6), of 16 females 54.55 mm (SD 1.2) (Howell and Robbins, 1995).

Wing of *sanchezi,* mean of 6 males, 87.3 mm (SD 1.4), mean of 5 females, 91.0 (SD 2.4); tail, mean of 5 males, 52.7 mm (SD 2.8), of 4 females, 55.0 mm (SD 0.5) (Howell and Robbins, 1995).

Wing of *griseiceps,* mean of 10 males, 87.2 mm (SD 2.7); tail, mean of 10 males, 47.8 mm (SD 1.8) (Howell and Robbins, 1995).

Binford (1989) reported that specimens he identified as *minutissimum occulatum* Moore from the Atlantic slope of Oaxaca match *palmarum* in wing and tail measurements (wing of males 79–85.1 mm, tail 51–55.5 mm), but are paler. No mensural information is available on the eggs of any of these three forms.

Weights

Males (unspecified sample sizes) of *sanchezi* 51–55 g; of *palmarum* ca. 50 g; of *griseiceps* 49.5–51.7 g, females of all races are heavier (König, Weick, and Becking, 1999). Howell and Robbins (1995) reported 19 male *palmarum* as averaging 44.9 g (SD 2.5), 2 *sanchezi* as averaging 53.5 g (SD 2.1), and 3 *griseiceps* as averaging 50.6 g (SD 1.1). Fifteen unsexed Mexican specimens of collective *minutissimum*-type owls averaged 47.8 (range 40.4–59 g) (Dunning, 1993). These weights place the birds as being only slightly larger than elf owls and thus among the smallest of the world's owls.

Identification

In the field. The pygmy-owls of Mexico are all allopatric with one another, which helps to distinguish them from one another in the field, although each overlaps with the northern pygmy-owl or the ferruginous pygmy-owl. All are separable from the ferruginous pygmy-owl by having variably spotted (not streaked) crowns, white-banded tails, and less contrasting whitish markings along the row of scapular feathers bordering the back. They are similar to the northern pygmy-owl in having similarly short tails and reduced scapular spotting but are not nearly so rufous overall, and all have single-noted rather than double-noted hooting calls. This series may be preceded by a quavering trill in all species. All pygmy-owls are prone to jerking the tail up and down or quickly

wagging it sideways when they are somewhat alarmed or agitated.

The Colima pygmy-owl of western Mexico is a thorn-forest (tropical deciduous forest) to dry oak woodland species with a short tail and a dull grayish brown to olive brown dorsal color, less rufous than the sometimes sympatric ferruginous pygmy-owl and with less conspicuous pale back spotting. The calls of the two are similar, but the Colima pygmy-owl tends to have a slower and more hollow hooting, the hoot phrases often uttered in a progressively increasing number of notes, up to 24 or more in a single series, with long pauses between. It occurs up to 1500 meters elevation.

The montane forest-dwelling Tamaulipas pygmy-owl occurs from 900 to 2100 meters in humid evergreen and subtropical forests, and varies from a rufous brown (in females) to more olive brown (in males) dorsally. Like the Colima it has a short tail, with only about four white bars (the basal one or two covered by coverts), a spotted crown, and an olive brown back with only slight pale spotting. It produces short series of from one to three slow-paced hoots, with fairly long and uniform intervals between the notes. The northern pygmy-owl is the only other pygmy-owl that might occur in cloud forest habitats, but it is more likely to be found in lower pine forests, and utters more rapidly placed double hoots in longer series.

The tropical evergreen forest–dwelling Central American pygmy-owl occurs from sea level to 1300 meters. It is generally darker brown on the back than the two preceding species, but has a somewhat contrasting grayish head with only slight white spotting. It is not as rufous as either the ferruginous pygmy-owl or the northern pygmy-owl, and utters short series of single, hollow hoots ranging from as few as 2 to as many as 18 notes, uttered at a uniform rate of about 3 notes per second, or with a break after the first few hoots, followed by a more rapid series of notes. No other pygmy-owl is likely to occur in tropical evergreen forests.

In the hand. These three species are readily separated from the ferruginous and northern pygmy-owls by their uniformly short tails (to 55 mm) with no more than four whitish cross-bars (plus one or two hidden by their coverts), variably spotted (not streaked) heads, and only slightly spotted scapulars. In addition to their geographic allopatry and distinctive habitat differences, the Central American species has the grayest head color (but juveniles of the others also have grayish, unspotted heads) and the darkest brown upperparts.

The Colima species has the most olive brown tints on the upperparts, and the Tamaulipas species is intermediate in upperpart color between the other two. It is paler overall than the Tamaulipas form, has more conspicuous white head spotting, and its underpart stripes are cinnamon brown. It is relatively long-tailed, with 6–7 pale bars.

The Tamaulipas pygmy-owl differs from the others in that the sexes are dichromatic, the male being grayish brown on the head with whitish spotting. The female is more cinnamon-toned on the head, with buffy cinnamon flecks or spots, and is more reddish throughout the upperparts. Males differ from the Central American pygmy-owl by their more olivaceous and less rufous upperparts, and females by their cinnamon brown head colors. It is relatively long-tailed, with six pale bars, and has fine crown flecking.

The Central American pygmy-owl has a distinctly grayish head, with white spotting, dark reddish underpart streaking, and five white or buffy white tail bars, rather than six as in the other two. It is relatively short-tailed as compared with the other two species.

Vocalizations

All three species utter single, rather than double, hooting notes as their primary songs. These notes are uniform in pitch and usually relatively uniformly spaced; they vary mainly in speed and number of notes per sequence. In all three species quavering trill notes my precede the hooting. As with other pygmy-owls, vocal imitation or tape broadcasting these calls tends to attract small songbirds to investigate their source.

The Colima pygmy-owl utters a series of from 2 to 24 hoots at the rate of about 3 per second, separated by long pauses. There is often a longer pause after the first note than the following ones. In a bout of hooting, the bird may gradually increase the number of notes progressively. The mean dominant frequency is 1416 Hz, the notes last an average of 0.1 seconds, and the internote intervals are 0.31–0.34 seconds, with average intersong intervals of 4.8 seconds (Howell and Robbins, 1995).

The Tamaulipas pygmy-owl utters 1–3 deliberately paced and rather high-pitched hoots, at the rare of about 2 per second with relatively long internote intervals. Two-note phrases are common, but 2-note and 3-note phrases may be alternated. The mean dominant frequency is 1485 Hz; the notes last an average of 0.26 seconds; and the internote intervals are 0.43 seconds, with average intersong intervals of 5.2 seconds (Howell and Robbins, 1995).

The Central American pygmy-owl often utters 2 to 4 well-spaced and bell-like notes, which are then sometimes followed by a series of 6–18 more rapid notes, at the rate of about 3–5 per second. Or it may begin with 2- and 3-note phrases, build up to longer phrases, and then vary these phrases lengths randomly. It may also utter a rather long series of notes grouped in phrases of from 2 to 6 or more notes, with short intervals between these phrases. In Panama a simple 4-note call is typical, low in tone, and with a pause after the first note. More northerly birds generally tend to have phrases with larger numbers of notes. The mean dominant frequency is 1385 Hz; the notes last an average of 0.14 seconds; and the internote intervals are 0.22–0.23 seconds, with average intersong intervals of 7.3 seconds (Howell and Robbins, 1995).

Habitats and Ecology

All the pygmy owls are diurnal hunters, and so are more likely to be seen during daylight, or at least heard during those hours, than the other Mexican owls. They also respond well to tape recordings during the day.

The Colima pygmy-owl ranges in Colima from the upper edge of the thorn forest zone (in which it encounters the ferruginous pygmy-owl) to the upper woodland zone of oaks and pines (where in nearby Jalisco it may encounter the northern pygmy owl), with a maximum abundance in the ravines and arroyos (creek drainages) of the upper part of the tropical deciduous forest zone (Schaldach, 1963). Its altitudinal range is from sea level to about 1500 meters.

The Tamaulipas pygmy-owl is a strictly montane species, occurring in Mexico from 900 to 2100 meters, in cloud forests and to a lesser degree in humid subtropical and pine-dominated evergreen forests.

The Central American pygmy-owl is a lowland evergreen forest species, ranging in Mexico from sea level to 1200–1300 meters in Guatemala, and to 800 meters in Costa Rica. It also occurs in second-growth woods and mature cacao plantations.

Movements

All these forms are apparently residential, although some vertical seasonal movements might occur in the montane-dwelling forms.

Foods and Foraging Behavior

Not much has been reported, but the Tamaulipas pygmy-owl is said to consume large insects and small reptiles such as lizards. The Central Ameri-

147

can pygmy-owl has been observed to take lizards, small mammals, and small birds such as tanagers and honeycreepers (Stiles and Skutch, 1989). Janzen and Pond (1976) have described the foods and feeding behavior of a captive bird in Costa Rica.

Social Behavior

These birds are said to be most active in calling just before dawn and again just after dusk, with much less calling during the day (Stiles and Skutch, 1989). However, the birds are active by day and night. They are sometimes attracted by an imitation of their call. Otherwise, their social behavior is probably very much like the better-studied species of pygmy-owls.

Breeding Biology

There is no reason to believe that the breeding biology of these species differs significantly from those of the better-studied pygmy-owls, but little specific information is available with which to compare them. The Colima pygmy-owl reportedly lays 2–4 eggs during May in Mexico, in tree cavities such as old woodpecker holes. Four clutches were obtained in Sonora between May 18 and June 3, and a female with a shelled egg in the oviduct was collected on May 6. Two of these nests were in woodpecker-made cavities in oaks (S. Russell and Monson, 1998). The Tamaulipas pygmy-owl's breeding biology is little known, but it is also said to use woodpecker holes and lay 2–4 eggs. The Central American pygmy-owl is reported to breed during April and May, and lay 2–4 eggs in old woodpecker cavities.

Evolutionary Relationships and Conservation Status

The North American and Old World species of *Glaucidium* diverged long ago (more than 7.8 million years ago in the judgment of Michael Wink and Petra Heidrich, 1999, p. 41), and *gnoma* and *passerinum* are thus clearly specifically distinct. Unfortunately only one of the Mexican forms described here (*griseiceps*) was part of their cytochrome b gene analysis, and it clustered with the South American *brasilianum* species group, rather than with the North American *gnoma* (and also *californicum*, which by their analysis is specifically separable from *gnoma*).

According to the work of Wink and Heidrich, the Central American pygmy-owl is part of a largely South American assemblage that includes the ferruginous pygmy-owl and its near relatives, specifically *peruanum* and *nanum*. This group in turn is part of a larger group that includes *bolivianum*, *jardinii*, and *hardyi*.

All three species discussed in this section are believed to be fairly common to common, although often rather local, across their Mexican ranges. The Colima pygmy-owl is the most tolerant of aridity and is found in habitats that at times are as dry as thorn forest. The other two species are associated with cloud forests and humid evergreen montane forests, both of which are declining rapidly in Mexico and elsewhere in their ranges. The Tamaulipas pygmy-owl has an especially small range, occurring at the northern end of cloud forests in Mexico, and is likely to be the most vulnerable of the three.

Ferruginous Pygmy-Owl   *Glaucidium brasilianum* (Gmelin) 1788

Other Vernacular Names:
cactus pygmy-owl (*cactorum*); ferruginous owl; streaked pygmy-owl.

Range

Resident from south-central Arizona (where listed as federally endangered), south through Sonora to Oaxaca, and southern Texas (where listed as state threatened) south through eastern Mexico to southern South America, reaching at least to northern Argentina. (See Figure 44).

North American Subspecies (Adapted from AOU, 1957.)

G. b. cactorum van Rossem. Southern Arizona in western Mexico south perhaps to Colima, and (probably) southern Texas south to Michoacan, Nuevo Leon, and Tamaulipas. Birds from Nayarit to Oaxaca have been separated as *G. b. intermedium* (Phillips), and those from the Pacific lowlands of Chiapas as *saturatum* (Brodkorb).

G. b. ridgwayi Sharpe. Mexico, south of *cactorum*, south to the Canal Zone. Ranges of this and the previous race are confused and need attention. Thus some authorities (e.g., König, Weick, and Becking, 1999; Proudfoot and Johnson, 2000) include the birds from Texas and eastern Mexico within *cactorum*.

Measurements

Wing, males 81–108 mm (ave. of 304, 94.4), females 89–113 mm (ave. of 194, 98.7). Tail, males 52–78 mm (ave. of 299, 60.6), females 52–79 mm (ave. of 188, 63.6) (Proudfoot and Johnson, 2000). The eggs average 28.5 × 23.3 mm (Bent, 1938).

Weights

Earhart and Johnson (1970) reported that 29 males averaged 61.4 g (range 46–74), and that 16 females averaged 75.1 g (range 64–87). Proudfoot and Johnson (2000) reported a mean weight of 64.3 g (range 53–79) and 77.2 g (range 66–102) for 68 males and 38 females, respectively. The estimated male:female mass ratio is 1:1.2. The estimated egg weight is 8.0 g; actual measured egg mass is 8.3 g (Proudfoot and Johnson, 2000). The estimated egg-to-female mass ratio is 10 percent.

Figure 44. Distribution of the ferruginous pygmy-owl. The inset map shows the South American distribution. Mexican distribution partly after Howell and Webb (1995).

Identification

In the field. Within the limits of the United
States, this species is found only in central and
southern Arizona and extreme southern Texas,
where its small size, long, narrow tail, and distinctive whistled or "popping" advertising call, an
indefinitely repeated *whoip* or *poip,* help to distinguish it. Its monotonous hooting song is uttered
at a rather constant rate about three times per
second, the notes at times speeding up and becoming more insistent (Howell and Webb, 1995).
In the United States it is the only pygmy-owl with
a narrowly buff-barred tail (rather than a broader
white-barred tail), but in Mexico it is best distin-
guished from several other similar pygmy-owls by
the narrow white streaks (not spots) on its crown
and nape. In Mexico, and generally in Latin
America, it occurs in tropical lowlands rather than
in montane forests, but it shares thorn forests, coffee plantations, and forest edges with the Colima
pygmy-owl in western Mexico. Like other pygmy-
owls it is active during the day and flies with rapid
wingbeats, often close to the ground, searching for
small lizards or ground-feeding birds. When excited, it often cocks or laterally flicks its tail momentarily, as do other pygmy-owls.

In the hand. The combination of small size (wing
under 120 mm), a rufous (in females) to brownish
(in males) crown that is narrowly streaked with
white, a generally rufous brown to wood brown
color over the upperparts and as breast streaking,
and a tail that has 7–8 whitish cross-bars (as seen
from above; up to 5 visible below) serves to distinguish this species. Adult measurements do not allow for certain sex identification, but females are
somewhat heavier and tend to be much more rufous brown than males, have tails of about the
same color as their backs (males have somewhat
darker tails), and have less distinct tail barring,
The downy plumage is white, like that of *gnoma.*
Juveniles are similar to adults of the corresponding plumage morph but initially have few or no
pale streaks on the crown (these are evident by 12
weeks of age), retain their juvenal flight feathers,
and perhaps have less distinct plumage patterning. Ageing criteria suggested for the northern
pygmy-owl might apply to this species, but verifi-
cation is needed.

Vocalizations

The vocalizations of this species are still only
poorly understood, but the territorial-advertising
call or "song" consists of a very prolonged series
of whistled notes (up to about 60), usually with a
slight jerk of the tail as each note is uttered.
These sequences are perhaps most often uttered
at dawn and dusk, but are also uttered during
daylight and at night, especially in spring and
summer. Stillwell and Stillwell (1954) described
the voice of this species as a clear and mellow series of notes, sounding something like *whah,* similar to the sound made by blowing across the opening of a partially filled bottle of water, and at the
approximate pitch of 1400 Hz. The call was uttered at the rate of about 150 times per minute,
in long sequences of 10–45, followed by 10-second
intervals of silence. The notes tended to be both
harsher and more rapidly delivered than those of
the northern pygmy-owl. Proudfoot and Johnson
(2000) stated that calling sessions may last from
5 minutes to 5 hours, and seasonally extend from
courtship to incubation. The advertisement call
may be heard for distances of up to 0.8 kilometers. The male's contact call is like its territorial
call but is lower in volume. The male's voice is
lower pitched and less wheezing than that of females, who may extend each note and sound "like
a dripping faucet" (Proudfoot and Johnson, 2000).

Other descriptions of the advertising song, as
summarized by Bent (1938), include a series of
chuck sounds repeated several times at the rate of
about 2 per second, a repeated *khiu* note, and a
series of repeated *chu* notes, most often heard in
the evening. Slud (1980) said that in Costa Rica
the call likewise consists of up to 30 *poo* notes, uttered at a rate of 2 or 3 per second, and Ffrench
(1973) described the major call in Trinidad as a
series of 5–30 musical notes, the pitch usually remaining constant. In contrast, Wetmore (1968)
described the species' call in Panama as a series
of 4–7 low, whistled notes, with a slight pause after the first. However, other accounts from South
America mention unbroken series of monotonous
single notes. Before, during, and after copulation
females utter a rapid chitter, comparable if not
identical to the food-begging call of nestlings, and
the female's alarm note is a short, sharp *pee-weet,*
given at irregular intervals (Proudfoot and Johnson, 2000). Low *chuck* notes have also been described, and high-pitched screams, presumably
distress calls, may be uttered by birds when re-
strained.

Scherzinger (1977) noted that the species'
vocalizations tend to be considerably softer and
rougher than in the Eurasian pygmy-owl; they
vary greatly in amplitude and length of sequence;
and do not ascend in pitch as occurs in the "scale
song" of the European pygmy-owl (Wolfgang
Scherzinger, pers. comm.).

Habitats and Ecology

In Arizona this species occurs in saguaro
(*Carnegiea*) desert habitats as well as in shady ri-

parian timberlands consisting of mesquite (*Prosopsis*) groves and cottonwood-mesquite habitats (Phillips, Marshall, and Monson, 1964). In Organ Pipe Cactus National Monument, where it is fairly regular, it occurs in saguaro desert "forests" and xeroriparian scrub along desert washes (R. Johnson and Haight, 1985). In Texas it is now mostly limited to edges and openings in subtropical riparian forests and woodlands of coastal live oak and mesquite; it also occurs on a few large ranches and in wildlife refuges and parks (Frederick Gehlbach, pers. comm.).

In general this species occurs in lower, more arid, and usually more open habitats than does the northern pygmy-owl. In Mexico it occurs from sea level to as high as about 1200 meters in western areas, and at least 300 meters in eastern areas. In Colima and Jalisco it is abundant in the desertlike lower thorn scrub and thorn forest zones, but is absent from the tropical deciduous forest zone and the higher vegetational zones, where it is replaced by two other species of *Glaucidium* (Schaldach, 1963). In Guatemala it extends from sea level to about 1850 meters and ranges from scrubby woodlands to the edge of the evergreen forest (Land, 1970). In Belize and Honduras it is apparently mostly a lowland species occupying relatively open habitats such as the transitional areas between open pine savanna and rain forest, second-growth forests, and arid to semihumid lowland habitats. Farther south, as in Panama, it seems to be mainly a tropical-zone bird, occupying rather restricted habitats.

Movements

This species is believed to be essentially sedentary, at least within its U.S. range. Generally, adult males defend territories some 700–1200 meters in diameter throughout the year, or perhaps for the length of their pair bond. In Arizona, year-around territories with radii of 160 meters have been found, and an unmated male held a territory with a somewhat smaller core area. Territorial boundaries are evidently established at the time of initial breeding (Proudfoot and Johnson, 2000).

When breeding, adult males had home ranges of 1.4–23.1 hectares, and 5 family groups had ranges of 9.3–59.5 hectares. Young birds disperse about 2 months after fledging, with 6 marked young moving 1.9–17.3 kilometers from natal sites. Ten radio-tracked adults in Texas expanded their home range area more than threefold after the dispersal of their young during early autumn relative to their breeding-season territorial activities. Two males had ranges of 117 and 125 hectares during October and November (Proudfoot and Johnson, 2000).

Foods and Foraging Behavior

Like the other pygmy-owls, this is a daring daytime predator, sometimes attacking prey as large as or even larger than itself. There are early accounts of it not only killing young domestic fowl (*Gallus domesticus*) but also attacking apparently captive guans (*Penelope*), the owl fastening itself firmly to the guan while tearing at it, eventually wearing it out and killing it. Wounded birds as large as American robins (*Turdus migratorius*) have also reportedly been pounced upon by ferruginous pygmy-owls (Bent, 1938). The largest prey reported from Arizona and Texas have been Gambel's quail (*Callipepla gambelii*) and hispid cotton rat (*Sigmodon hispidus*). Studies of prey remains in Texas have found that on a numerical basis insects comprised 58 percent of identified prey remains around eight nests, and represented 66–90 percent of observed or videotaped prey deliveries. Additionally, reptiles comprised 7.1–22.5 percent, birds 2.3–10.5 percent, and mammals 1–9 percent. Similar studies in Arizona produced rather different results: reptiles 47 percent, birds 21 percent, mammals 9 percent, and insects 5 percent (Proudfoot and Beasom, 1997; Cartron, Richardson, and Proudfoot, 1999; Proudfoot and Johnson, 2000).

Social Behavior

Like the other pygmy-owls these are nonsocial birds, occurring solitarily or at most in pairs except when caring for dependent young. Adults are sedentary and almost permanently territorial. Scherzinger (1977) noted that a captive pair began presumed territorial calling in mid-January, about two months before the first mating was observed, and almost three months prior to the laying of the first eggs. Likewise in both Arizona and Texas calling begins in February, two to three months prior to egg laying. In Texas, three pairs maintained monogamous pair bonds during four years of continuous study. Pair bonds are maintained by the male's territorial-advertisement calls and food-presentation; receptive females utter chitter calls. Allopreening may occur both before and after copulation, and prey presentation usually precedes copulation (Proudfoot and Johnson, 2000). Some representative postures of the ferruginous pygmy-owl are shown in Figure 45.

Breeding Biology

Egg records from Texas are primarily for May; Oberholser (1974) reported a spread of records (number unstated) from April 9 to May 28; Bent (1938) lists 8 for the period March 28 to May 28, and 5 records from the National Museum of Nat-

Figure 45. Ferruginous pygmy-owl behavior, including (*A*) rear view of perched adult (showing "eye-spots"), (*B*) same posture, with head reversed and plumage somewhat fluffed, and (*C*) alert posture, with intermittent tail-cocking. After photos by Art Wolfe (*A* and *B*) and the author (*C*).

ural History and the Western Foundation of Vertebrate Zoology are from May 11 to 28. A much larger series of 29 Mexican records from the same sources plus Bent (1938) are from April 4 to June 17; of these 19 are within the period April 26 to May 15.

Of 30 nest site records known to me, 17 were natural cavities of trees, stumps, or snags; 10 were in holes of various medium-sized woodpeckers (*Centurus, Melanerpes, Dryocopus*); 2 were in tree forks or depressions; and 1 was in a hole in a sandbank. The specific trees or vegetational supports of 12 nests included 6 oaks, plus 1 each in mesquite, pine, cypress (*Cupressus*), palmetto (*Sabal*), cottonwood, and an unidentified evergreen. Surprisingly, none of these was in a saguaro cactus, where, like elf owls, they are also known to nest (Karalus and Eckert, 1974; Millsap and Johnson, 1988). Giant cacti have also been used as nest sites in Sonora (S. Russell and Monson, 1998). The average height of 26 nest sites was 4.9 meters (range 3.3–9 m). Nest boxes with entrances of 4.5–5.8 centimeters have been used, but others with entrances of 6.4 and 8.3 centimeters were not. Among 99 sites, cavity heights ranging from 2 to 12 meters aboveground have been observed. Among 29 Texas cavity nests, the entrances were 4.9–6.2 centimeters in diameter, and cavity depths were 24–60 centimeters. Cavity height and directional orientation are apparently not significant factors in nest selection (Proudfoot and Johnson, 2000).

The average clutch size of 43 nests from Mexico and the United States was 3.3 eggs, with a range of 2–5 eggs. Of the total, the most common clutches were of 4 and 3 eggs, represented by 17 and 16 nests, respectively. Another sample of 58 clutches had a mean of 4.9 eggs (Proudfoot and Johnson, 2000). Incubation has been estimated to last 21–23 days, and fledging at 26–28 days (Proudfoot, 1996). In Texas, eggs laid in June are presumed to be replacement clutches; such clutches are laid 20–24 days after loss of the initial clutch. The earliest observed hatching date in Texas is May 12, the first fledging there June 14, and the first dispersal of juveniles August 14. Corresponding Arizona dates are 2–3 weeks earlier (Proudfoot and Johnson, 2000).

Scherzinger (1977) reported that a captive female laid a clutch of 5 eggs starting on April 10. A second clutch was begun 20 days after the first unsuccessful one was removed. He determined an incubation period of 30 days and noted that the owlets had opened their eyes by the 7th day following hatching. By the 17th day they were able to perch readily, and initial fledging occurred 28 days following the hatching of the first egg. After fledging, the young sometimes returned to the nesting box to sleep.

These small owls are sensitive to nest predation, with snakes being likely but unproved culprits. In a Texas study, 42 percent of 28 natural nest cavities were depredated, as well as 20 percent of 30 nest boxes (Proudfoot and Johnson, 2000).

Evolutionary Relationships and Conservation Status

Seemingly this is a moderately close relative of *gnoma*, but there are several other species of strictly New World and more tropical pygmy-owls that are more likely candidates. Coats (1979) judged from vocal evidence that the Andean pygmy-owl (*G. jardinii*) is a distinct species, but part of a species group that also includes *brasilianum* and the Austral pygmy-owl (*G. nanum*). Molecular studies summarized by Wink and Heidrich (1999) do not support a close relationship between *jardinii* and *brasilianum* but do suggest an affinity of the ferruginous pygmy-owl with the Chaco pygmy-owl (*G. tucumanum*) of Argentina and Paraguay. The Austral pygmy-owl does not appear to be a close relative of the ferruginous on this basis and is certainly not conspecific, as has at times been suggested.

The current U.S. range of the ferruginous pygmy-owl is very small and still retracting. In Texas it was apparently limited to remnant mesquite thickets by the middle of the twentieth century (Oberholser, 1974). Recent Texas observations are confined to the area extending from Zapata and Falcon Reservoir southeast along the Rio Grande to Rio Grande City, north to Falfurrias, northeast to Baffin Bay, and south coastally to Brownsville. The population of Brooks and Kenedy counties was estimated at 745–1823 birds in the mid-1990s (Mays, 1996). Another recent estimate was of 1308 birds in Brooks, Kenedy, and Willacy counties. In Kenedy County, 99 nests were found and monitored in 1994–1999 (Proudfoot and Johnson, 2000).

Recent Arizona records are known only from Organ Pipe National Monument, near Ajo, and near Tucson, where nesting occurred in the 1990s (Monson, 1998). In the Tucson Basin 16 nests were located between 1994 and 1999. Little or nothing is known of its status or population trends in Mexico and farther south in Latin America (Proudfoot and Johnson, 2000). The Arizona population of this species was declared endangered nationally in 1997. In terms of all localities where the species was detected during the 1997–2000 Christmas Bird Counts, the ferruginous pygmy-owl was second only to the elf owl as to relative rarity of all North American owl species then encountered. It was seen at an average of only 2.25 U.S. and Mexican localities, for an average of 8.75 total birds. Single birds were reported from 3 U.S. localities, but 18 were seen at Alamos, Sonora, Mexico, in 2000. At lest 280 birds of the *cactorum* race were recently reported from the Sonoran desert of Mexico, helping to balance the precarious status of that race on the U.S. side of the border.

Elf Owl *Micrathene whitneyi* (Cooper) 1861

Other Vernacular Names:
Sanford's elf owl (*sanfordi*); Texas elf owl (*idonea*); Whitney's elf owl (*whitneyi*).

Range

Breeds from extreme southern Nevada, southeastern California, central Arizona, northwestern New Mexico, western Texas, Coahuila, Nuevo Leon, and southern Texas south at least to Sinaloa, Coahuila, and Tamaulipas, and perhaps sometimes or locally to Guanajuato, Distrito Federal, and Oaxaca; also resident in Baja California Sur and (historically) on Socorro Island and Revillagigedo Island. Winters in central and southern Mexico, perhaps from Sinaloa south to Guerrero and Puebla. (See Figure 46.)

Subspecies (Adapted from AOU, 1957; Howell and Webb, 1995; and Henry and Gehlbach, 1999.)

M. w. idonea (Ridgway). Resident from Rio Grande Valley of Texas south at least to Coahuila and Tamaulipas. Migratory in part or all of breeding range. Validity of this subspecies is questionable.

M. w. whitneyi (Cooper). Lower Colorado River Valley, southern Arizona, southwestern New Mexico, and southwestern Texas south probably to Sinaloa. Migratory in part or all of the indicated range, which is poorly documented and may be largely nonbreeding.

M. w. sanfordi (Ridgway). Resident in Baja California Sur.

M. w. graysoni Ridgway. Resident historically on Socorro and Revillagigedo islands; possibly extinct.

Measurements

Wing, males 105–115 mm (ave. of 7, 110.7), females 106.5–112 mm (ave. of 10, 108.9); tail, males 46.5–53.5 mm (ave. of 7, 49.7), females 45–51 mm (ave. of 10, 47.2) (Ridgway, 1914). The eggs of *whitneyi* average 26.8 × 23.2 mm (Bent, 1938).

Weights

Walters (1981) reported that 20 adults (sex not specified) averaged 41 g (range 35.9–44.1). Ligon (1968) reported the weight of breeding females as 41–48 g. Three breeding males averaged 43 g (range 41–46), 3 nesting females 46 g (range 45–47) (Henry and Gehlbach, 1999). The estimated male:female mass ratio is 1:1.07, comparable to the flammulated owl's corresponding ratio. The

estimated egg mass is 7.5 g; the actual mass of 7 partly incubated eggs was 6.9 g (Ligon, 1968). The estimated egg-to-female proportional mass ratio is16 percent, a very high percentage that is also comparable to that of the flammulated owl. These relative sexual and egg-mass ratios represent extremes for North American owls.

Identification

In the field. The tiny size of this species is distinctive, as is its desert habitat, where it typically can be found nesting or roosting in old woodpecker holes of about 5–7 centimeters in diameter (which are sometimes also used by pygmy-owls). It lacks ear tufts, and its diverse high-pitched notes are generally reminiscent of a puppy's yelping calls. Separated from the similar-sized ferruginous pygmy-owl by its shorter tail and much more grayish brown rather than rufous brown upperpart coloration. Unlike pygmy-owls, it is normally inactive during daylight hours.

In the hand. No other owl species is as small (wing under 116 mm) as this one; the somewhat larger pygmy-owls have tails that are longer (more than 50 mm) and more definitely barred, and underparts that are more distinctly streaked with rufous. Adults cannot reliably be sexed by plumage or measurements, although the slightly larger females average somewhat redder and would have brood patches when breeding. Chicks are initially covered with pure white down, and have a grayish yellow iris and a horn gray bill. Juveniles are similar to adults, but the crown is nearly immaculate deep brownish gray and lacks any cinnamon buff on face or throat, or buffy brown on underparts. The latter is irregularly marbled or clouded with white and light brownish gray, narrowly barred with darker gray. Reliable criteria for ageing older juveniles are still undeveloped, but the leading edges of the outer primary coverts may have a notched pattern of whitish markings rather than a uniformly broad whitish edging. Immature birds retain their juvenal flight feathers, so relative wear might be useful in ageing (Pyle, 1997b). It is likely that the iris color also gradually becomes a brighter yellow with age.

Vocalizations

Ligon (1968) determined that elf owls have about a dozen distinct vocalizations, some of which are

Figure 46. Distribution of the elf owl, showing breeding ranges of the races *graysoni* (gr), *idonea* (id), *sanfordi* (sa), and *whitneyi* (wh); indicated racial limits are only approximations, and the breeding range in northern Mexico is speculative. Wintering areas of *whitneyi* and *idonea* are shown by stippling.

limited to a single sex. The advertising song (also called the "chatter" or "song B" type), uttered only by males, varies in length but usually has 5–7 descending notes uttered in slightly more than a second. The notes sound surprisingly like a puppy's yelping. The first note of the chatter song may be fainter than the rest, and the final note may be more emphatic or lower-pitched. The song's main energy is centered at about 1.5 kHz, and thus is higher-pitched than most North American owls except pygmy owls. The chatter song is used for territorial and nest site advertisement, as well as to attract females. The intensity of the song is influenced by time of day (highest during early evening and at daybreak), season (highest during the breeding season, from spring to June), degree of moonlight (higher on moonlit nights), as well as temperature and wind conditions. Similar sequences, with 8–9 notes per second (song B type) are sometimes uttered for longer periods of a minute or more. The song B version stimulates the female to approach and enter the cavities found by her mate, and it increases in volume and intensity as the female responds. As the female approaches the nest the male gradually descends slowly into the cavity, simultaneously decreasing the song's volume. Paired birds also sometimes utter duets, with the female's call resembling the male's primary advertising song, but shorter and softer.

A call uttered by the female, which assists the male to locate her and is uttered both before and after the start of incubation, is a single *peeu* or *seeu* note. When the male flies from a cavity that he has been showing to his mate he sometimes utters a short flight song, with the song rate and volume varying from one series of notes to another, sounding like a repeated *CHU-ur-ur-ur.* When a bird of either sex is disturbed it utters a scolding or "bark" note, a single sharp *cheeur* that is often repeated. Another single-note call uttered by both sexes is the "station call," a whistled and slurred *peeu,* mainly uttered by pair members when feeding dependent young (Henry and Gehlbach, 1999).

The male's precopulatory call is an excited and repeated *che-o,* uttered as it flies toward its mate. During copulation the female utters a shrill *sheee....* Some females also utter a drawn-out *rrrr* ... when being fed by their mate, and young utter a high-pitched trill when being fed. When hungry they utter a repetitive rasping call.

Habitats and Ecology

Traditional descriptions of elf owl breeding habitats have overemphasized their association with giant saguaro (*Carnegiea gigantea*) cacti, for although they are extremely common in the dry, upland Sonoran deserts that are dominated by these plants, they also occur on the low plains of river bottoms and the adjacent tablelands, where if giant cacti are not present they nest in woodpecker holes in cottonwoods (*Populus*), sycamores (*Plantanus*), and probably almost any other tree species within this general habitat type. Woodpecker holes made by several genera of medium-sized woodpeckers (*Picoides, Melanerpes, Colaptes*) are regularly used for nesting. The birds are able to survive indefinitely in the absence of surface water.

In Texas, where the saguaro is lacking, the birds occur in thorny desert scrub (where they often nest in agave flower stalk cavities), riverside cottonwood groves, mesquite (*Prosopsis*) groves on flood plains, and mixed juniper-piñon-oak woodlands, thus occurring in nearly every type of xeric woody vegetation but avoiding solid stands of pines (Oberholser, 1974).

Ligon (1968) described the breeding habitat of the Cave Creek area of the Chiricahua Mountains in southeastern Arizona as being characterized by supporting sycamores (*P. wrightii*) along streambeds, junipers (*Juniperus*) in overgrazed areas, and locally various oaks (*Quercus*) as well as pines (*Pinus*), willows (*Salix*), box elders (*Acer*), ashes (*Fraxinus*), and cypresses (*Cupressus*). In such canyon habitats nesting most often occurs in sycamore cavities; the abundance of the owls is correlated with the abundance of such cavities and also that of western screech-owls, which perhaps are unable to use such small holes as those chosen by elf owls (Henry and Gehlbach, 1999).

In most parts of their winter range that have been studied, the birds occur where the vegetation varies from sparse to dense, but without large trees or columnar cacti to provide roosting cavities, and they roost in bushes or shrubby trees during the day (Ligon, 1968). Columnar cacti may be a regular component of wintering habitats, and may perhaps also be used for nesting sites in Mexico (Henry and Gehlbach, 1999).

There are few good estimates of breeding densities, but Walker (1974) mentioned finding 11 nests in an approximate square-mile area of saguaro desert, indicating a minimum density of 4.6 nesting pairs per square kilometer. He thought that even greater densities might occur in woodlands on adjacent mountain slopes. Goad (1985) found that elf owl density was positively correlated with the presence of the largest size class of saguaro cacti present in her study areas in Saguaro National Monument but concluded that any saguaro with a cavity present was a potential nest site. Ligon (1968) mapped the nest sites and territories of several pairs, which sometimes had

nest sites as close as only 9 meters apart (estimated average of 4 mapped distances, 40 m). One
mapped area of approximately 1.1 hectares supported at least three territories, or a density of
about a pair per 0.3 hectares. Evidently the territory centers on a nest site, and males probably establish a territory to encompass it only after locating a suitable nesting cavity. Each territory
also includes a foraging area, but inasmuch as
food is sometimes apparently superabundant, the
foraging areas of adjoining territories sometimes
partially overlap. Territories are apparently established by the males shortly after they arrive on
their breeding grounds, before the females arrive.
Territories appeared to Ligon to be stable in size
from the time his study was initiated in mid-May
until it was terminated in early August; although
he judged that the intensity of territorial defense
might wane in midsummer. Males may defend
more than one potential nest cavity but not most
of the space between these cavities. So far only a
single case of possible polygyny has been reported
(P. Hardy, 1997; Henry and Gehlbach, 1999).

Movements

This species is one of only two (the other being
flammulated) highly migratory owls in North
America, and both species are unusually small
and highly insectivorous. Ligon (1968) judged
that the elf owls breeding in southern Arizona
may be forced to winter farther south than if they
were diurnal, inasmuch as there are not insects
active during cold winter nights in that area. On
their Mexican wintering grounds there is an
abundance of insects and other arthropods at
night. Little is known of their migration, but the
earliest seasonal records for elf owls in Arizona
are for mid-February, and the last fall record is
for mid-October. They are generally present in
Arizona from March or April through September
(Phillips, Marshall, and Monson, 1964; Henry and
Gehlbach, 1999). In Texas the birds are reportedly resident throughout the year (Oberholser,
1974), although Ligon (1968) was unable to locate
them during searches in February. It seems likely
that the entire northern population is migratory,
with only infirm birds likely to persist in winter,
although climatic amelioration may be changing
that situation (Henry and Gehlbach, 1999).

During spring migration the males apparently precede the females, and probably the first
males to return to Arizona establish territories at
lower altitudes, forcing the tardier males to move
into higher canyons as the lower-altitude habitats
become saturated. Once on their territories the
birds are seemingly quite sedentary; the maximum distance of observed foraging away from a

nest site in the territories of five pairs mapped by
Ligon (1968) was about 70 meters, and the two
smallest foraging territories were no larger than
about 45 meters in maximum diameter. Gamel
(1997) found that nine radio-tagged adults in
Texas had summer home ranges of 0.2–2.6
hectares, averaging 1.0.

Foods and Foraging Behavior

The best analyses of foods of the elf owl are those
of Ligon (1968) and Henry and Gehlbach (1999),
both based on original data and a review of the
literature. Apart from a few records of lizards
(*Sceloperus, Comphosaurus*), blind-snake (*Leptotyphlops*), and young kangaroo rats (*Dipodomys*),
nearly all the reported prey are of arthropods, including insects, centipedes, spiders, and scorpions. Centipedes are taken by adults and fed to
the young, after their stingers have been removed. Muller (1970) reported that arachnids
may be an important component of the food of
nestlings, perhaps comprising as much as half of
the total.

The majority of food items taken by elf owls
are certainly insects. Henry and Gehlbach (1999)
tabulated 18 families of insects as reported prey
in Texas and Arizona, with 7 insect families common to both regions, as well as spiders, scorpions,
and two lizard families. Orthopterans, beetles,
and lepidopterans were especially well-represented
insect groups. Ligon (1968) found that during
early July crickets (Gryllidae) and moths (mainly
Noctuidae, some Sphingidae) were the primary
dietary items, but by mid-July scarab beetles comprised most of the food brought to the young, reflecting the onset of the summer rains and corresponding emergence of the adult beetles. Various
other orthopterans, coleopterans, homopterans,
hemipterans, hymenopterans, and neuropterans
have also been reported as foods. The average
prey length of arthropods is only 1.8 centimeters,
and at least in Arizona there is a high level of dietary overlap between the elf owl and the coexisting western screech-owl (Henry and Gehlbach,
1999).

Foraging is often done by flying over open
ground, sometimes hovering above prey, and also
pursuing prey on the ground. Additionally the
birds sometimes fly out from perches to hawk flying insects, typically capturing their prey with the
feet. Hovering above insects on the ground and
capturing them as they take flight is another
common method of foraging. Unlike most owls,
elf owls are not completely silent fliers, which
probably does not matter for most of their invertebrate prey. Peak feeding periods, at least while
the young are being fed, are around dusk and

dawn, with some foraging throughout the night (Ligon, 1968). However, the birds exhibit little or no eye shine (Walker, 1974), suggesting that their nocturnal vision may be rather poor.

Social Behavior

Little is known of pair-bonding patterns or pair-forming behaviors of this species. It is likely that males arrive on their breeding grounds in advance of the females and quickly establish territories containing one or more potential nest sites. The male then advertises his location, attracts a female, and shows her his potential nesting cavities. Ligon (1968) judged that the pair bond lasts at least three months in Arizona, with birds pairing shortly after arrival. By early April the territorial males are singing almost continuously on moonlit nights. On one occasion Ligon heard two males calling from about 50 meters apart, with a female between. She responded with some scold-like calls and then flew to a tree halfway between them and called several times. One of the males flew to a nearby tree and continued to sing; the female then began to utter the typical call of mated females. The other male also continued to sing, but did not approach the female. The first male eventually flew some distance and began his second (shorter) song type from a cavity. As the female approached he flew out of the cavity while singing his flight song. He later entered a second cavity and also sang from it. The female's calls indicated an apparent willingness to be fed by him, suggesting that pair bonding had begun. In Ligon's (1968) view pair bonding is probably completed when a male actually begins feeding a female. Some representative postures of the elf owl are shown in Figure 47.

Following pair formation the two birds usually remain close together, with nest sites the focus of their activities. Singing periods of the male alternate with periods of foraging and of feeding his mate. After pairing the male sings from his potential nesting cavities, stimulating the female to investigate them and presumably choose one of them as an actual nest site. The cavity is also used as a daytime roost by the female prior to egg laying. The female may begin roosting in the nesting cavity a week or two before laying the first egg, perhaps in part to prevent its occupancy by other hole-nesting birds (Ligon, 1968). If any nesting materials left by an earlier occupant are present, these may be removed by the owl (Henry and Gehlbach, 1999).

Copulation often occurs in the nest tree and may be preceded by the male flying to his mate and feeding her. The pair may also bill for some time. The male utters an excited precopulatory call before mounting the female, who perches crosswise on a horizontal limb and utters a copulation call. Males typically fly silently away following treading. Copulation may be repeated several times in a single night but apparently occurs only infrequently after egg laying is completed. During the incubation period the female forages independently at dusk, but after the young hatch she remains in the nest cavity for much of the time, with the male providing food for her and the nestlings. The increasing food demands of the growing young may determine when the female begins to leave the nest and forage independently again.

Breeding Biology

Evidently elf owls are wholly dependent upon woodpeckers for their nest sites. Of nearly 30 nest sites analyzed by Ligon (1968), the average height aboveground was 10.3 meters; the average cavity depth was 24.5 centimeters; its inside diameter averaged about 10 × 11 centimeters; and its entrance averaged about 5.0 centimeters in diameter (range 3.8–6.4 cm). Woodpeckers responsible were mainly gilded flickers (*Colaptes auritus chrysoides*), Gila woodpeckers (*Melanerpes uropygialis*) in saguaro desert habitats, or acorn woodpeckers (*Melanerpes formicivorus*) and Strickland's woodpeckers (*Picoides stricklandi*) in wooded habitats. In Texas, holes excavated by the golden-fronted woodpecker (*Melanerpes aurifrons*) and ladder-backed woodpecker (*Picoides scalaris*) are known to be used. The vegetational substrates are highly variable and probably in part depend on what is locally available to woodpeckers. Ligon (1968) reported 26 nests in sycamores, 4 in pines, and 2 in walnuts. This apparent preference for sycamores probably reflects the fact that this tree has fairly soft wood, often with rotten or dead limbs present. Among a sample of 46 clutches, mostly from the Western Foundation of Vertebrate Zoology, all but 3 were in giant saguaro cacti, with 2 of the remainder in sycamores and 1 in an oak. The cavity entrances averaged 5.5 meters aboveground (range 3–9 m). When nesting occurs in saguaro cacti, both the main stem and larger branches may be used, and in the warmest parts of southwestern Arizona the nest openings are likely to face north or northwest. Historically, egg dates for assumed first nestings in Arizona are from early May to early June, but in recent years nesting has begun as early as late April, in apparent correlation with warming climatic trends (Henry and Gehlbach, 1999). In Texas, the birds favor nesting in thorn woodland near riparian habitats, perhaps because of reduced competition from screech-owls and ferruginous pygmy-owls

Figure 47. Elf owl behavior, including slimmed concealment posture (*A*) and normal posture (*B*). After sketches in Ligon (1968) and Henry and Gehlbach (1999). Also shown (*C*) is a forward-bowing (probably aggressive) posture, after a sketch in Voous (1988).

(Gamel, 1997; Henry and Gehlbach, 1999). Although many other hole-nesting birds occur in the same region, holes and natural cavities are typically so abundant that their numbers are unlikely to limit nesting densities of elf owls. Nest site openings of gilded flickers average 8.3 × 7.0 centimeters; those of Gila woodpeckers, 6.3 × 5.7 centimeters; and acorn woodpeckers, 4.7 × 5.0 centimeters. All are commonly used by elf owls. In a recent Sonoran desert study, 68 elf owls nested only in woodpecker cavities in saguaro cacti (P. Hardy and Morrison, 2001).

The eggs are usually laid on alternate days, with three days sometimes separating successive eggs. Of 54 clutch records known to me, the average clutch was of 3.04 eggs, and the most com-

mon clutch size was 3 eggs (70 percent of total). Among 90 clutches summarized by Ligon (1968), the most common clutch (56 percent of total) was of 3 eggs, and the observed range was 1–5 eggs. The average of all 90 clutches was 2.98 eggs, with clutches averaging slightly higher in desert habitats than in wooded canyons. Lost clutches may be replaced, and brood losses may be followed by second nesting efforts. These replacement clutches usually have only 1 or 2 eggs (Henry and Gehlbach, 1999).

In a sample of 29 intensively studied nests the clutch averaged 2.55 eggs, the average initial brood size of 23 successful nests was 2.4 young, and the average number of young fledged from these successful nests was 2.3 young, a 94 percent rearing success rate (Ligon, 1968). The overall breeding success (90 percent of all eggs producing fledged young) was one of the highest reported for owls, a remarkable fact considering the vulnerability of the tiny adults and a probable reflection of the difficulties large avian or mammalian predators have in gaining entrance to the nest. In a Texas study, half the nests studied (16–28 per season) over a four-year period were successful, and 90 percent of all nestlings fledged. The overall breeding success (percent of eggs producing fledged young) was 55 percent (Henry and Gehlbach, 1999).

The incubation period is from 21 (Muller, 1970) to 24 (Ligon, 1968) days, with the first 2 eggs usually hatching about the same time and the third a day later. This would suggest that incubation normally begins with the laying of the second egg. Incubation is entirely by the female. Hatching success rates of 60 percent (in Texas) and 95 percent (in Arizona) have been recorded; the latter figure seems extraordinarily high for such small and relatively defenseless owls (Henry and Gehlbach, 1999). After the young hatch the female spends most of the night in the nest passing food on to the chicks as it is brought by the male. There is a feeding peak in early evening and another before dawn, with feeding terminating about 5:00 A.M. and starting again about 7:30 P.M. Evidently the male alone is able to provide enough food for the young and his mate, bringing in insects at the rate of almost a trip per minute during peak periods around dusk. Some of the prey brought to the nest are not immediately consumed, but may be incapacitated to prevent their escape before finally being eaten.

The nestling period lasts for 28–33 days, and the young are able to fly (weakly) at 27–28 days, apparently being able to capture insects such as crickets as soon as they are able to fly efficiently. The period of postfledging dependency of the young on their parents is still unreported. The maximum known longevity under natural conditions is 4 years and 11 months (Patuxent Wildlife Research Center banding data).

Evolutionary Relationships and Conservation Status

In his osteological study, Ford (1967) placed *Microthene* between *Glaucidium* and *Athene* in his taxonomic sequence, all within the tribe Surniini. This treatment coincides with traditional classifications that place *Micrathene* and *Athene* in close taxonomic proximity. Molecular information on its taxonomic position is not yet available.

The population status of elf owls is minimally dependent upon the availability of nesting holes made by various woodpeckers, as well as adequate insect supplies during the breeding season. In California, at the extreme northwestern edge of its range, the elf owl is considered state endangered. It is extremely rare in the few desert riparian habitats it occupies, and only 15–18 pairs were found in a 1987 survey of the lower Colorado River (Halterman, Laymon, and Whitfield, 1989). On the Arizona side of that river it is also very rare, with 10 known pairs in the Havasu National Wildlife Refuge being the major concentration. Environmental degradation (clearing and burning of native trees; channelization and damming of the river) in that area has been severe (Henry, 1998). There may have been a general statewide population decline in Arizona; on the other hand the species may have increased its range in north-central Arizona, and in the Sonoran uplands it may locally be the most common raptor (Henry, 1998). A three-year study in the Sonoran desert of Arizona revealed a stable population of elf owls, compared with a declining trend in western screech-owls. A positive correlation has been found between elf owl abundance and amounts of mesquite cover, overstory perennial vegetation cover, and the density of saguaro cacti (P. Hardy, 1997; P. Hardy, Morrison, and Barry, 1999). There have also been breedings since 1976 in the Magdalena Mountains of western New Mexico, which are outside the known historic range of the species in that state (Stacey et al., 1983). The species' status and population trend in Texas is unknown but probably fairly secure; Gamel (1997) estimated a maximum of 802 elf owls at Santa Ana National Wildlife Refuge alone. The species has also recently been reported from east of the Pecos River for the first time. R. Johnson, Haight, and Simpson (1979) were unable to judge the species' status trends in the Southwest as a whole, and Breeding Bird Survey data are also inadequate for judging such trends. The elf owl was reported on only two Christmas Bird Counts in 1999 and 2000, and in each case was represented by only a single bird, and in Mexican locations. It was the least common species and was seen at the fewest average number of localities of all 18 owl species then reported.

Burrowing Owl   *Athene cunicularia* (Molina) 1782

Other Vernacular Names:
billy owl; Florida burrowing owl (*floridana*); ground owl; howdy owl; prairie dog owl; western burrowing owl (*hypugaea*).

North American Range (Adapted in part from AOU, 1983.)

Breeds, at least historically, from southern interior British Columbia (extirpated since 1979, with some attempted reintroductions), southern Alberta, southern Saskatchewan, and southern Manitoba south through eastern Washington, central Oregon, and California to Baja California, east to western Minnesota (now very rare or extirpated), northwestern Iowa (last nested in 1980s), central (previously eastern) Nebraska, central Kansas, central Oklahoma, and central Texas, and south to northern Sinaloa and the Mexican Plateau; also resident in Florida, the West Indies (Bahamas, Hispaniola), and on Clarion and Guadalupe islands. Locally distributed in South America south to northern Tierra del Fuego. Resident generally in the southernmost parts of the U.S. breeding range (Florida, southern California), but variably migratory in the Great Basin and Great Plains. Migrates south regularly to Honduras, and a winter vagrant occurs south to western Panama. Local breeding may occur within the winter range. Declining generally in the United States, with retracting ranges in many areas, but locally expanding in northern Florida. (See Figure 48.)

North and Central American Subspecies (Adapted from AOU, 1957; J. Peters, 1940, exclusive of extinct and Lesser Antilles forms.)

A. c. hypugaea (Bonaparte). Range as described for mainland North America, exclusive of Florida, south to central Mexico.

A. c. floridana (Ridgway). Florida and the Bahama Islands.

A. c. troglodytes (Wetmore and Swales). Hispaniola.

A. c. rostrata (Townsend). Clarion Island and others of Revillagigedo group.

Measurements

Wing, males (of *hypugaea*) 164.5–178 mm (ave. of 26, 172.3), females 162.5–181 mm (ave. of 33, 170.3); tail, males 74.5–86 mm (ave. of 26, 81.6), females 71.5–85.5 mm (ave. of 33, 79.0) (Ridgway, 1914).The eggs average 31 × 25.5 mm (Bent, 1938).

Weights

Thirty-eight Colorado males averaged 146.3 g, 31 females 156.1 g. A sample of 111 Florida males averaged 148.8 g, 162 females 149.7 g (Haug, Millsap, and Martell, 1993). The estimated male:female mass ratio is 1:1.01–1.07, one of the lowest such ratios of all North American owls. The estimated egg weight is 10.5 g, the same as the actual mean of 10 eggs (Haug, Millsap, and Martell, 1993). The estimated egg-to-female proportional mass ratio is thus 7.0 percent.

Identification

In the field. This small owl is almost always found in open, low-grass fields, usually among prairie dog "towns" or other available rodent-dug cavities in which nesting occurs; rarely, in natural earthen or rock cavities. The birds often perch on low prominences or fence posts, when their unusually long legs are evident (Figure 49). They lack ear tufts, and the underparts tend to be crossbarred rather than streaked as in most small owls. The male's advertisement call is a repeated, dovelike *cu-coo,* uttered mainly at night during the pair-forming period in spring.

In the hand. The very long legs (tarsus more than twice as long as the middle toe) and generally small body (wing under 185 mm) serve to separate this species from all other North American owls. Adults cannot be reliably sexed by measurements (males of this species are as large or larger than females), but breeding females tend to have darker underparts during breeding (a result of differential feather wear), as well as brood patches. Chicks are initially scantily covered with grayish white down, which after about 14 days is gradually replaced by contour feathers. The crown, hind neck, and back of juveniles are mostly plain light grayish brown to buffy brown; the wing coverts, mostly light buff. The underparts and upper tail coverts are immaculate buff, the sides of chest shaded with brown, and the throat band uniform brown. First-year birds tend to have narrower rectrices and remiges than older age-classes, outer primaries that are initially rather broad-tipped but gradually become more tapered, and somewhat more distinct white barring on the central rectrices (Pyle, 1997b).

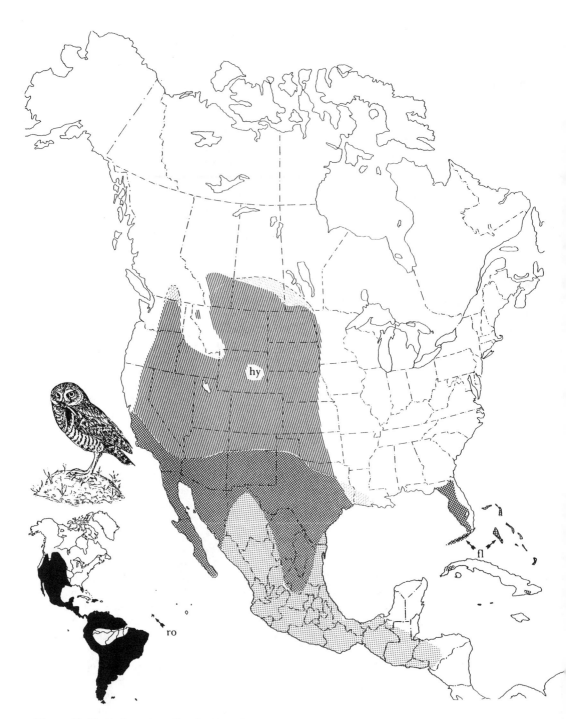

Figure 48. North American distribution of the burrowing owl, showing breeding (hatched), residential (cross-hatched), and wintering (stippled) ranges of *hypugaea* (hy), and residential ranges of *floridana* (fl) and *rostrata* (ro). Recent retractions of Canadian range indicated by stippling; extralimital distribution shown in inset.

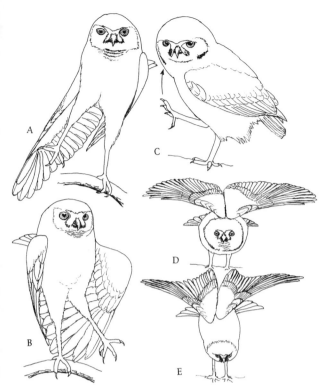

Figure 49. Burrowing owl comfort behavior, including early (*A*) and late (*B*) stages of wing-and-leg stretching, (*C*) self-scratching, and early (*D*) and late (*E*) stages of two-wing stretch. *C* and *D* after photos by Art Wolfe; others after photos by author.

Vocalizations

According to D. Martin (1973b), the burrowing owl has a repertoire of at least 17 vocalizations (including 3 by young), making it one of the most accomplished vocalists of all North American owls and comparable to the common barn owl in this respect. The male's primary song, used in pair formation, precopulatory behavior, and territorial defense, consists of a double-noted *coo-coo,* the second note longer than the first and the total duration about 0.6 seconds, with a similar interval between calls. There is little frequency variation; the fundamental frequency is at about 1000 Hz (1.0 kHz), and harmonics are variably developed, producing a rich and musical timbre. The song is uttered only when the male is near its burrow, and as it is uttered the bird expands its throat and dips the anterior part of its body downward in a "bowing" display (Figure 50). The song is fairly variable in its frequency characteristics, but its temporal components are less variable, especially the intervals between notes. It is uttered at all times of the year, but least frequently from September to December (Thomsen, 1971). During the spring pair-forming period singing may continue from sunset throughout the night. Males utter one or two primary songs during copulation, terminating with a multinoted tweeter call. Females utter an extended "smack" or infrequently a warbling call during copulation. A similar warbling termination to the male's song may also be added during copulation (D. Martin, 1973b).

Females produce a rasping call when distressed, when begging for food, when receiving food from the male, or when passing it on to the young. The rasp call may also be used by the adults as an "all's-clear" signal to the young (Thomsen, 1971). Females also respond to the male's primary song by uttering an "eep" call that may grade into this rasping note, and they may utter an irregular warble when defending the nest burrow against other females. A rattling note is produced by females when playbacks of a male's primary song are made, and both sexes utter various chucking, chattering, and screaming calls as warning signals or during mobbing behavior toward possible nest predators. When alarmed the bird usually stands quite erect, calling as she bobs up and down.

The young owls have three distinct calls plus defensive bill snapping. These include an "eep" used by young birds as a low-intensity alarm, distress, or hunger call, a rasping call that is produced when being fed or as an indication of hunger, and a "rattlesnake rasp," uttered by the chick when severely distressed or cornered by a predator. This call closely resembles that of a rattlesnake's (*Crotalis*) rattle. Bill snapping and the rattling call are often uttered by cornered adults

163

Figure 50. Burrowing owl social behavior, including (*A*) male bowing display, (*B*) singing postures of male (after Martin, 1973a) with sonogram of song shown above, (*C–D*) variations in exposure of white throat and facial feathers (after Thomsen, 1971), (*E*) reciprocal preening by paired birds, and (*F–G*) nest-defense postures (after Bent, 1938, and photo by author).

as the feathers are fluffed out, the wings are opened and rotated forward, and the bird weaves back and forth while crouching (Figure 50).

Habitats and Ecology

The usual habitat of this species is level, open, and dry vegetation that is typically of heavily grazed or low-stature grassland or desert vegetation, with available burrows. These are primarily those of colonial rodents (mainly *Citellus* or *Cynomys*) but occasionally are produced by other animals including even tortoises (*Gopherus*). Rarely they may be dug by the birds themselves, as has been reported in Florida. Nesting areas always have available perching sites, such as fences, utility poles, or raised rodent mounds (R. Grant, 1965). The distribution and abundance of burrowing rodents are thus central to the species' ecology, providing not only nest sites but also sites for perching, food storage, escape from enemies, and ameliorated temperature changes during both extreme cold and hot conditions (Coulombe, 1971; Thomsen, 1971). An exhaustive review of the habitat needs of this species and the effects of management practices on it was provided by Dechant et al. (1999a).

MacCracken, Uresk, and Hansen (1985) reported that in South Dakota the birds select sites that are in early stages of grassland succession,

offering an abundance of annual forbs and relatively low average vegetation height, thus providing hiding sites for young owls while not obscuring the owls' vision. Low vegetation around the nesting site may also increase hunting efficiency. Burrows located in relatively sandy sites may also be favored there, perhaps because they are more easily modified by enlarging passageways and may also drain more rapidly following rainfall. Too sandy soil, however, may cause the burrow to collapse, and burrowing owls are rare throughout the Nebraska sandhills. On the other hand, Toombs (1997) found that soils with a high amount of coarse materials are avoided by prairie dogs. He also noted that active prairie dog colonies are much more likely to support burrowing owls than inactive ones, that larger colonies have more nests than small ones, and that the areas with a high density of active burrows were those favored by burrowing owls. Somewhat similar results were obtained by Desmond and Savidge (1996) and Ekstein (1999) in Nebraska, where owls selected nests in areas of high prairie dog density, but closer to the edge of the colony, even through owl breeding success was often higher in areas of lower burrow density and in sites farther from the colony edge.

Burrows used by the owls vary considerably as to their origins. R. Rich (1986) found in Oregon that the birds preferred to nest in small rock outcrop cavities, perhaps for reasons of increased safety against badgers (*Taxadea taxus*) and canid predators. In this area as well as in Nebraska, badger excavations are sometimes used as nest sites, but at generally low rates. Those studied by D. Martin (1973a) were all made by rock squirrels (*Citellus variegatus*), and those with entrances at the bottom of arroyo gullies or other vertical cuts were favored over those at the lip or on the vertical face of a cut. Most of the utilized burrows slanted gradually down from the entrance and invariably had a sharp right- or left-hand turn present. The entrance size varied greatly, but averaged 32 × 24 centimeters; the inner tunnel dimensions were somewhat more uniform, averaging 11 × 20 centimeters. Based on studies of use of artificial burrows, the inner dimensions and composition may not be critical but the tunnel should be sufficiently convoluted as to maintain the nest chamber in darkness (C. Collins and Landry, 1977; C. Collins, 1979).

Probably nesting density is strongly influenced by local burrow abundance and condition, but in optimal California habitats it may reach about 8 per square kilometer (Coulombe, 1971). In a New Mexico study the observed population densities were evidently considerably lower (15 pairs located along 3.7 linear km of habitat) and territory sizes correspondingly higher, in spite of available and apparently suitable burrow sites in the area; there the average distance between occupied nests was 166 meters (D. Martin, 1973a). R. Grant (1965) estimated that 5–9 pairs nested in an area of 32 hectares in North Dakota, suggesting a nesting density of 3.5–6 hectares per pair, and Wedgwood (1976) also found a colony of 6–8 active burrows in a Saskatchewan pasture of about 100 hectares, or about 13–16 hectares per pair. In areas where the birds nest in badger (*Taxidea*) burrows or natural rock cavities, the density may be as low as a pair per 58 square kilometers (Gleason and Johnson, 1985).

Home ranges were estimated by Haug (1985) as ranging from 0.14 to 4.81 square kilometers (averaging 2.41 sq km), using radio-telemetry. Territoriality in the burrowing owl is seemingly limited to defending the immediate vicinity of the burrows, with mutually shared foraging areas by adjacent pairs commonly observed (Coulombe, 1971). Thomsen (1971) estimated six territories as ranging from 0.04 to 1.6 hectares (averaging 0.8 ha), and the distances to the nearest neighbor ranged from 9 to 120 meters. D. Martin (1973a) believed that the territory centers on the burrow and includes an undefined and unmeasured area in both directions from it, but not the foraging grounds. R. Grant (1965) estimated the foraging ranges of two pairs to be 4.8 and 6.5

hectares. Flight distances of as great as 2700 meters from the nest have been measured for radio-tagged males (Haug and Oliphant, 1990).

Movements

Evidently burrowing owls are migratory only near the northern end of their range, with occasional birds overwintering even as far north as British Columbia, Washington, Nevada, Colorado, and Nebraska, and regularly overwintering in Kansas and Oklahoma. However, some birds banded in South Dakota have been recovered in Oklahoma and Texas (Bent, 1938), and vagrants of unknown origin have appeared in various parts of Central America during winter. Burrowing owls have been documented in nearly all the Mexican states (Enriquez-Rocha, 1997).

The Florida population is apparently sedentary, as is the population breeding in southern California, although the latter varies seasonally in numbers, perhaps supplemented by some immigration from more northern areas (Coulombe, 1971). In New Mexico the birds become infrequent during winter and possibly wander about to some degree during that period, especially in the case of young birds (D. Martin, 1973a). It is probably advantageous for adult males to overwinter on their nesting areas, wherever and whenever local conditions permit, in order to retain possession of their burrows and be able to keep them in good repair. During the winter months the birds are inclined to remain within their burrows during daylight hours and become more strictly nocturnal, presumably preying on nocturnal mammals.

Migrants in a New Mexico colony arrived between March 15 and April 3 (D. Martin, 1973a), while at the northern end of their breeding ranges birds arrived about a month later, from April 12 to May 8, with a peak around April 21 (Wedgwood, 1976). In a Manitoba study, adult males returned to their nesting colony in subsequent years more often than did females (40.2 percent vs. 24.4 percent), and males correspondingly dispersed shorter mean distances between years. Average dispersal distances for juveniles were 29.5 kilometers for nine males and 33.7 kilometers for nine females (De Smet, 1997).

Foods and Foraging Behavior

A very large literature on the foods of the burrowing owl has developed, much of which has been summarized by Bent (1938), Zarn (1974a), and Haug, Millsap, and Martell (1993). At least on the basis of relative numerical abundance, this species' major summer foods consist of insects, especially larger species such as beetles, grasshoppers, crickets, locusts, dragonflies, and the like, many of which it captures on the wing or by hovering above the prey and quickly dropping down on them. Scarab beetles, such as dung beetles, which are often abundant in grassland areas, are among the commonest of the insect prey. Among the mammals an equally broad array have been identified in pellets, including mice, rats, ground squirrels, gophers, chipmunks, shrews, young prairie dogs, cottontails, and even bats (Bent, 1938). Various birds, apparently especially horned larks (*Eremophila alpestris*), are sometimes taken. These include species occasionally weighing almost as much as the owls themselves, such as adult mourning doves (*Zenaida macroura*), weighing nearly 130 grams on average (C. Collins, 1979). In spite of the owl's rather small, nonreflective eyes and unspecialized hearing morphology, nocturnal foraging activity is well documented (D. Martin, 1973a; Butts, 1973). One advantage of nocturnal hunting in this and other variably desert-adapted owls is that more prey animals are typically active at night than during the hot daytime hours, and there are fewer associated dangers of overheating and evaporative water loss (Cook, 1997).

Prey weights have been analyzed by Schlatter et al. (1980) in Chile, and at least there it is typical that about three-fourths of the total diet is made up of mammals, with anurans of secondary importance and other vertebrates used very little. The most frequently taken rodents have adult average weights of less than 100 grams, but substantial numbers of juveniles of a species of *Octodon* averaging 230 grams as adults were also consumed. The authors estimated that the upper limit for mammal kills is about 115 grams. Among the arthropods in the prey, beetles were much the most frequent, followed by dragonflies, orthopterans, and caterpillars. Most of the arthropod prey taken were of ground-dwelling forms, and the agile fliers such as dragonflies are probably captured mainly when they are torpid.

In similar biomass analyses, Gleason and Craig (1979) and Gleason and Johnson (1985) estimated that about 61–68 percent of the burrowing owl's prey in southern Idaho is of mammalian origin, mainly consisting of kangaroo rats (*Dipodomys*) and voles (*Microtus*). Mammals as large as pocket gophers (*Thomomys*), with an average adult weight of 150 grams, were apparently killed. Insects comprised about 29–32 percent of the remaining biomass total and were primarily represented by Jerusalem crickets. *Microtus* was also reported as an important summer food in North Dakota (Konrad and Gilmer, 1984), while in Arizona scorpions and scarab beetles were

most important on a numerical basis (Glover, 1953), as were beetles in Iowa (Errington and Bennett, 1935) and in North Dakota (James and Seabloom, 1968). Observations by R. Grant (1965) on a pair with two young in Minnesota indicated that about 85 percent of the biomass (or about 300 g) consumed during a 24-hour period were comprised of vertebrates (mostly *Peromyscus* and *Microtus,* some frogs or toads) averaging about 25 grams each, and the remainder comprised of various insects (mostly beetles). In an Alberta pellet study, Schmutz, Wood, and Wood (1991) tallied 136 cricetine rodents (*Peromyscus* and *Microtus*), 247 insects, 2 birds, and a ground squirrel. It is quite possible that insects are especially important to young and relatively inexperienced birds, while adult owls continue to prey on larger vertebrate foods during the summer months, as suggested by Errington and Bennett (1935). However, the incidence of insect foods remained rather constant (as measured by frequency of occurrence in pellets) throughout all seasons of the year in Thomsen's (1971) study.

Marti (1974) reported that the burrowing owl in Colorado fed heavily on insects throughout his period of study (April to September), but perhaps depended more strongly on vertebrates during times when insects were less abundant. The mean prey weight for all species was only 3 grams, with more than 90 percent of the identified prey individuals weighing no more than a gram. Nevertheless, the majority of their food biomass was of mammalian origin. The owls hunted by direct aerial chases, by hovering flights, by running down prey on the ground, and by hawking insects during short sorties from perches. Most vertebrates were captured during low light levels, when the owls had a visual advantage. However, the birds hunted actively all day long, with activity peaks around sunrise, near midday, and again around sunset. R. Grant (1965) reported a distinct peak in hunting activities between 8 and 11 P.M. in Minnesota (sunset at 9:20 P.M.) and almost no midday hunting. Most prey were captured during fairly short (under 100 m) glides or darting flights out from a perch.

Social Behavior

Pair formation in this species is still little studied, but from information on banded birds it is clear that pair bonding is not permanent. Of 9 males and 9 females banded as breeding birds one year, 6 males and 2 females returned the next (D. Martin, 1973a). All the males selected the same burrow they had occupied the year before (if it were still usable), but none of the 6 pairs in which both members had been banded was reunited. In one case both members of an original pair were present but had acquired new mates. Thomsen (1971) reported that, of 9 pairs studied one year, 5 were reestablished the following year; 3 lost one or both members; and 1 pair dissolved and its members established bonds with new mates. However, in the residential Florida population pair fidelity is well developed, with 92 percent of pairs remaining intact among those birds who both survived to the next breeding season (Millsap and Bear, 1992). Some cases of polygyny have also been reported (Haug, Millsap, and Martell, 1993).

Birds arrive on the breeding area either already paired or singly. Immediately after arrival the males begin to occupy burrows and prepare them for new use. This involves excavation by scratching, as well as transporting mammalian feces to the nest entrance, which are shredded and partially used to line the nest chamber. They also begin singing through the nighttime hours and performing courtship during the early evening hours (D. Martin, 1973a). Mutual billing and preening are among the owl's pair-bonding activities. Some representative egocentric and social postures of the burrowing owl are shown in Figures 49 and 50.

Thomsen (1971) and D. Martin (1973a) observed a variety of apparent courtship displays near the nesting site, including billing, mutual nibbling of the head and facial feathers, food presentation, and copulation. While the male sings, a female may stand near him or in the burrow entrance, uttering rasping or "eep" calls. This in turn may stimulate the male to forage and, on returning, to present the female with his food. Following this, mutual billing may occur, followed by renewed calling, more food presentation, and finally copulation. Copulation was observed by Martin to occur mainly during the first hour after sunset, and was seen as often as 8 times in 35 minutes. Exposure of the white feather patches of the throat and "eyebrows" is an apparently precopulatory display in both sexes; the male simultaneously standing very tall while raising his body feathers, but the female keeping hers more sleeked. While treading, the male utters a copulation song, and the female may also utter a special copulation warble.

Although not observed by D. Martin (1973a), R. Grant (1965) twice observed an apparent display flight that exceeded in its duration, distance, and height all other observed flights. It was characterized by an ascent up to about 30 meters, hovering for 5–10 seconds, steeply diving down 7–15 meters, and repeating the sequence (for at least 8 min on one occasion). This description by Grant is rather similar to the described display flights of

snowy owls but has apparently not yet been reported for burrowing owls by others except perhaps Thomsen (1971), who described rare "circular flights," performed mainly by males.

Territorial behavior is most evident during the period of pair formation. It seemingly involves primarily defense of the burrow itself, mainly by standing erect and exposing the white feather patches on the face and throat. In one case a female was observed attacking an intruding female, but otherwise defense of the territory was mainly performed by the somewhat larger male. Territorial defense declined once egg laying began, after which females remained in their burrows throughout the day and the males ceased their singing (D. Martin, 1973a). However, Thomsen (1971) reported vigorous defense of territory until fledging of the young. Bowen (2000) reported that nonbreeding adults represented 8.2 percent of a Florida population of 1757 adults, and among 946 territories possible polygynous matings were observed at 4 territories. In one study area 66.7 percent of the territories were reoccupied and the same burrows used during a second year of study; these reoccupied territories produced a higher mean number of young than those that were not reoccupied.

The unique entrance- and nest-lining behavior of burrowing owls with dried feces and various other materials such as food remains is noteworthy and is generally believed to function in providing insulation, camouflage for the owls' scent from mammalian predators, or both. That it is not simply an incidental accumulation of debris is indicated by the fact that if the feces are removed from the nest opening and tunnel, they are replaced within one day (D. Martin, 1973a). Frederick Gehlbach (pers. comm.) has observed owls bringing in fresh dog feces to their burrow entrances daily, sometimes flying in from more than 300 meters with these materials.

Breeding Biology

Because of the great difficulties in excavating nests, relatively little statistical data on clutch sizes are available for this species. Bendire (1892) estimated that clutches of from 7 to 9 eggs are most common, but they range from 6 to 11, or rarely 12, eggs. Murray (1976) reported that 439 clutches from the species' entire range had from 1 to 11 eggs, averaging 6.49. Other reported means include 9.9 eggs for 30 clutches in Idaho; 7.0 eggs for 32 California clutches; and a median of 4 eggs in 14 Florida clutches (Haug, Millsap, and Martell, 1993). Counts of young surviving long enough to reach the surface have varied from 3.4 to 5.2 (D. Martin, 1973a; Wedgwood,

1976), but by that age a substantial amount of egg and owlet mortality must certainly have occurred and these figures must be somewhat below actual clutch sizes. Replacement clutches, usually smaller than the original ones, are frequently laid following the loss of the first clutch (Bent, 1938). In favorable years renesting is frequent, and some second broods may even be produced (Gervais and Rosenberg, 1999).

As with other owls, a small proportion of territorial birds, perhaps yearlings, evidently fail to nest each year. Wedgwood (1976), reviewing his own and other available data, reported that this estimated incidence varied from 8 to 41 percent of the adult population (weighted average of all studies, 15 percent). The egg-laying interval is not known with certainty, but laying may be done on an approximate daily basis (Henny and Blus, 1981). Judging from the staggered sizes of young in families, incubation must begin with the laying of the first egg. A sample of 41 California records are from April 1 to June 17, with half from April 14 to May 2. Colorado and Kansas records are from March 29 to July 1, and a few egg records from the Dakotas are from May 1 to June 13 (Bent, 1938). A larger sample of 32 active nests in North Dakota range from May 15 to August 23 (R. Stewart, 1975). Egg dates for Florida are mostly from March to May with half of 52 records from April 4 to 23, but there are some fall-to-winter egg records (October, December, January) as well (Stevenson and Anderson, 1994). Renesting occurs following clutch loss, and a few cases of double-brooding have been recorded in Florida (Haug, Millsap, and Martell, 1993).

The incubation period is from 27 to 30 days (Henny and Blus, 1981). Females become highly secretive during the incubation and brooding periods, apparently incubating all day, as well as most if not all night, and being provided food by the male. Temporary food depots may be produced by the male almost anywhere within his territory, but usually within 100 feet of the burrow (R. Grant, 1965). In spite of some early assertions to the contrary, there is no evidence that the male assists in incubation. Apparently the materials brought in to line the nest prior to incubation are removed at some time during the incubation or brooding periods (Thomsen, 1971).

The owlets weigh 8.9 grams on average at hatching, and their eyes begin to open when they are 4 days old. Hatching success in one study was 55 percent (Landry, 1979), but up to 90 percent in another (Olenick, 1990). By 14 days, when their contour feathers begin to break out of their sheaths, they may begin to appear at the mouth of the burrow. By 30 days their average weight approaches its adult limit, and they fledge at 40–45

days (Landry, 1979). Assuming an incubation period of 30 days and a minimum fledging period of 40 days, about 70 days would be required between egg laying and fledging. If a clutch of 5 eggs, laid over a period of 10 days, is considered typical, perhaps nearly 80 days would be needed to complete a reproductive cycle. Nearly half a year would be needed for double-brooding birds.

Wedgwood (1976) reported that among all 45 breeding pairs studied by him the number of aboveground young totaled 172, or an average of 3.8 young per family. If only counts made before the dispersal of young began are used, the average brood size was 4.4 young. With additional statistical adjustments, an estimated 4.6 juveniles per breeding pair were raised to independence, out of an average of 5.1 chicks surviving long enough to reach the surface. Among four other studies summarized by Wedgwood the initial observed brood size varied from 3.4 to 5.2 young, and the average number of young surviving to independence from 1.9 to 4.6 (fledging success rates of 56–90 percent, averaging about 78 percent). This relatively high estimated fledging success rate in a small, otherwise vulnerable owl species suggests that there is a strong selective advantage for it in breeding in the relatively protected environment provided by rodent burrows. In a Florida study, 77 percent of 406 nestings produced at least 1 fledged chick, and 3 chicks were the median number fledging from successful nests (Haug, Millsap, and Martell, 1993). In a sample of 926 Florida territories, the mean number of young produced per breeding territory was 2.4 (Bowen, 2000). Wellicome (2000) determined that supplemental foods provided during the prehatching period did not influence clutch or egg size, nor did it affect the seasonal decline of clutch size. However, birds provided with supplemental foods during the posthatching period had larger and more massive offspring and raised 47 percent more fledglings than did controls.

In California annual estimated survival rates for immatures and adults, respectively, are 30 percent and 81 percent (Thomsen, 1971). In Florida minimum adult survival rates are 68 percent for males and 59 percent for females (Millsap and Bear, 1992). Lower rates have been estimated for migratory populations, based on return rates of marked birds. Maximum known longevity under natural conditions is 8 years and 8 months (Patuxent Wildlife Research Center banding data).

Evolutionary Relationships and Conservation Status

The American Ornithologists' Union (1983) has merged *Speotyto* with *Athene,* thus following the recommendations of various investigators such as Ford (1967). Presumably *A. noctua* is the nearest living relative of *cunicularia,* judging from their apparent zoogeographical affinities and essentially replacement ranges through the Holarctic; other similar but forest-adapted species of *Athene* occur in India and southeastern Asia. Wink and Heidrich (1999) judged from molecular data that *Athene* and *"Speotyto"* may have separated 6 million years ago, but nevertheless are a monophyletic group and can be merged as a single genus. Desmond (1997) similarly judged that the North and South American populations of the burrowing owl split some 2 million years ago, with the North American birds then disseminating southward via the isthmian land bridge of Panama.

The Committee on the Status of Endangered Wildlife in Canada has classified the burrowing owl as threatened in Canada, collectively, and it has been provincially listed as threatened in Saskatchewan and Alberta, and endangered in British Columbia and Manitoba. Its precarious status in British Columbia was discussed by Howie (1980a), and in Manitoba by Ratcliffe (1986). In the early 1990s the Canadian population may have numbered only about 2000 pairs (Haug, Millsap, and Martell, 1993). Some reintroduction efforts have been made in British Columbia. In Manitoba, the known population dropped from 34 nesting pairs to only 1 between 1987 and 1996 (De Smet, 1997).

In 1966 the U.S. Fish and Wildlife Service classified the burrowing owl as a rare species, a classification that was dropped in 1968, and further changed to undetermined in 1973. Among states, it is listed as endangered in Minnesota and Iowa, and of special concern in 12 other states. The overall U.S. status of the burrowing owl has been reviewed by Evans (1982) and more recently by Haug, Millsap, and Martell (1993). Regional reviews of its range and status have also been provided by a variety of other authors. Ligon (1963) noted that the species' breeding range locally increased in northern Florida, in conjunction with increased cattle grazing, as well as the establishment of golf courses, airports, and similar developments (Wesemann, 1986). As of 1990 the estimated population of this race was 3000–10,000 adults (Haug, Millsap, and Martell, 1993), and Christmas Bird Counts and Breeding Bird Surveys data suggest a possibly declining population (Stevenson and Anderson, 1994). Bowen (2000) estimated the densities of burrowing owls in South Florida to be 10 times greater than those in North Florida, with those in the south mainly occurring coastally and those in the north mainly in the interior. Vacant residential lots, airfields, parks, ball fields, and schoolyards all provide significant Florida breeding habitats, in addition to agricultural lands.

At the eastern edge of its range in Minnesota it has been reduced from a locally common resident to virtually extirpated in about 50 years (Martell, 1985). It has also been essentially eliminated from western Iowa and eastern Nebraska. There is a statistically nonsignificant annual population trend of +5.0 percent, based on annual Breeding Bird Surveys done between 1966 and 1993 (Price, Droege, and Price, 1995). A more recent summary of these data suggest a 1.2 percent annual decline from 1966 to 2000, which is also not statistically significant (http://www.mbr-pwrc. usgs.gov/bbs/bbs.html). During the 1999 and 2000 Christmas Bird Counts the species was detected on an average of 88 counts, or about 5 percent of all U.S. and Canadian counts. In terms of all localities where the species was detected, the burrowing owl was the tenth most widely reported of the 18 North American owl species then encountered. The maximum number ever reported on a single CBC was 161, at Fort Meyers, Florida, in 1991. The average number of total counted was 385 birds, or about 4 percent of all owls reported in the tabulated summaries of counts for the United States and Canada during those years. The Canadian population has recently been estimated at 1000 pairs (Kirk and Hyslop, 1998) and is believed to be endangered and declining.

The proceedings of an important symposium on the burrowing owl were published after this book went to press (Wellicome and Holroyd, 2001). They include five papers on biology, nine on population status and trends, and five on conservation and management, all of which should be consulted for the most recent information on these subjects.

By and large the species has suffered over most of its historic range in western North America, as colonial rodent populations have been controlled or eliminated by poisons; insecticides have reduced its food supplies and have perhaps directly poisoned it; and traditional rangelands have been converted to agricultural purposes through irrigation (Desmond, Savidge, and Eskridge, 2000). In most western states the familiar "howdy owl" is saying a long, sad farewell.

Mottled Owl   *Ciccaba virgata* Cassin, 1849

Other Vernacular Names:
Cassin's wood owl; squamulated owl (*squamulata*), Tamaulipas owl (*squamulata*), Búho café;
Lechuza café; Mochuelo llanero (Spanish)

Range

Ranges residentially in Mexico on both slopes from southern Sonora and central Nuevo Leon south through Belize, Guatemala, and the rest of Central America (except the unforested arid regions) to Colombia and western Ecuador; also occurs over most of forested tropical South America east of the Andes to northeastern Argentina. (See Figure 51.)

North American Subspecies

C. v. squamulata Bonaparte. Northern and western Mexico from Sonora and Tamaulipas south to Guerrero.

C. v. centralis (Griscom). Southern Mexico (Oaxaca) south to western Panama.

Measurements

Wing (of *squamulata*), males 239–254 mm (ave. of 9, 247.1), females 240–265 mm (ave. of 10, 251.6); tail, males 139–151 mm (ave. of 9, 143.9), females 147–163 mm (ave. of 10, 154.1) (Ridgway, 1914). Four Trinidad eggs averaged 39.2 × 32.0 mm (Ffrench, 1973), and 16 eggs from Guatemala averaged 42.2 × 36.1 mm (Gerhardt et al., 1994a). Schönwetter (1967–1984) reported a mean of 41.8 × 33.4 mm, and an associated estimated weight of 25.7 g for 17 eggs from Trinidad.

Weights

Eleven females averaged 334.9 g (SD 22.0), seven males averaged 239.7 (SD 13.3) (Gerhardt and Gerhardt, 1997). This mean sexual size difference amounts to about 40 percent, or a good deal more than the 5–10 percent difference estimated earlier by Voous (1988). Three males 177–242 g (ave. 216.3), 2 females 251 and 345 g (ave. 298) (S. Russell, 1964). Smithe (1966) reported a female weight of 300 g, and a male weight of 240 g. A female of *centralis* 356.2 g (Field Museum of Natural History). Range of both sexes (unspecified sample size) 176–305 g (König, Weick, and Becking, 1999). Stiles and Skutch (1989) reported a sex-unspecified weight of 275 g; Gerhardt and Gerhardt (1997) reported a mean weight of 28.2 g for 16 Guatemalan eggs; the estimated weight of eggs from Trinidad is 22 g. This would suggest that the eggs represent about 7–8 percent of the adult female's mass.

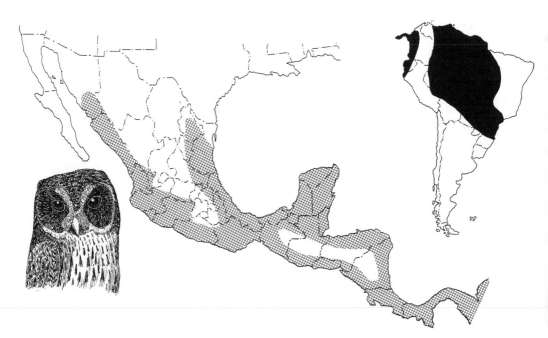

Figure 51. Distribution of the mottled owl. Partly after Howell and Webb (1995). The inset map shows the South American distribution.

170

Identification

In the field. This woodland owl is somewhat similar to a barred owl, with comparable dark brown eyes, and underparts that are vertically streaked with dark brown marking on a paler background. However, the mottled owl is vertically streaked from its throat to its under tail coverts, whereas the barred owl is horizontally barred on the breast and vertically streaked on the lower underparts. Both are highly nocturnal birds of similar size, and both are more likely to be heard than seen. See also "Vocalizations."

In the hand. This is a medium-sized (wing under 275 mm) owl, with no ear tufts, and with naked toes, a rather small bill, a long but rather weak tarsus, and dark brown eyes. Unlike the similar barred owl, its external ears lack opercular flaps in front of the auditory meatus, but the right ear opening is 1.4–1.5 times longer than the left one. There is only a rudimentary dermal flap behind the meatus. The structure of this species seems to be somewhat transitional between the highly developed external ear structure of *Asio* and the much simpler structure typical of *Ciccaba* (Voous, 1988). Adults occur in two plumage morphs, the darker one having buffy cinnamon underparts and the lighter one being more whitish below. Linear measurements do not allow for sex identification, and the plumages of the two sexes are also apparently alike. Nestlings have a cinnamon buff downy plumage, with a paler buff face. Later plumage stages remain undescribed.

Vocalizations

The usual territorial or advertisement song is a series of short, guttural or barking hoots that initially becomes faster and louder, and then fade away, the series having up to about 10 notes. The notes are uttered at approximate half-second intervals, so a 5-note phrase lasts about 2.5 seconds. Shorter series of up to 3 notes have also been heard. Tape recordings of the species' song from Mexico were provided by J. Hardy, Coffey, and Reynard (1999). They include a series of repeated "woof" notes, as a second bird, likely the female, utters more whining calls. There is also a prolonged song sequence of hoots, in which 2 series of 9 gradually louder hoots are uttered in about 20 seconds. A third sequence consists of 2 "bouncing-ball" phrases of up to about 20 rapid hoot notes, much like that of a screech-owl but lower in pitch. This no doubt is the "bouncing ball" sequence, with increasing acceleration but the final notes becoming softer and lower, that was mentioned by Burton (1973). The birds are highly vocal, much like barred owls, and no doubt many additional call types exist. As for screech-owls, whinny-like notes have been described for the female. Wailing screams, especially of young birds, have also been heard, and duetting by adults of both sexes occurs. The call of the males is noticeably higher-pitched than the females'.

Gerhardt (1991) found that mottled owls in Guatemala were quite responsive to playbacks of conspecific calls, and this method may make a useful censusing tool. There the usual call is of 4–6 notes, with 1 or 2 preliminary and muffled notes, followed by 3 higher and booming ones with fundamental frequencies of about 250–750 Hz. The female produced a catlike yowl as an apparent food-solicitation call, but no "bouncing-ball" notes were mentioned. In spite of the rather small size of this owl, the proximal part of their bronchi are much enlarged, and their vibratory membranes are unusually large, which is likely to be the reason why the birds can produce such loud, barking calls seemingly typical of larger owl species (A. Miller, 1963).

Habitats and Ecology

This is an ecologically highly tolerant species, ranging in Mexico from sea level to about 2500 meters, and in habitats from thorn forest and coffee plantations to second-growth forest edges, pine-oak forests, and humid evergreen forests. It is especially abundant around coffee and cacao plantations. It also favors patchy forest and tolerates limited human presence very well.

In Costa Rica the species has a prey base (mainly invertebrates) similar to the Middle American ("vermiculated") owl and crested owl, and all three responded to playback of recorded calls of the other two species, as well to others to which they were exposed (Enriquez-Rocha, 1997).

Movements

Apparently residential and sedentary.

Foods and Foraging Behavior

Wetmore (1968) stated that the stomachs of birds he collected in Panama usually contained large beetles and orthopterans, but one had eaten a snake and yet another contained a salamander. Elsewhere in its broad range it has been known to take a variety of rat-sized rodents such as rice rats (*Oryzomys*), cotton rats (*Sigmodon*), and climbing rats (*Ototylomys*), bats, tree frogs (*Hyla*), and snakes and lizards (*Anolis*). Gerhardt et al. (1994b) examined 52 mottled owl pellets from Guatemala, determining that almost half contained only insect fragments, and only one lacked any insect remains. Half the pellets contained rodent remains, and a few lizard and bat remains were also present. Among prey items brought to nests were beetles, cockroaches, raniid and hylid frogs, various native rats (*Oryzomys, Ototylomys*), and a small bird. This owl is a nocturnal hunter but sometimes exploits the presence of artificial lights while hunting prey. Most of the prey taken are probably terrestrial, and hunting is done by swooping down from low perches.

Social Behavior

Among the most nocturnal of all strigid owls, these apparently solitary birds remain little studied as to their social behavior. Like other nocturnal owls such as barred owls, they are highly vocal, and probably most of their important signaling is done in this manner. Gerhardt et al. (1994a), working in Tikal National Park, Guatemala, radio-tagged 7 males and 4 females. The estimated home ranges for the 4 males were 20.8 hectares. The breeding density for their study area was estimated at 7 adults per square kilometer, or 8.8 adults if some pairs that didn't attempt to build nests are included. Roost sites were found to average 5.3 meters high, and roosts were typically in dense wooded swamps.

In tropical dry forest habitat of Tamaulipas Frederick Gehlbach (pers. comm.) estimated a density of 1 breeding pair per square kilometer, but in tropical evergreen forest of Belize the estimated density was 5 pairs per square kilometer. In Tamaulipas the birds were vocal from April to June, whereas in Belize singing occurred from January to March.

Breeding Biology

Good information on this species' breeding biology comes from the work of Gerhardt et al. (1994a) in Tikal National Park, Guatemala. They located 13 nests, with eggs being laid in March and early April, and fledging occurring as late as the end of May. All the nests were in live trees and typically were in natural cavities. These cavities were usually formed by rotting wood, but one was in a branch depression overhung by a *Philodendron*, and another was in a crotch. The mean nest height was 12.9 meters aboveground, and the mean clutch-size was 2.2 eggs. Incubation began with the first egg, and required at least 28 days. Three young fledged at 27–29 days, and another at 32 or 33 days. Nine of the 13 nests fledged at

least 1 chick. The chicks remained within their parents' home range for at least 3 months after fledging and were still being fed at that time.

Less information on breeding in Mexico is available, but a nest with 1 apparently abandoned egg was found in Oaxaca during June (Rowley, 1984), and 2 very young birds were collected there in early May (Binford, 1989). Three 2-egg clutches have been obtained between mid-May and mid-June in Sonora, from nests placed 3–8 meters high in oaks (S. Russell and Monson, 1998). Five sets of eggs from Mexico in the Western Foundation of Vertebrate Zoology are all of 2 eggs, with dates ranging from April 5 to June 11. These nests were all in natural tree cavities, stubs, or at the tops of tree stumps, from about 3 to 5 meters aboveground.

Farther south in Central America (Costa Rica) the birds begin breeding activity (egg laying) between February and April. In South America the egg-laying period is widely spread, from as early as February in Colombia to as late as November in Argentina. Typically a natural tree cavity is used, or the birds may lay in old nests of other species such as hawks. Gerhardt and Gerhardt (1997) reported a mean clutch size of 2.2 eggs for 13 clutches.

Evolutionary Relationships and Conservation Status

The generic separation of *Ciccaba* from *Strix* has been questioned by Voous (1988), who found that previously described external ear differences fail to separate all the species of these two reputed genera. The mottled owl does exhibit the specialized asymmetry of the outer ears, if not the complex structure that is also supposedly typical of *Strix*, and so at least it should probably be merged with *Strix*. Relevant DNA studies comparing *Ciccaba* with *Strix* have yet to be done. Voous suggested that, rather than being the ancestors of *Strix*, the *Ciccaba* owls of Latin America are perhaps their tropical offshoots.

The mottled owl, like the spectacled owl, is believed to still be common to fairly common in Mexico. It is also probably one of the most common owls of Latin America generally, although also one of the more nocturnal and thus elusive ones. Besides occurring in heavy to open forest it can also survive in medium to tall second growth, coffee plantations, and even tall thorn forest.

Black-and-White Owl   *Ciccaba nigrolineata* Sclater, 1859

Other Vernacular Names:
Búho Blanquinegro, Lechuza Listada (Spanish)

Range

Ranges residentially in Mexico along the Atlantic slope from southeastern San Luis Potosí (and Veracruz), and along the Pacific slope from eastern Oaxaca southeastward through all of Central America (except the arid interior) to South America, where it extends from western Ecuador northeast to Venezuela. (See Figure 52.)

North American Subspecies

None recognized by König, Weick, and Becking (1999). A South American race *Ciccaba nigrolineata spilonota* has been described and is sometimes recognized, making the North and Central American form the nominate one.

Measurements

Wing, males 272–285 mm (ave. of 4, 277.2), females 255–293 mm (ave. of 8, 274.4); tail, males 161–171.5 mm (ave. of 4, 164.6), females 154–179.5 mm (ave. of 8, 164.5). (Ridgway, 1914). These and a few other linear measurements from

Central America would suggest that very little sexual dimorphism occurs in this species, which is difficult to explain and may simply reflect small sample sizes. The eggs measure 46.4 × 38.4 mm (ave. of 4) (Gerhardt et al., 1994a).

Weights

One adult male weighed 436 g, a female 535 g (Gerhardt and Gerhardt, 1997). A male from Belize, 405 g (S. Russell, 1964). One female from Mexico, 468 g (Field Museum of Natural History). Weight (sex unspecified) 350 g (König, Weick, and Becking, 1999; Stiles and Skutch, 1989); also 458 g (W. Robinson, Brown, and Robinson, 2000). Gerhardt and Gerhardt (1997) reported a mean weight of 33.8 g for 4 eggs, or 6.3 percent of adult female body mass.

Identification

In the field. This is the darkest of the earless owls of Mexico, with a blackish facial disk, a black back, and a black crown. Scalloped black and white feathers cover the entire underparts, and

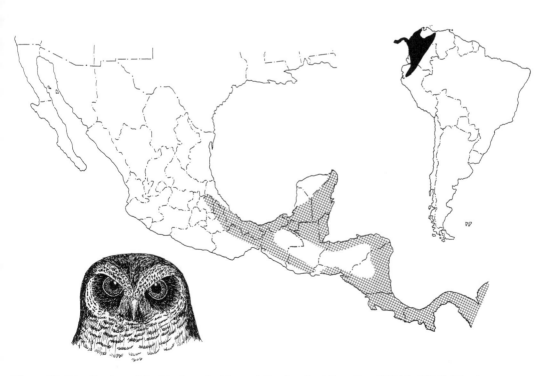

Figure 52. Distribution of the black-and-white owl. Partly after Howell and Webb (1995). The inset map shows the South American distribution.

extend up around the facial disk to form inconspicuous eyebrows. The tail and flight feathers
are also regularly barred with whitish. The eyes
are (usually?) dark reddish brown but have also
been reliably described as yellow, so some age or
regional variation is possibly present. The species
is usually found in dense, lowland forests. See also
"Vocalizations."

In the hand. This is a medium-sized (wing length
under 300 mm), mostly blackish owl, without ear
tufts, and with naked toes but feathered tarsi, a
relatively small bill, and small external ears that
lack opercular flaps. No other New World owl is
so uniformly barred with black and white on the
underparts. A combination of linear measurements and weights should allow for external sex
identification of adults. Nestlings are evidently
silvery white on the upperparts and more yellowish white below, with scattered blackish feathers
of the juvenal plumage, especially on the wing
coverts. Other immature plumage stages are undescribed.

Vocalizations

This species' distinctive song consists of a sequence of low, rapid and single hoots that gradually increase in loudness and end, after a brief
pause, with an explosive and higher-pitched *wow!*
note. Such series often consist of up to about nine
notes; sometimes it is the penultimate rather
than the final note that is the loudest. This sequence in turn may be followed immediately by
an extended series of much fainter and rapid *wio*
sounds. Tape recordings of the species' song from
Venezuela were provided by J. Hardy, Coffey, and
Reynard (1999). In a Venezuelan recording six sequences of the song are uttered in 30 seconds,
and in each case the eighth note is much the
loudest and most explosive, while in two of them
an "afterthought" softer ninth note is added.
Loud, prolonged, and catlike calls have also been
described; these might be female or defensive
calls.

Habitats and Ecology

This is preferentially a lowland evergreen forest
species, ranging in Mexico from sea level to 1200
meters, but it also extends to some degree into
semideciduous forest gallery forests, tall mangroves, plantations, and mature second-growth
forest. At times it occurs fairly close to human
habitations. Farther south in Central and South
America it occurs at progressively higher elevations, up to 2400 meters in Colombia.

Movements

Apparently residential and sedentary.

Foods and Foraging Behavior

This is a nocturnal hunter, taking a wide variety
of insects and a variety of small vertebrates, especially bats. The insects include beetles such as
scarabaeids, curculionids, and cerambycids, orthopterans including locusts, and roaches and cicadas. The mammals include a wide variety of
fruit-eating and insectivorous bats (*Eumops, Sac-
copteryx, Phyllostomus, Glossophaga, Chiroderma, Lasiu-
rus, Myotis, Artibeus, Molossus, Centurio,* and *Uro-
derma*) and rodents such as rice rats (*Oryzomys*)
and introduced rats (*Rattus*). Some birds
(thrushes, tanagers, sparrows) are also taken,
perhaps on their nocturnal roosts. In a sample of
pellets from Venezuela, Ibanez, Ramo, and Busto
(1992) found the remains of 14 species of bats
(representing 10 genera), 2 rodents, 5 bird
species, and a few hylid tree frogs. The majority
of the estimated prey biomass (64.2 percent) was
of bats, with rodents comprising 14.2 percent and
birds 15.4 percent. The diversity of bats taken indicates varied prey capture strategies are employed. Vertebrate prey ranged in size from 10 to
150 gm, and about half of the prey taxa were classified as aerial rather than terrestrial. Gerhardt
et al. (1994b) found evidence of similar prey in
Guatemala, with most of the 73 pellets they examined containing bat remains, and all some insect matter. All foraging occurred at night, and
there was little apparent overlap in prey types
taken by black-and-white owls and mottled owls
that occurred in the same study area. Bats and insects are often captured in flight, sometimes using any available artificial light.

Social Behavior

Almost nothing is known of the social behavior of
this species. Gerhardt et al. (1994a) studied a
radio-tagged male in Tikal National Park, Guatemala, and found it had a mean home range of
437.3 hectares. This home range included a nest
site, several diurnal roosts, and two foraging areas. Many of the roost sites were in dense vegetation with hanging vines present, and the roosts
averaged 14 meters high (range 3.5–26 m). A
pair studied over three consecutive years was
found to be monogamous and wholly sedentary.

Breeding Biology

Little is known of this species' nesting in Mexico, but it is said to breed during the latter part

of the dry season (March to May in Central America) and to lay in tree cavities. One nest record in the Western Foundation of Vertebrate Zoology is a single egg, placed in a tree cavity about 7 meters aboveground, the tree near a running creek and the nest in deep shade. The late date (June 25) would suggest a second nesting effort. The work by Gerhardt et al. (1994a) in Tikal National Park, Guatemala, offers the best available breeding information to date. They located four nests, three of which were used in different years by the same female. They were placed in three different tree species, the heights of which averaged 26.3 meters. In all cases the eggs were placed among epiphytic vegetation, and each nest had a single egg. Nesting began in late March, and hatching occurred in late April. None of the four nests were successful.

The black-and-white owl reportedly may also lay its eggs in abandoned stick nests of some other bird species such as raptors. The young are said to pass through an extended sequence of plumages before becoming fully adult.

Evolutionary Relationships and Conservation Status

The generic separation of *Ciccaba* from *Strix* is questionable on morphological grounds; at least *Strix* would seem to be this group's nearest relatives (Voous, 1988).

The black-and-white owl is thought to still be uncommon to fairly common in Mexico (Howell and Webb, 1995), although others have said it is rare throughout its entire range (Enriquez-Rocha, 1995). Stotz et al. (1996) listed it as occurring in at least 12 Latin American countries, and fairly common. Besides occurring in humid primary forests it can also survive in plantations and along forest edges. However, over much of its range it is believed to be fairly rare, at least as compared with its congener the mottled owl. This in part relates to its larger body mass, greater dependence on ver-tebrate prey, and its correspondingly greater territorial and home range requirements, its home ranges averaging about 20 times larger than those of mottled owls, according to Gerhardt et al. (1994a).

Spotted Owl   *Strix occidentalis* (Xantus de Vesey) 1860

Other Vernacular Names:
Arizona spotted owl (*lucida*); California spotted owl (*occidentalis*); Mexican spotted owl
(*lucida*); northern spotted owl (*caurina*); western barred owl; wood owl.

Range (Adapted from AOU, 1983.)

Resident in the mountains and in humid coastal
forests from southwestern British Columbia
south through western Washington and western
Oregon to southern California and, very locally,
northern Baja California; and in the Rocky Mountain region from southern Utah and southwestern
Colorado south locally through the mountains of
Arizona, New Mexico, extreme western Texas
and Mexico, south to Michoacan and San Luis Potosí. (See Figure 53.)

Subspecies (Adapted from AOU, 1957, and
J. Peters, 1940.)

S. o. caurina (Merriam). Southwestern British Columbia south to central California in the coastal
ranges.

S. o. occidentalis (Xantus). Coastal ranges and
western slopes of Sierra Nevadas in California
from Tehama to San Diego counties.

S. o. lucida (Nelson). From northern Arizona,
southeastern Utah, and central Colorado south
through western Texas, New Mexico, and Arizona
to the Mexican Plateau, and in eastern Mexico
from Coahuila south to San Luis Potosí.

Measurements

Wing (of *occidentalis*), males 310–326 mm (ave. of
6, 320.5), female 328 mm; tail, males 210–220
mm (ave. of 6, 215.8), female 225 mm (Ridgway,
1914). Measurements of all 3 races presented by
Gutiérrez, Franklin, and LaHaye (1995) show
only minor racial differences in wing length and
tail length, with no consistent clinal trends, although northern birds average largest and Mexican ones the smallest. The eggs of *occidentalis* average 49.9 × 41.3 mm (Bent, 1938).

Weights

Earhart and Johnson (1970) reported that 10
males averaged 582 g (range 518–694), and that
10 females averaged 637 g (range 548–760).
Mean masses of *caurina*: 68 males 579 g, 65 females 663 g; of *occidentalis*: 218 males 566, 195 females 646; of *lucida*: 68 males 509, 68 females 569
(Gutiérrez, Franklin, and LaHaye, 1995). The estimated male:female mass ratio is 1:1.12–1.15.
The estimated egg weight is 44 g. The estimated

egg-to-female proportional mass ratio is thus 6–8
percent.

Identification

In the field. This species is associated with mature coniferous forests of the West, where it is
likely to sit quietly in shady tree roosts throughout the day. It closely resembles the barred owl
(and the two species now are slightly sympatric
in Washington and Oregon and possibly also in
northern California), but the entire underparts
(rather than just the breast) are barred, and its
advertisement call typically consists of only 4–5
notes ("Whooo . . . are you, you-all?") rather than
the distinctive 9-noted sequence of the barred
owl. Both species show strong orange-red eye
shine when illuminated by direct light (Gutiérrez,
Franklin, and LaHaye, 1995).

In the hand. This is a medium-sized owl (wing
210–330 mm) that lacks ear tufts, has dark
brown eyes, and has distinct white spots on the
rich brown neck, upperparts, and flanks. The similar barred owl has brown and whitish barring on
the neck and breast, abruptly changing to brown
streaking on the underparts, without such clear
white spotting evident either above or below. As
adults, the sexes are nearly identical, but females
have darker head and face color, and breeding females have brood patches. Footpad length in females averages larger, and females have a greater
average number of tail bars, but specific measure-
ment differences may vary among subspecies. A
discriminant function analysis, using a variety of
measurements, may separate the sexes of about
80 percent of adults (Blakesley, Franklin, and
Gutiérrez, 1990). The downy chicks are initially
pure white, as in *S. varia*. The juvenal body
plumage is pale brownish buff, broadly barred
with light brown, the bars widest and most distinct on scapulars, which are tipped with white.
The head is mostly pale brownish buff, the feathers dark brown basally; the legs immaculate buff
or dull buffy white. First-year birds retain their
juvenal flight-feathers, which tend to be uniform
in color and wear, and have relatively narrow rectrices that have more pointed tips with clear
whitish terminal bands. The light bars on the
flight feathers and rectrices are relatively larger
and more closely spaced than in adults (Pyle,
1997b).

Figure 53. Distribution of the spotted owl, showing residential ranges of the races *caurina* (ca), *lucida* (lu), and *occidentalis* (oc).

Vocalizations

The most complete summary of vocalizations in this species is that of Forsman, Meslow, and Wight (1984). They called the advertisement call the "four-note location call" and described it as a single introductory hoot (sometimes omitted) followed by a short pause, 2 closely spaced hoots, and a final note that trails off at the end. This call is uttered by both sexes in various situations, serving both as a territorial challenge and as a general location call. It is thus often uttered during vocal territorial exchanges and also by members of a pair calling back and forth to one another on their territory. Males also utter the call when arriving near the nest with food and both before and after copulation; in the latter situation and during territorial disputes other calls are commonly alternated with this one.

A similar call, or a variant of the first, is the "agitated location call," which is more intense than the just-described call, and its last note more emphatically uttered. As in the other, the first note may be omitted, at least after the first call in a sequence. Both sexes use the call in territorial disputes and the males sometimes utter it in association with copulation.

Three kinds of calls are given in long series. The first, the "series location call," consists of 7–15 hoots in series, which may be evenly spaced, or more commonly of 5–7 evenly spaced notes followed by single note or paired notes at longer intervals. These may be uttered during territorial disputes or as signals between paired birds. The "bark series" is a rapidly uttered series of 3–7 loud barking notes uttered at the rate of 2–3 notes per second. It is typically uttered by females during territorial disputes but may also be used as a long-distance contact call between pairs. The third call given in series is the "nest call," uttered by both sexes as they call from the nesting tree during prenesting activities. All three of these calls may last for several minutes, with only minor breaks, the individual notes uttered at about 3 per second.

Two contact call types were described by Forsman, Meslow, and Wight (1984), both sounding like *co-weep!* and mainly uttered by females. The typical call, uttered at 15–45 second intervals, apparently serves to inform the male or young of the female's location, thus facilitating food exchange, copulation, and other pair activities. The "agitated contact call" is louder and shriller, and is often associated with territorial disputes. In the same situation a loud, grating, two-syllable *wraaak!* note may be uttered. During copulation the female utters a series of chittering notes, and a similar call is produced by both adult and young owls when being handled or, sometimes, when being preened by another. The male produces a series of hoots while copulating, and both sexes utter terse, single-syllable alarm grunts and groans in response to approaching predators. Various cooing calls are produced during close-range encounters among associated birds, such as during roosting or mutual preening. Finally, nestlings utter two calls similar to those of adults (chittering and alarm grunting) as well as a specific begging call that gradually differentiates over time into the adult contact calls.

Habitats and Ecology

Forsman, Meslow, and Wight (1984) reported that 98 percent of the 636 sites found supporting spotted owls were in areas of old-growth forests, or mixtures of mature and old-growth timber, usually with an uneven-aged and multilevel canopy. The birds were found from nearly sea level to the upper edge of the ecotone separating the midlevel mesic forests from the subalpine forest zone. The authors also analyzed the habitat requirements of spotted owls in terms of foraging, roosting, and nesting needs, based on trackings of radio-tagged birds. All these birds exhibited a strong preference for foraging in old-growth conifer forests (forests at least 200 years old), with younger forest types used progressively less, and forests cleared or burned in the last 20 years used rarely or not at all. This preference for forest foraging occurred in all seasons, but was strongest in midwinter. All of more than 1600 roost sites were in forests, and 90 percent were in old-growth forests. During hot and warm weather these sites were often close to the ground, but in cold or wet weather they were typically higher, in tall conifers, where the owls would roost close to the tree trunk and have woody vegetation or foliage directly above to serve as a shelter. Of 47 nests studied, 90 percent were in multilayered, old-growth forests, and the remainder were in stands of 70- to 140-year growth, with scattered older trees. Most nest sites had a high degree of vegetational canopy closure overhead, were on the lower halves of moderate slopes, and were within 250 meters of a source of water. Most roosts also had southern exposures, for unknown reasons.

The pairs studied by Forsman, Meslow, and Wight (1984) in Oregon had average nearest-neighbor distances of 2.6–3.3 kilometers for extreme western Oregon and the eastern Cascades, respectively, with a minimum observed distance of 1.9 kilometers. No estimates of territory size were made by these authors, who were unsure whether the male-female interactions observed between birds having adjacent home ranges should be considered as territorial or not. However, Gould (1974) estimated territories in the Sierra Nevadas of California to average about 93 hectares, compared with overall home ranges averaging 182 hectares. Territorial disputes between long-term neighbors appear to be rare, but the vocalizations of unrecognized intruders produce aggressive responses. Apparently individual adults occupy the same home range for long periods of time, probably for life, and the mean home range estimates by Forsman, Meslow, and Wight for all radio-tagged owls on two different study areas averaged about 1100–1900 hectares. Females had somewhat larger estimated average home ranges than males, although these differences were not statistically significant. Home ranges of paired individuals exhibited substantial overlap (averaging 68 percent), and home ranges of adjacent pairs overlapped slightly (averaging 12 percent). Spotted owl density and distribution in Oregon are strongly affected by the distribution of federal lands that have historically been protected from logging, and thus any large-scale estimates of population density are likely to vary greatly, depending on the age and foresting history of the area concerned. Forsman and Meslow (1985) estimated that, based on home range studies, a minimum of about 400 hectares of old-growth habitat is needed to support a single pair of birds, and as the proportion of old growth decreases within a pair's home range their overall home range tends to increase correspondingly (Forsman, Meslow, and Wight, 1984). Possible reasons for this species' high dependency on old growth include adequate nest-platform needs, broad temperature gradients and widely diverse roosting sites associated with multilayered canopy trees, associated prey-abundance or prey-availability variables, and other possible adaptations (Carey, 1985).

In California, Gould (1974, 1979) located spotted owls at 404 sites, and documented habitat parameters at 192 of these. Dominant trees in these areas were larger than 0.83 meters in diameter (breast height) at 85 percent of the sites and were classified as moderately decadent to decadent. The degree of canopy closure was at least 40 percent at 90 percent of the sites. The average adult home range for 5 pairs of radio-tagged owls in Washington was estimated at 2776 hectares, with individual adults having average winter home ranges of 1663 hectares, compared with an average of 870 hectares during summer. Within the pairs' home ranges was an average of 951 hectares of old forest growth (Brewer and Allen, 1985). Available home range information has been recently tabulated for all 3 races by (Gutiér-

rez, Franklin, and LaHaye, 1995).Home ranges of pairs typically range from about 10 to 20 square kilometers (1000–2000 ha), and of individual birds from 5 to 15 square kilometers, but with much variation. Generally, home ranges tend to increase from the breeding season to winter, and also vary with dominant prey species, topography, and the amount of old-growth habitat available. Home ranges for members of a pair may overlap almost completely, and those of adjacent pairs up to nearly 50 percent.

Half of 20 pairs found by Garcia (1979) in Washington were in forests more than 200 years old, and only 6 were in stands less than 100 years old; a population density of 0.09 birds per square kilometer of mature and old-growth forest stands was estimated. This estimate is smaller than that of 0.20 per square kilometer given as an average in suitable habitats in California (Gould, 1974); 0.24 per square kilometer, in northwestern California (unpublished report of B. G. Marcot cited by Garcia); and 0.36 per square kilometer, in the Coast Range of Oregon (Forsman, Meslow, and Wight, 1984).

The Rocky Mountain race of the spotted owl lives in a quite different climatic environment from that of the coastal race, and its distribution may be influenced by the presence or absence of wood rats (*Neotoma*). These rats are apparently its primary prey in New Mexico, Arizona, and Utah (Webb, 1983), although in the northern Pacific coastal region the flying squirrel (*Glaucomys*) is at least as important a prey species (Carey, 1985). In New Mexico the birds once were distributed along many lowland riparian forests but now occur as small, relatively isolated populations in higher elevations of scattered mountain ranges (Arsenault, Hodgson and Stacey, 1997). In eastern Utah and southwestern Colorado the birds are strongly associated with moist and cool canyon bottoms of canyon-mesa topography, where shady microclimatic conditions may prevent possible heat stress (Barrows, 1981) in otherwise relatively hot environments. In such "mountain islands" the birds nest in cliff crevices of shady canyons down to about 1500 meters and in trees where about 10 or more square kilometers of forest exist (Gehlbach, 1995a). In north-central Colorado, at the extreme northern edge of this race's range, only a few records of apparent stragglers (possibly young birds) seem to be available for the region (Webb, 1983).

Although it has been suggested that spotted owls tend to avoid areas occupied by the great horned owl, presumably because of potential predation dangers, they have occasionally been found using the same general habitats in some areas. Interspecific conflicts and occasional hybridization have also resulted from the recent expansion of barred owls into spotted owl habitats in western Washington, Oregon, and northern California (Hamer et al., 1994; Gutiérrez, Franklin, and La-Haye, 1995).

Movements

It is generally believed that spotted owls are relatively sedentary, with movements limited to the usual dispersion of young birds during their first fall of life. However, Laymon (1985, 1988) reported rather substantial seasonal movements of 4 adults in the central Sierra Nevadas of California. The birds moved downslope during autumn a distance of from 19 to 32 kilometers, descending an average of 70 meters in elevation. The 4 birds occupied winter home ranges of 300–2000 hectares until at least late February and returned by mid-April to their nesting sites. Movements of up to 65 kilometers horizontally and 1500 meters vertically have been detected (Verner, Gutiérrez, and Gould, 1992). Similar vertical (up to 1000 m or more) and horizontal (20–50 km) movements have been found in Mexican spotted owls (Gutiérrez, Franklin, and LaHaye, 1995).

Several studies of movements in juvenile spotted owls have been performed. Gutiérrez et al. (1985) followed the movements of 11 fledged owlets between early September and late October, during which time the birds dispersed a distance of 30–156 kilometers from their nests, averaging 78 kilometers. Major ridges, rivers, and similar topographic barriers did not noticeably affect the direction of dispersal, which was primarily southerly. Typically, after an initial rapid movement out of the natal area, the birds attempted to settle. Three such settled young had measured home ranges of 146–186 hectares. Of the 11 banded owls that survived to disperse, 7 died during this process from starvation, predation, or unknown causes, suggesting that this is a highly risky period during the bird's life. Similarly, Forsman, Meslow, and Wight (1984) reported a 35 percent mortality rate of 29 young birds between fledging and the end of August.

In New Mexico, Arsenault, Hodgson, and Stacey (1997) followed 12 juvenile and 3 yearling spotted owls during fall dispersal. The juveniles exhibited two types of dispersal, namely extensive local exploration and rather rapid movements across the landscape, at rates as rapid as 11.3 kilometers per night. During the latter type, the birds crossed grassland and savanna habitats between mountain ranges, and at times roosted in open pine or even piñon-juniper woodlands. The maximum observed straight-line dispersal distance was 57.6 kilometers, which occurred within

52 days, but some birds disappeared from tracking monitors and perhaps moved even greater distances. Three yearling females temporarily paired with males during their first summer, but separated from then during fall, perhaps dispersing even farther. Breeding incidence of the entire adult population varied greatly, from 8 to 80 percent of the known pairs during this study.

Similar studies of juvenile dispersal by radio-tagged owls have been undertaken in Oregon. Allen and Brewer (1985) reported that 6 radio-tagged juveniles dispersed in September and October over distances of 48 kilometers in some cases, in seemingly random directions. Four of the 6 birds had died by the following June. G. Miller and Meslow (1985) noted that radio-tagged juvenile birds dispersed as far as 76.8 kilometers from their natal sites, but survival of such long-distance dispersers was apparently low. Eleven owls that dispersed for average distances of 33 kilometers all died before settling in. Two more juveniles dispersed an average of 18 kilometers and survived to settle in. Three that dispersed an average of 26 kilometers settled in and survived to the next breeding season. Survival of radio-tagged juveniles was low, with first-year mortality 60–95 percent. Like adults, dispersing juveniles exhibited habitat selection (18 of 19 cases) for mature and old-growth forest stands (G. Miller and Meslow, 1985).

Foods and Foraging Behavior

One of the most complete analyses of the northern spotted owl's foods is that of Forsman, Meslow, and Wight (1984), whose data were based on more than 4500 identified prey items from 62 pairs of owls in various parts of Oregon. On both a frequency of occurrence and biomass basis the most important single prey species there is the northern flying squirrel (*Glaucomys sabrinus*), which comprised at least half total estimated biomass of food intake. An additional 30 species of mammals were represented, with wood rats (*Neotoma*) and hares or rabbits of special significance, plus 23 species of birds, 2 of reptiles, and various invertebrates. Mammals comprised more than 90 percent of the biomass in all areas, with mean prey weights ranging from 54 to 150 grams, and with no significant differences between the sexes as to prey weight or prey identity. In wet coniferous forests the flying squirrel was the principal prey species, while in mixed conifers wood rats were of primary importance. Flying squirrels were also seasonally most important during fall and winter, whereas during spring and summer a greater variety of foods was consumed, especially deer mice (*Peromyscus*), voles, and other small mammals. A more recent summary of the foods of this race, based on more than 6000 prey remains in pellets from the northern parts of the range and more than 5000 from more southern areas, was provided by Thomas et al. (1990). On a numerical basis, flying squirrels comprised 17.7–31.7 percent of the total, wood rats 6.5–27.4 percent, *Phenacomys* voles 12.3–14.9 percent, *Peromyscus* mice 9.8–10.2 percent, *Microtus/Clethrionomys* 8.8–9.5 percent, other mammals including lagomorphs 8.2–20.0 percent, plus small numbers of birds and other taxa, mostly invertebrates. Varied degrees of concentration on wood rats and flying squirrels depended on prey ranges, with northern birds concentrating on flying squirrels and southern ones on wood rats.

A similar spectrum of foods of the California race was found typical by Laymon (1985) in smaller samples (based on 8 pairs) for the Sierra Nevadas of California, while Barrows (1985) reported that among 4 pairs of successfully breeding and unsuccessful owls in California the successfully breeding birds concentrated on large (more than 100-g) prey species such as wood rats and flying squirrels. During years of low breeding success the incidence of smaller prey species was higher (the mean prey weights of the successful and unsuccessful pairs being 115 and 79 g, respectively). Barrows thus suggested that the presence and availability of large prey such as wood rats and flying squirrels are important to attainment of breeding success in spotted owls, which might also have implications in habitat selection of the species. In a larger sample of prey remains from this race, Gutiérrez, Franklin, and LaHaye (1995) tabulated data on more than 11,000 prey items from pellets obtained in two regions. Again, flying squirrels (2.0–19.2 percent) and wood rats (9.8–39.7 percent) dominated the prey taxa, with *Peromyscus* mice (7.7–12.2 percent) also well represented. Other mammals (7.3–19.7 percent), birds (3.5–15.1 percent), and various other taxa (21.9–38 percent) were relatively well represented. Gutiérrez, Franklin, and LaHaye (1995) also summarized available data for the Mexican race, based on more than 14,000 prey items from across its range. Of these, wood rats were most significant (26.3–35.7 percent of remains), followed by *Peromyscus* mice (27–30 percent). Other mammals, including bats, voles, and rabbits, made up much of the remainder, plus birds (4.7–7.8 percent) and other miscellaneous prey, mainly invertebrates (15.1–19.1 percent). At least in terms of biomass, wood rats were the most important prey item found.

A comparative study of barred owls and spotted owls foods taken in an area of sympatry (Hamer et al., 2001) indicated that in both species nocturnal mammals predominated, but the barred owl took a more diverse and more evenly divided array of prey. The estimated degree of dietary overlap was 76 percent, with the most important prey item of the barred owl being snowshoe hares (35 percent of prey biomass), whereas the northern flying squirrel comprised the majority (57 percent) of prey biomass for the spotted owl.

Forsman, Meslow, and Wight (1984) observed nine predation attempts by spotted owls, seven of which were on squirrels or birds in trees and two on mammals at ground level. In either case the method of attack was to dive upon the prey from an elevated perch. When such a dive was unsuccessful the owl might fly or hop after the fleeing animal. Prey that was not eaten immediately was sometimes cached in various places on the ground, on large rocks, or in trees. On average the birds left their roosts 14 minutes after sunset and stopped foraging 21 minutes before sunrise. During the day most of the time was spent roosting, but sometimes the owls would dive down to capture prey beneath roost trees, make flights to retrieve cached prey, or fly to nearby streams to drink or bathe. Caching of excess prey is often done, the sites chosen including on mossy limbs or among mossy rocks, on broken stumps, under fallen logs, and in tree hollows (Gutiérrez, Franklin, and LaHaye, 1995).

Social Behavior

Breeding is known to occur and perhaps is regular in 2-year-old and older females (G. Miller, Nelson, and Wright, 1985), but quite possibly few if any first-year birds breed even though they may form pair bonds. There is also a fairly high incidence of nonbreeding among wild birds, which has been estimated by Forsman, Meslow, and Wight (1984) at 38 percent and by Gutiérrez and Carey (1985) at 54 percent, with some repetitively nonnesting pairs tracked for as long as 5 years. Additionally, in all three races, birds 2 years old or less breed less frequently than do older birds (10–40 percent vs. 58–76 percent) and also fledge young less frequently (7–28 percent vs. 45–51 percent). Among all nesting females, the percent of such birds 3 years old or older has ranged from 85.4 to 92.1 percent in various studies (Gutiérrez, Franklin, and LaHaye, 1995).

Barrows (1985) followed three occupied territories for 4, 6, and 7 years, although the birds were apparently not individually marked and thus might not have been composed of exactly the same pair members throughout the entire period of study. Miller (in Walker, 1974) described watching a remarkably tame pair of breeding

181

spotted owls for 8 successive years. After the eighth year the female of the original pair was replaced by a new and warier bird, bringing an end to Miller's detailed observations.

Because of the high level of nesting site fidelity typical of this species, it is highly likely that pair bonds, once formed, hold indefinitely. So far, no cases of polygyny or extra-pair copulations have been reported, and "divorce" appears to be infrequent (Gutiérrez, Franklin, and LaHaye, 1995). Forsman, Meslow, and Wight (1984) were unable to determine whether mate constancy occurred between breeding seasons, but if so they considered it to be more a function of the birds' attachment to a traditional home range than of attachment to a particular mate. They stated that between October and January, adult owls in Oregon lived a largely solitary existence, the birds calling only infrequently but roosting together. In February or early March the resident pair in each territory began roosting near their eventual nest site, which was usually the same site as in previous years. The birds also began to call almost every night, especially at dusk prior to foraging and again at dawn as they rejoined near the nest.

Both members continued to forage nightly until about 12 days before the eggs were laid, when males presumably began to feed their mates and the latter became increasingly sedentary, rarely moving more than a few hundred meters from the nest. During the last 5–7 days prior to egg laying the females spent most of the nighttime hours near the nesting tree, waiting for their mates to arrive with food.

Miller (in Walker, 1974) observed daytime hunting by a female while she was raising a brood of three young, although the male ceased hunting at dawn during that period. Mutual preening also was observed by Miller among paired adult spotted owls, as well as between mothers and their young, and almost certainly is a regular part of their social behavior. Some representative examples of spotted owl behavior are shown in Figure 54.

Sexual behavior was observed at two nests by Forsman, Meslow, and Wight (1984) during the two weeks prior to egg laying. Copulations were observed on the first night each of the two pairs was watched and continued until a few days after clutch completion. Copulations usually occurred

Figure 54. Spotted owl behavior, including adults in (*A*) alert posture, (*B*) slimmed posture, (*C*) relaxed posture, (*D*) young nestling calling, and (*E*) fledgling, showing head movements associated with investigatory peering. After photos by Lewis Walker (*A, D, E*), A. and S. Carey (*C*), and the author (*B*).

at dusk, just after the birds left their roosts, and were preceded by preliminary vocalizations such as the four-note location calls for several minutes by both pair members. The male then flew to the female, who perched sideways on a limb and uttered the copulation call during treading as the male fluttered his wings and also called. Copulation was a nearly nightly event during the two-week period prior to egg laying, and as many as two copulations were seen in a single evening.

Breeding Biology

Of 47 nests studied by Forsman, Meslow, and Wight (1984), 64 percent were in cavities and 36 percent on stick platforms or other debris on tree limbs. All 30 of the cavity nests were in old-growth trees, and the majority of the platform nests were also in old-growth trees. All but 2 of the nests were in living trees, all of which were in conifers. The nest heights in this sample ranged from 10 to 50.3 meters, averaging 27.3 meters. Of the 17 platform nests, most were constructed by other raptors, wood rats, or squirrels, while others were in dense clusters of dwarf mistletoe (*Arceuthobium*) or simply natural accumulations of debris that had become caught by the limbs. Most of the platform nests were located directly against the tree trunks. Of the 25 nests checked for 2 or more years, 17 were used more than once, and one was used 6 times during an 8-year period. Platform nests are more typical of younger forests, and cavity nests of older stands, where they are more generally available. In Arizona, both types of sites are used, and ledges on cliffs or pothole sites are commonly selected, as are mistletoe clusters (Gutiérrez, Franklin, and La-Haye, 1995).

A sample of 38 clutches included 27 with 2 eggs (71 percent), 9 with 3 eggs, and 2 with a single egg (Gutiérrez, Franklin, and LaHaye, 1995). Of 40 nest records from various U.S. locations available to me (mainly obtained from the Western Foundation of Vertebrate Zoology), clutch sizes averaged 2.5 (range 2–4; 2 eggs in 62 percent of the nests). Of 13 for which location was specified, 6 were on rock ledges or cavities, 4 on tree forks, 2 in tree cavities, and 1 was on a pigeon coop. The average nest height was 9.4 meters. By comparison, 47 tree nests measured by Forsman, Meslow, and Wight (1984) averaged 22 meters (platform nests) to 30.1 meters (cavity nests) aboveground. The clutch size of 4 nests was 2 eggs in all cases, and the egg-laying interval as determined from a captive bird was always between 66 and 78 hours.

Twenty-five egg records from California (counting those listed by Bent, 1938) are from March 1 to May 10, with 12 of these from March 27 to April 1. A few records from Arizona and New Mexico are from April 4 to 17, and both actual records and indirect evidence (Forsman, Meslow, and Wight, 1984) suggest that eggs are laid in Oregon during early April, with birds nesting on the western slope of the Cascades and on the Coast Ranges laying somewhat earlier than those on the eastern slope.

Incubation begins soon after the first egg is laid, and is performed entirely by the female, judging from observations by Miller (in Walker, 1974) and by Forsman, Meslow, and Wight (1984). The latter authors estimated an incubation period of 28–32 days, based on observations of a nest abandoned shortly before hatching would have occurred. For 8–10 days after hatching the young are brooded almost constantly by the female, but when they are 14–21 days old the females begin foraging for progressively longer periods during the night.

Miller (in Walker, 1974) described the nest life of young spotted owls. She noted that only during 1 year did 3 owlets survive to fledging, and these required nearly constant hunting by the parents. The male would typically bring in prey that he had eaten the heads of, passing the rest of the food on to the female to feed the young. Prey that the female caught were similarly decapitated; when the young were very small it was also torn to bits before being fed to the young. At times the female would whistle to the male, apparently to wake him up and stimulate him to search for food. After the young had been out of the nest for about 2 weeks he would sometimes deliver prey directly to the owlets.

Most owlets observed by Forsman, Meslow, and Wight (1984) left the nest when they were 32–36 days old, when the juvenal plumage was nearly complete but the remiges were not more than two-thirds grown. Thus, the young often fell or fluttered to the ground, but they soon usually climbed up into nearby tree perches. Most owlets were able to fly or climb into such perches within 3 days of leaving the nest, using their talons and beak and fluttering the wings as required. Although persistent if clumsy climbers, some individuals that left the nest prematurely spent as long as 10 days on the ground. Within a week of leaving the nest, or at about 40–45 days of age, most owlets could fly for short distances, and within 3 more weeks were able to hold and tear up prey by themselves. Shortly after that they were starting to capture insect prey independently.

Forsman, Meslow, and Wight (1984) estimated an overall nesting success rate of 81 percent for 81 nests observed over a 5-year period, and an average of 1.4 young raised per successful nesting attempt in 46 cases. Of 130 pairs surveyed (including nonbreeders) at least 68 young were raised, or 0.52 young per pair, a relatively low rate of reproductive success. Similar rather low rates of fledging success have been reported elsewhere and may be fairly typical of spotted owls, although marked yearly variations in success also seem characteristic (Barrows, 1985).

Of 2113 northern spotted owl broods, 42.2 percent fledged single chicks; 55.9 percent fledged two chicks; and 1.9 percent fledged three. Of 354 California spotted owl broods, 42.9 percent fledged single chicks; 51.7 percent, 2 chicks; 5.1 percent, 3; and 0.3 percent, 4. The percentage of all females fledging one or more young in various studies has ranged from 35 to 45 percent in the northern spotted owl, 33 to 90 percent in the California spotted owl, and 21 to 92 percent in the Mexican spotted owl. In all races the annual survival rates of juveniles have proved quite low (14–31 percent) but those of adults relatively high (83–90 percent) (Gutiérrez, Franklin, and LaHaye, 1995). Maximum known longevity under natural conditions is 13 years and 9 months (Patuxent Wildlife Research Center banding data).

Evolutionary Relationships and Conservation Status

The status of the spotted owl is of great concern at present, as a result of its dependence on old-growth forests, a habitat type that is rapidly disappearing from western North America (Bart and Forsman, 1992). Because of the National Forest Management Act, the U.S. Forest Service is legally obligated to maintain minimum viable populations of all vertebrate species on its lands, which has brought the spotted owl into direct conflict with national demands for lumber and the Forest Service's tendency to bow to the timber industry's strong political influences (Doak, 1989; Yaffee, 1994). In various physiographic provinces of the Pacific Northwest the loss of forest habitats critical to the spotted owl has ranged from 54 percent to nearly 100 percent (Gutiérrez, 1994), and the species' population has correspondingly suffered.

The U.S. Fish and Wildlife Service proposed the spotted owl's listing as a nationally threatened species in 1973, a classification that did not become fully effective until the 1990s. The northern race of the spotted owl was listed as threatened by the U.S. government in 1990, and the Mexican race was added under the same category in 1993. The northern spotted owl was also listed in 1994 as endangered in Canada, and the Mexi-

can spotted owl was listed as threatened by the Mexican government in 1995. The California race has not yet received a similar federal listing but has been classified as a species of special concern by the state of California and is considered as sensitive by the federal government. The spotted owl and barred owl are closely related species that represent ecological replacement forms in mature forests of western and eastern North America, respectively, and that in the historic past have been allopatric in distribution. The first known breeding of barred owls in Washington was in 1974, but by 1985 there were 130 state records for the species, 70 percent of which were within the range of the spotted owl. The barred owl first entered Oregon in 1974, and by the 1990s it had spread through the entire length of the Cascades, and into the Blue and Wallowa mountains, as well as coastally. Similar expansion and secondary contacts between the species have occurred in southwestern British Columbia (Dunbar et al., 1991). Hybrids with spotted owls have since been found in several locations in Washington and Oregon (Gilligan et al., 1994). Barred owls not only sometimes hybridize with spotted owls but also are slightly larger and may displace them from their territories (Allen, Hamer, and Brewer, 1985; Hamer, 1988; Hamer et al., 1994; Herter and Hicks, 2000).

As a result of these and related concerns, a great deal of research on the spotted owl's status, ecology, and management has occurred (Zarn, 1974b; Gutiérrez and Carey, 1985; Gutiérrez, Franklin, and LaHaye, 1995), and a bibliography of spotted owl research has been published (Campbell, Forsman, and van der Ray, 1984). Besides surveys and ecological studies of the species in Oregon (Forsman, Meslow, and Wight, 1984; Carleson and Haight, 1985; Carey, Reed, and Horton, 1990; Meyer, Irwin, and Boyce, 1998; Swindle et al., 1999), surveys and studies have been conducted in Washington (Garcia, 1979; Seaman, 1997), California (Gould, 1985; Laymon, 1988; Mills, Fredrickson, and Moorhead, 1993; Franklin et al., 1996; Franklin, 1997), Arizona (Ganey and Balda, 1985; Ganey, 1988), New Mexico (Arsenault, Hodgson, and Stacey, 1997), and Colorado (Webb, 1983). In Arizona it is still fairly widespread in coniferous and mixed-forest habitats between 1500 and 3000 meters, and in the 1990s there may have been 600–1200 pairs present in the state (Ganey, 1998). In Colorado it is a very rare breeding species, breeding in narrow and steep canyons in the vicinity of Mesa Verde National Park, and it is equally rare in the Central Rockies. The highest numbers counted in recent Colorado surveys produced (in 1993) a total of only 7 known pairs (Kingery, 1998). It is also locally present in moist canyons of Utah, such as at Capitol Reef and Canyonlands National Parks and at Zion National Park (Kertell, 1977; Marti, 1979; Rinkevich and Gutiérrez, 1996). Its numerical status in Mexico is unknown but probably unfavorable, given the high rate of logging activities in recent years.

It is nearly impossible to estimate total recent populations for this elusive species, but the northern spotted owl had a 1987–1992 population of at least 3778 pairs, and about 1000 additional single territorial owls were detected. Most of the pairs were in Oregon (1971), followed by California (1111), Washington (671) and British Columbia (25). The Canadian population has recently been estimated at 10–100 pairs (Kirk and Hyslop, 1998) and is considered endangered. Between 1970 and 1992 a minimum of 3050 pairs of the California race were detected; 777–1554 birds of the Mexican race were estimated for the period 1990–1993 (Gutiérrez, Franklin, and LaHaye, 1995). During the 1999 and 2000 Christmas Bird Counts the species was detected on an average of 17 counts, or about 1 percent of all U.S. and Canadian counts. In terms of all localities where the species was detected, the spotted owl was the 14th most widely reported of the 18 North American owl species then encountered. The maximum number ever reported on a single CBC was 13, in western Sonoma County, California, in 1994. The average total number counted was 30 birds, or about 0.3 percent of all owls reported in the tabulated summaries of counts for the United States and Canada during 1999–2000.

A variety of studies (Forsman et al., 1996) have indicated an increasing rate of decline in female survival rates of the northern spotted owl as well as the California race. In spite of these recent state and federal efforts at improved protection, the time is running out, with probably less than 5 percent of the Pacific Northwest's old-growth forests still intact and the rest likely to disappear in the next few decades. The U.S. Forest Service has been notable in its unflagging and long-standing efforts to satisfy local and regional logging interests, in spite of massive evidence of spotted owl declines and old-growth forest destruction. When our ancient western forests are entirely gone, along with all of their specialized constituent plants and animals, the antiquated timber industry of the Pacific Northwest will also have effectively undergone self-destruction, and our chance to study, and appreciate this wonderful and unique ecosystem will be lost forever. Also gone will be valuable plants, such as trees that provide the basis for cancer-treating drugs like taxol, and ancient firs that were already present at the time of Christ, trees that have withstood countless earthquakes and climatic changes but that have no evolutionary means of dealing with chain saws.

Barred Owl   *Strix varia* Barton 1799

Other Vernacular Names:
Florida barred owl (*georgica*); northern barred owl (*varia*); swamp owl; Texas barred owl (*helveola*); wood owl.

Range (Modified from AOU, 1983.)

Resident from southeastern Alaska, southeastern Yukon, British Columbia, western and northern Washington, eastern and central Oregon, and northern California, east through northern Idaho, northwestern Montana, central Alberta, and central Saskatchewan, and from southern Manitoba, central Ontario, southern Quebec, New Brunswick, Prince Edward Island, and Nova Scotia south through riparian woodlands of the Great Plains (west locally to southeastern South Dakota, central Nebraska, central Kansas, and central Oklahoma) to central and southern Texas, the Gulf coast, and southern Florida; also resident in Mexico, from Durango south locally to Oaxaca. Recently (1999) reported from northeastern Colorado, perhaps bridging the Central Plains grasslands via riverine forests. (See Figure 55.)

Subspecies (Adapted from AOU, 1957.)

S. v. varia Barton. From southeastern Alaska (Skagway to Ketchikan), British Columbia and southeastern Yukon east to Nova Scotia, south through the eastern Great Plains to Oklahoma and east to Virginia.

S. v. georgica Latham. From central Arkansas east to North Carolina and south to eastern Texas and southern Florida.

S. v. helveola (Bangs). Endemic to south-central Texas.

S. v. sartorii (Ridgway). Western and southwestern Mexico, from Durango to Oaxaca.

Measurements

Wing (of *varia*), males 320–340 mm (ave. of 11, 332.8), females 330–352 mm (ave. of 7, 338.3); tail, males 215–230 mm (ave. of 11, 225.4), females 224–257 mm (ave. of 7, 230.3) (Ridgway, 1914). The eggs of *varia* average 49 × 42 mm (Bent, 1938).

Weights

Earhart and Johnson (1970) reported that 20 males averaged 632 g (range 468–774), and that 24 females averaged 801 g (range 610–1051). Twelve males averaged 621.9 g (range 483–812), 14 females averaged 872.6 g (range 650–1020)

(Mazur and James, 2000). The estimated male:female mass ratio is 1:1.3–1.4. The estimated egg weight is 45 g; 25 eggs averaged 45.5 g (Elderkin, 1987), and the estimated egg-to-female proportional mass ratio is thus 5.2–5.6 percent.

Identification

In the field. The highly distinctive "Who cooks for you; who cooks for you-all?" advertisement call (sometimes uttered antiphonally as a duet or chorus) is the most convenient means of identifying this owl. It hides inconspicuously in heavy woodland vegetation during the daylight hours. If seen, the very large head is "earless," and it has dark brown eyes. On the breast there is a sharp break between the lateral barring of the throat and upper breast and the vertical streaking of the lower breast and flanks.

In the hand. This medium-sized owl (wing 215–340 mm) lacks ear tufts, has dark brown eyes, and has distinct dark streaks on the flanks and lower breast, whereas the upper breast is strongly barred. The nape lacks the definite white spots found in the similar but slightly smaller spotted owl, and instead is barred like the breast. Adult differences in linear measurements may be used in discriminant function analysis to distinguish the sex of about 75 percent of birds (Carpenter, 1992). Chicks are initially covered with pure white down, which in a few weeks is replaced by a second longer downy coat that is buffy basally and white terminally. The head, neck, and entire underparts of this plumage are broadly barred with rather light brown and pale buffy and whitish; the scapulars and wing coverts are similarly barred, but the bars are broader, the brown ones of a deeper shade, and each feather is broadly tipped with white. The remiges and rectrices of young birds are as described for the spotted owl (Pyle, 1997b). The first winter plumage is adultlike, but buff tones tend to replace white on the body feathers (Bent, 1938).

Vocalizations

Bent (1938) described this species' "ordinary" call (presumably the territorial advertisement song, but uttered by both sexes) as consisting of 2 groups or phrases of 4 or 5 syllables each, uttered rhythmically and strongly accented, as well as loud, wild, and strenuous *hoo'-hoo-to-hoo'-ooo, hoo-hoo-hoo-to-whooo'-ooo.* The first 2 syllables of the

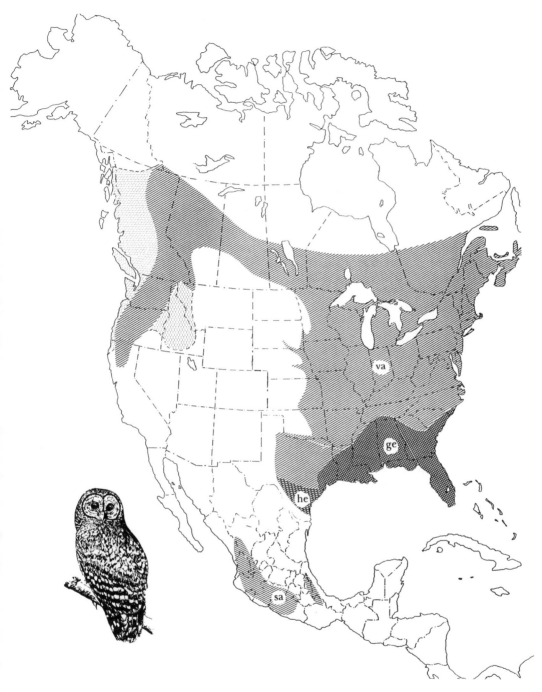

Figure 55. Distribution of the barred owl, showing residential ranges of the races *georgica* (ge), *helveola* (he), *sartorii* (sa), and *varia* (va); indicated racial limits are only approximations. Recent range extensions in western North America are indicated by light stippling.

first phrase and the first 3 of the latter one are distinct, deliberate, and low-toned. The last 2 are run together, with a strong accent on the penultimate one, which is loudest of all, first rising in pitch and then descending and diminishing in volume as the final note terminates. In some cases the series ends on a loud, harsh note, and occasionally each phrase may be reduced to only 2 or 3 syllables, or only 1 instead of 2 phrases will be uttered. Both sexes utter variations of this call, the female having a higher-pitched voice than the male. A variant of this sequence is the "ascending hoot," consisting of 6–9 evenly spaced and equally accentuated notes of gradually ascending frequency, followed by a downwardly inflected *hoo-aw* (Mazur and James, 2000).

A characteristic barred owl vocalization is "caterwauling," which has sometimes been described as loud and prolonged outbursts of maniacal cackling, laughing, and whooping calls. This exchange, uttered by duetting pairs, may last up to two minutes and is sometimes stimulated by call playbacks. It is sometimes also uttered as large prey are being subdued (Mazur and James, 2000).

McGarigal and Fraser (1985) recognized 6 distinct vocalizations during studies involving playbacks of prerecorded songs. The most common was the typical 9-syllable, 2-phrase hooting, which was heard at 80 percent of the sites tested. The ascending hoot, was noted at 56 percent of the stands. Both of these call types were uttered by both sexes and perhaps serve as location calls between members of a pair as well as territorial challenge calls. A third call, heard at 36 percent of the stands, was the 2-syllable *hoo-aw* note uttered independently. Caterwauling was associated only with duetting. A single-syllable sharply ascending wail-like note, uttered by one bird when near its mate, was an apparent contact call. Finally, an irregular and patternless assemblage of hoots was heard at one stand.

A phrase sounding "angry" to Bent (1938) is a *whah-whah-whah-to-hooo'* with the notes of loud, nasal, and rasping quality, as in derisive laughter. Or 2 or 3 soft, hooting notes of uniform rhythm and little accent, similar to that of the great horned owl, may also be uttered. A loud, tremulous call, similar to that of the screech-owl, but much louder, was noted twice by Bent, and once this note had a whining quality. A husky, almost humanlike whistling call, as well as a doglike barking note, was also noted by Bent. Some unpublished studies on adult barred owl vocalizations and their possible functions also exist (DeSimone, Root, and Roddy, 1985).

Dunstan and Sample (1972) reported that calling by barred owls occurs in Minnesota during all months of the year but is most frequent in February and early March (prior to egg laying) and again during late summer and fall, which probably corresponds to the dispersal period of the young, when they are presumably trying to establish territories. In southern Canada there is a similar peak in calling during late March and April, just prior to egg laying (Elderkin, 1987; Mazur and James, 2000). Most calling occurs at night, and probably serves to advertise territories and to attract mates, as in other owls.

Habitats and Ecology

Most observers have characterized the typical breeding habitats of the barred owl as consisting of relatively heavy, mature woods, varying from upland woods to lowland swamps, often with nearby open country for foraging but with densely foliaged trees for daytime roosting (including conifers or deciduous trees with persistent leaves for winter roosts) and the presence of enough large trees (roughly 50 cm in breast-height diameter or larger) with suitable cavities to allow for nesting. Old-growth woods seem to be favored owing to larger numbers of potential nest sites, greater ease of hunting in the more open understory vegetation, and closed-canopy woods that provide both improved protection from mobbing and easier thermoregulation. General prey availability may also be higher in older and vegetationally complex forests (Mazur and James, 2000). Barred owls are highly territorial, and the birds actively defend their entire home range. Chases and fights may occur at territorial boundaries, and these boundaries are defended throughout the year (Nicholls and Fuller, 1987; Mazur, Frith, and James, 1998).

In an early application of radiotelemetry to owl ecology, Nicholls and Warner (1972) radiotracked 10 barred owls in Minnesota during a period of more than 1100 days, obtaining nearly 27,000 habitat locations. These birds showed strong avoidance and preferences for particular habitat types, with the general decreasing order of preference being oak (*Quercus*) woods, stands of mixed hardwoods and conifers, white cedar (*Thuja occidentalis*) swamps, oak savannas, alder (*Alnus*) swamps, marshes, and open fields. There were no marked sexual, seasonal, or weather-related variations in habitat preference nor any major year-to-year differences. The two habitat types having the greatest apparent year-round use were oak woods and mixed woods. Both of these typically were located in upland areas, were free of dense understories, and had few herbaceous plants on the forest floor, probably providing ideal hunting conditions. The lack of brush made for excellent visibility, and the many dead or dying trees provided habitats for prey such as mice and squirrels, and probably also nesting sites.

In a similar radio-tracking study, Elody and Sloan (1985) used seven radio-equipped owls to establish habitat use in Michigan. There, the most-used habitats were old-growth stands of hemlock (*Tsuga canadensis*) and maples (*Acer* spp.), which singly or in combination were the dominant cover types throughout the study areas. Mixed pine stands and marshes were used according to their relative availability. It was suggested that these old-growth stands offered a combination of dense forest cover for daytime roosting and a supply of natural cavities in dead timber for nesting sites. Most of the owls' territories were less than half a kilometer from water, and both marshes and swamps are known to be commonly used by barred owls for hunting purposes. Dunstan and Sample (1972) found that many of the Minnesota barred owl nests they studied were situated close to lakes. Similarly, Bosakowski, Speiser, and Benzinger (1987) found that swamp habitats, especially those with hemlocks that were well removed from humans, were preferred habitats.

Estimated annual home ranges of 9 radio-tagged barred owls studied by Nicholls and Warner (1972) averaged 231 hectares, and varied from 86 to 370 hectares. The estimated annual home ranges of 7 owls similarly tracked by Elody and Sloan (1985) averaged 282 hectares, but during the summer months only about 118 hectares of the overall home range were used. The increased area used during winter was believed to be most likely a result of relative unavailability of prey at that season, which caused some of the males to vacate the study area, leaving the females to utilize and defend their territories alone. A similar seasonal increase in home ranges of 8–13 birds was reported in Saskatchewan (mean breeding, nonbreeding, and annual home ranges 149, 1234, and 971 ha, respectively) (Mazur, Frith, and James, 1998). Similarly, the mean breeding and annual home ranges of 8–10 birds in Washington were 321 and 644 hectares (Hamer, 1988). The breeding home ranges of 15 barred owls studied by B. Olson (1999) averaged 337.9 hectares and consisted of relatively old coniferous and mixed forests as compared with the general surrounding landscape. According to Nicholls and Fuller (1987), the birds exhibit territorial defense and a high level of exclusive use of their entire home ranges, the boundaries of which often remain quite stable from year to year and even from generation to generation.

Population densities of barred owls are relatively low. J. Craighead and Craighead (1956) estimated that an area of 93.2 square kilometers in

Michigan with extensive deciduous woodlots supported as many as 3 pairs, an overall density of 0.03 pairs per square kilometer, while an area of about 462 hectares of lowland forest in Maryland supported approximately 1 pair per square kilometer (R. Stewart and Robbins, 1958). Low densities (0.07 pairs per sq km) were reported by Bosakowski, Speiser, and Benzinger (1987) in a study area of 120 square kilometers in northern New Jersey.

Movements

Little information exists relative to local movements of barred owls. The birds individually appear to be highly sedentary, using the same nest site year after year. The same territory may be occupied for as long as three decades, presumably by several generations of owls. The longest documented individual is a young bird banded in Nova Scotia and recovered less than a year later 1,600 kilometers west in Ontario (Mazur and James, 2000). Nevertheless, during the past century the barred owl has gradually expanded its western limits to encompass essentially all of southern Canada's boreal forests, as well as reaching southeastern Alaska (north to Skagway) and moving south to northern California. First reported in southern Manitoba in 1886, it had reached Saskatchewan by 1948 and was nesting in Alberta by 1966. It was first documented in British Columbia in 1946 and had reached the Pacific coast by the 1960s. It then moved southward, reaching Washington by 1965, Idaho by 1968, and Oregon by 1974. Although first reported from northern California in 1981, nesting was not documented there until 1991 (Mazur and James, 2000).

Foods and Foraging Behavior

The early studies of barred owls summarized by Fisher (1893) and Bent (1938) indicate that a wide variety of small mammals, especially rodents, are consumed, along with an equally wide or wider array of birds, rarely up to the size of grouse and domestic fowl. Frogs, lizards, small snakes, salamanders, fish, and some invertebrates (mollusks and insects) have also been reported among the foods consumed. Errington (1932a) noted that mammals comprised 47–76 percent of prey items identified in various study areas, birds 7–40 percent and miscellaneous prey 11–38 percent. Thus it is clear that barred owls are largely opportunistic foragers, taking what is available within their power to subdue. This ordinarily consists of birds up to about the size of flickers (*Colaptes*) and mammals as large as moles and partly grown cottontails (*Sylvilagus*). However, they have been known to kill considerably larger or more formidable birds, including eastern screech-owls, and there is even a record of long-eared owl remains present in the stomachs of two barred owls.

Wilson (1938) reported that remains of *Microtus* voles comprised about 83 percent of the 777 prey items found in barred owl pellets in Michigan, with progressively smaller numbers of short-tailed shrews (*Blarina brevicauda)*, *Peromyscus* mice, and other small mammals, plus a very few remains of birds, amphibians, and insects. Similarly, in Montana *Microtus* spp. comprised more than 90 percent of the total 107 prey items in pellets from winter roosts, although some bird remains of gray partridge (*Perdix perdix*) and ring-necked pheasant (*Phasianus colchicus*) were also present. In a small sample of pellets from Illinois *Microtus* was again the most numerous prey item, comprising about 30 percent of the total individuals identified (Cahn and Kemp, 1930).

In a survey of barred owl foods across its range from Nova Scotia to Montana, Mazur and James (2000) summarized data based on 1637 prey items. Mice, voles, and shrews comprised from 32 to 97.3 percent of the identified items, lower vertebrates 0–24.5 percent, squirrels 0–12.1 percent, birds 2.7–25.1 percent, and invertebrates 0–24.5 percent. A few other prey were present in very small numbers. Seasonally, small mammals may be most important during winter, whereas lower vertebrates and invertebrates may increase in importance during summer. These figures are generally similar to those of Snyder and Wiley (1976), based on 2234 prey items, with mammals comprising 76 percent, invertebrates 15.8 percent, birds 5.8 percent, and lower vertebrates 2.5 percent.

The barred owl is essentially a seminocturnal to nocturnal hunter (although birds with broods may also hunt during the day), with a hunting technique and prey preferences that are apparently virtually identical to those described for the somewhat smaller spotted owl. This size difference places the latter at a competitive disadvantage where the two species might come into contact. However, the barred owl is in turn dominated by the much larger great horned owl, so that in areas where woodlands are relatively small the barred is seemingly excluded, but in places where the forest habitats are more extensive the barred owl can survive in the presence of the great horned owl, apparently using habitats less frequented by the great horned owl (J. Craighead and Craighead, 1956).

Social Behavior

The barred owl is believed to be quite sedentary, and thus might be expected to exhibit fairly per-

manent pair bonds and a high degree of both territoriality and nest site tenacity. Bent (1938) reported that one nest site was occupied by various barred owls over a period of 10 years, and the same woodlot was occupied for at least 33 years. Another area had a barred owl occupancy record of 34 years. A third site mentioned by Bent had a record of 26 years of occupancy, during which time the birds nested in five different pine groves owing to disturbance by forest cutting. These long occupancies were not necessarily by the same birds throughout, and both home range location and size may remain fairly constant from year to year despite changes in individual occupants (Nicholls, 1970; Nicholls and Fuller, 1987). Such data suggest that a very high level of site tenacity may be present in barred owls. D. Johnson (1987) noted that none of 158 barred owl band recoveries in North America occurred more than 10 kilometers from the point of banding, which also suggests that a high degree of sedentary behavior is typical of barred owls. However, in Nova Scotia, five banded chicks dispersed up to 64 kilometers from their nest site, and one chick moved from Nova Scotia 1600 kilometers west to Ontario in less than a year (Elderkin, 1987; Mazur and James, 2000).

Little has been written on pair bonding in this species, but it is likely to be similar to that of its near relative, the tawny owl (*Strix aluco*). In that species pair bonds are permanent. Courtship begins in winter, becoming progressively centered on the future nest site, with exchanges of hooting by the male, and contact and other calls by the female. As nesting approaches the female becomes seemingly lethargic, remaining near the nest and uttering food-begging calls to its mate. The male also utters a wild variety of calls when pursuing the female, and may perch near her, swaying from side to side and then vertically, raising each wing in turn and then both simultaneously. The plumage may be fluffed out, and then slimmed down, as the male sidles along the branch toward the female and back again (Mikkola, 1983). Courtship feeding and mutual preening are also important parts of social behavior in the closely related tawny owl, and the evident pleasure shown by even wild-caught barred owls in having their heads scratched by humans suggests that the same may apply to this species. Such mutual preening has apparently not yet been seen. Some representative postures of the barred owl are shown in Figure 56.

Nest sites are typically used year after year by barred owls, so long as they remain usable. The birds often select a natural tree cavity or an old hawk, squirrel, or crow nest, with few if any repairs or modifications being made. They often

Figure 56. Barred owl behavior, including (A) adult calling, (B) young juvenile leaning against tree trunk, and (C) older juvenile performing defensive wing-spreading. A and B after photos by Elaine Bachel; C after photo by author.

nest in very close proximity to red-shouldered hawks (*Buteo lineatus*), without evident conflict, and Bent (1938) mentioned some cases of mixed clutches of the two species being found in the same nest. Both living and dead trees are selected for nest sites. Peck and James (1983) reported that 8 of 10 nests in Ontario were in natural tree cavities, and the others in a squirrel drey and a stick nest. Five nests were in balsam poplar (*Populus balsamifera*); 2 each were in beeches (*Fagus*) and birches (*Betula papynfera*); and 1 site was in an unspecified tree. The heights ranged from 4.5 to 10.5 meters (half between 7.7 and 10.5 m), and the tree diameters of two cavity nests were 61 and 76 centimeters. In three other studies, mean tree diameter at breast heights ranged from 47.4 to 61 centimeters, and mean nest height from 6.8 to 13.4 meters. Natural cavity interior diameters average 2533 centimeters, and cavity depths 45–54 centimeters (Mazur and James, 2000). B. Olson (1999) found similar characteristics for tree cavity nests, and further noted that trees were chosen that were closer to other larger-diameter trees than those typical of the owl's overall home range. Postupalsky, Papp, and Scheller (1997) regarded barred owls as obligate cavity nesters.

In the central Appalachians, Devereux and Moser (1984) found that 8 nest sites were significantly closer to forest openings than would be randomly expected, and tended to have well-developed understories and fewer but larger overstory trees. Six of the nests were in the tops of hollow tree stubs, and all nesting trees were in declining or dead and decomposing stages. In a sample of 33 nests from throughout the species' range, 88 percent were in deciduous trees, with elms (*Ulmus*), beeches, and maples (*Acer*) predominating. Trees chosen for nests tend to be larger than random snags, suggesting a preference for nesting in taller and larger trees. Of 37 nests, 24 were in deciduous forests, 10 in mixed woods, and 3 in coniferous forests (Apfelbaum and Seelbach, 1983). Bent (1938) reported finding 38 nests, with 21 among white pines (*Pinus strobus*), 6 in deciduous woods, and the rest in mixed woods. Of the 38 nests, 18 were in old hawks' nests; 15 were in tree hollows; and 5 were in apparently old squirrel nests. A higher mean level of breeding success in birds using cavity rather than platform nests has been found in Michigan (Postupalsky, Papp, and Scheller, 1997). Reuse of the same site in subsequent years is common; rarely such sequences have extended for as long as 10 consecutive years.

Breeding Biology

The breeding season of this species is fairly long, and renesting is common following egg or brood loss. Bent (1938) noted that a second clutch is normally laid 3–4 weeks later, and sometimes a third set may even be laid. Peck and James (1983) noted that in two Ontario cases replacement clutches were laid after the first was collected, suggesting that even at the northern limits of its range renesting may be fairly common. In southern New England the egg dates are from March 31 to May 18, with half of 63 records from April 2 to 21. A total of 41 records from New Jersey are from February 28 to April 14, with half from March 17 to 29. Twenty-three records from Illinois and Iowa are from February 25 to April 30, with half from March 6 to April 13. A sample of 22 Florida records are from January 11 to March 10, with half between January 28 and February 20. Twenty-two Texas records are from February 17 to June 4, with half between February 27 and March 25 (Bent, 1938). Six active nests from Ontario range from April 4 to May 18 (Peck and James, 1983). The overall seasonal span mentioned here is from late January to mid-May, or nearly five months. December records exist for Florida, as well as June records for Texas (Stevenson and Anderson, 1994; Mazur and James, 2000).

Murray (1976) reported that among 315 clutches from across the barred owl's range the average clutch was 2.41 eggs, with slight but significant increases in clutch size with increasing latitude in 2 of 3 regions. Bent (1938) reported that among 61 sets of eggs of the race *varia*, 41 were clutches of 2, 18 were of 3, and only 2 of 4, the average being 2.36 eggs. Forty-eight Nova Scotia nests averaged 2.31 eggs (Elderkin, 1987). One 5-egg clutch in the National Museum of Natural History may have been the work of 2 females.

Incubation begins with the laying of the first egg, resulting in staggered hatching. All incubation is by the female, and it requires 28–33 days. By the end of a week, the young begin to open their eyes, but they continue to be brooded extensively until they are about 3 weeks old. Beak clapping, hissing, and food-begging calls begin after 11 days of age and continue through fledging (Dunstan and Varchmin, 1985). When about 4 or 5 weeks old they begin to leave the nest and clamber about on nearby branches, but do not fledge until they are about 6 weeks old. However, even when they are as old as 4 months they may continue to receive some food from their parents (Bent, 1938).

There is little information on nesting success, but Apfelbaum and Seelbach (1983) calculated the average number of nestlings from 55 broods as 2.02 young. Devereux and Moser (1984) reported 1.9 nestlings per nest in 7 active nests, and 1 young successfully fledged in each of 2 successful nesting attempts. In Michigan, 49 cavity nests averaged 1.9 fledged young, whereas 13 open platform nests fledged an average of 1.0 young. Tree cavities and nest boxes were both much more successful sites for nests than were hawk nests or other open sites. All told, of 114 nesting attempts, 85 were successful, with an average of 1.97 young per successful nest, and 1.48 young per breeding attempt. Most often (57 percent of nests with known brood sizes) 2 young were reared to advanced stages of development, and no more than 4 were raised (Postupalsky, Papp, and Scheller, 1997). These results are similar to those of D. Johnson (1987), in which 86 percent of 22 nests were successful, and the mean number of young produced per successful nest was 2.42, and 2.09 young per nesting attempt. Maximum known longevity under natural conditions is 18 years and 2 months (Patuxent Wildlife Research Center banding data).

Evolutionary Relationships and Conservation Status

The spotted and barred owls are certainly close relatives, and so too is the tawny owl of Europe. Even more closely related is the fulvous owl (*Strix fulvescens*), which ranges from southern Mexico (Oaxaca) south to Honduras and has at times been considered conspecific with *varia* (AOU, 1983). The fulvous owl is associated with pine-oak woodlands and humid montane cloud forests, and has hooting calls that are very similar to those of the barred owl.

The barred owl is a forest-dependent species, requiring at least some old-growth trees for nesting. As such, it has probably suffered in the eastern and southeastern parts of its range, as large stands of old-growth forests have disappeared as a result of lumbering. However, the recent expansion of the barred owl into the spotted owl's range in the Pacific Northwest deserves special attention. This expansion was first documented in the early 1960s, when the birds began moving into southeastern British Columbia (J. Grant, 1966). Shortly thereafter the first records for Washington and Oregon were obtained (A. Taylor and Forsman, 1976; Rohweder, 1978). By 1975 nesting in Washington had been documented (Leder and

Walters, 1980), and the first California record occurred in 1982. By 1985 the birds had extended their range to northern California and southeastern Alaska, and had become common in southern British Columbia, northern Idaho, and northeastern Washington (Hamer and Allen, 1985). By the late 1980s the species had occupied most or all of British Columbia (Campbell et al., 1990) and much of Idaho (Stevens and Sturts, 1997). Parts of northwestern Montana have also been recently colonized (Bergeron et al., 1992). The first known California breeding occurred in 1991, and by the late 1990s the species had reached south to Sonoma County (Dark, Gutiérrez, and Gould, 1998).

In Olympic National Park barred owls apparently displaced two spotted owl pairs from their territories between 1985 and 1986 (Sisco and Sharp, 1986). However, extensive hybridization may not be occurring where barred owls have become firmly established within the range of spotted owls (Herter and Hicks, 2000). There is also the possibility that barred owls will soon have crossed the Central Great Plains via human tree plantings or the increasingly forested Plains rivers and will begin to occupy forests along the Front Range of the Rocky Mountains. There have been two recent reports of barred owls from eastern Colorado (*North American Birds* 1999, 53:82; 2000, 54:304).

There is a statistically highly significant upward national population trend based on annual Breeding Bird Surveys done between 1966 and 1993 (Price, Droege, and Price, 1995). A more recent Internet summary of these data suggest an overall $+2.6$ percent annual change from 1966 to 2000, which is also highly significant statistically. The greatest increases have occurred in the vicinity of the Northeast and the Great Lakes, the Fish and Wildlife Service's regions 5 (5.4 percent) and 3 (4.6 percent) (http://www.mbrpwrc.usgs.gov/bbs/bbs.html). During the 1999 and 2000 Christmas Bird Counts the species was detected on an average of 623 counts, or about 17 percent of all U.S. and Canadian counts. The maximum number ever reported on a single CBC was 105, at Pine Prairie, Louisiana, in 1999. In terms of all localities where the species was detected, the barred owl was the third most widely reported of the 18 North American owl species then encountered. The average total number counted among the 100 counts having the highest species totals was 944 birds, or about 10 percent of all owls reported in the tabulated summaries of counts for the United States and Canada during 1999–2000. The Canadian population has recently been estimated at 10,000–50,000 pairs (Kirk and Hyslop, 1998) and is believed to be stable or increasing.

Great Gray Owl   *Strix nebulosa* Forster 1772

Other Vernacular Names:
cinereous owl, Lapland owl, sooty owl, speckled owl, spectral owl.

North American Range (Adapted from AOU, 1983.)

Breeds in North America from central Alaska, northern Yukon, Nunavut, northern Manitoba, and northern Ontario south locally in the interior along the Cascades and Sierra Nevadas to central California; in the Rockies from northern Idaho and Montana to western Wyoming, and to central Alberta, central Saskatchewan, southern Manitoba, northern Minnesota, and south-central Ontario (rarely to northern Wisconsin and northern Michigan). Winters generally through the breeding range, but wanders south irregularly to southern Montana, North Dakota, southern Minnesota, southern Wisconsin, central Michigan, southern Ontario, and central New York, casually as far as southern Idaho, Nebraska, Iowa, Indiana, Ohio, and from southern and eastern Quebec, New Brunswick, and Nova Scotia south to Pennsylvania and New Jersey. Also distributed widely in northern Eurasia. (See Figure 57.)

North American Subspecies (Adapted from AOU, 1957.)

S. n. nebulosa Forster. Range in North America as described above.

Measurements

Wing (of *nebulosa*), males 410–447 mm (ave. of 5, 433), females 430–465 mm (ave. of 7, 446); tail, males 300–323 mm (ave. of 5, 313.6), females 310–347 mm (ave. of 7, 323.3) (Ridgway, 1914). The eggs of *nebulosa* average 54.2 × 43.4 mm (Bent, 1938).

Weights

Earhart and Johnson (1970) reported that 7 males averaged 935 g (range 790–1030), and that 6 females averaged 1296 g (range 1144–1454). J. Craighead and Craighead (1956) noted that 7 females averaged 1084 g. Bull and Duncan (1993) reported a mean of 890.5 g (range 825–1050) for 21 breeding males, and 1267 g (1025–1700) for 63 breeding females. Mikkola (1983) stated that 24 males and 31 females of the Eurasian population averaged 871 and 1242 g, respectively. The estimated male:female mass ratio is 1:1.39–1.42. The estimated egg weight is 53 g. The estimated egg to adult female proportional mass ratio is thus 4.2–4.8 percent.

Identification

In the field. This enormous owl is almost instantly recognizable by its very large and "earless" head, and by a generally dark body plumage except for a white "moustache" that is variably broken in the middle by a black "bow tie." The usual call is a deliberate series of soft and low-pitched single- or double-syllable hoots that gradually drop in frequency and decelerate toward the end of the series.

In the hand. The large size (wing more than 410 mm) and large but "earless" head, with yellow eyes that are surrounded by a series of 6–8 dark concentric rings in a distinct and circular facial disk, instantly identify this species. The wing is broad, with the sixth primary the longest, and the inner webs of the outer five primaries emarginated. The tarsus and heavily feathered toes are both relatively short, but the claws are long and slender. As adults, not only do the sexes differ in linear and weight measurements but females also tend to be darker throughout, with underwing coverts that are rich buff to tawny, rather than silvery, whitish, or pale buff, as typical of males (Pyle, 1997b). Newly hatched birds have grayish down dorsally and white down below, with yellowish legs and yellowish gray iris color. Juveniles are olive brown, darkly barred and spotted with white above, barred below, with broad black facial markings. The definitive body plumage is attained in less than five months (Mikkola, 1983), but first-year birds have gray-tipped flight feathers. These remiges are also shorter and narrower than in adults, and are uniform in color and degree of wear; the central rectrices and tertials are narrower and more tapered than in adults (Pyle, 1997b). Some first-year remiges may be retained as long as three or four years (Bull and Duncan, 1993).

Vocalizations

The vocalizations of this species have only been carefully studied in Europe (Berggren and Wahlstedt, 1977). Those of the North American race have been described by Oeming (1955), Nero (1980), K. Collins (1980), and Winter (1980, 1986), and sonograms have been produced by K. Collins (1980). The vocal repertoire of the species seems rather sparse, at least as compared with some of the more highly nocturnal owls.

In the Sierra Nevadas of California the birds

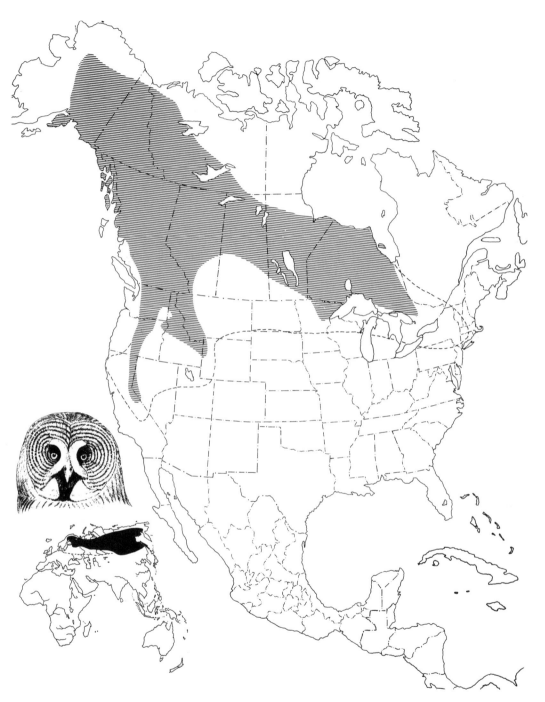

Figure 57. North American breeding distribution of the great gray owl. The dashed line indicates usual limits of wintering vagrants. Extralimital distribution shown in inset.

are vocal throughout the year, responding to tape-recorded calls at virtually any time, but primarily uttering territorial calls between March and mid-May. Typically there the calling begins late in the evening, with a premidnight peak, followed by a sharp decline around midnight but a second peak shortly thereafter, and then gradually declining. Each call phrase of the territorial hooting lasts 6–8 seconds, the individual soft hooting notes uttered at the rate of about 3 per 2 seconds, and with an average interval of 33 seconds between calls (Winter, 1980). Under ideal conditions the call can be heard for up to 800 meters, but it often carries only about 500 meters (Mikkola, 1983). In Scandinavia, the males begin their territorial calling in January or February, often during the first period of mild weather, with a peak in calling activity during the nesting period. Territorial calling there may also be heard late in the breeding season, during June or July, and again sometimes in autumn (Mikkola, 1983). There is also an increase in vocal activity during autumn in North America (Bull and Duncan, 1993).

Although the female sometimes also utters the territorial call (at a higher pitch than males) prior to the egg-laying period in spring, her most common note is a single soft and mellow hoot, described by Nero (1980) as a *whoop*, which serves largely as a food-begging call. It also serves as a contact call between pair members and between the pair and their young (Bull and Duncan, 1993). A similar hoot that can be heard for up to about 300 meters is used by the male at the nest. A double, excited *ooh-uh* is uttered by the female when the male is arriving with food. As a defensive or warning cry both sexes produce an extended series of double notes, uttered in groups of up to 100 in sequence and at the rate of up to 3 notes per second. The female's typical alarm call is a deep growling, together with bill snapping. During intense alarm, as when performing nest-distraction or injury-feigning displays, she may produce a series of wails, squeaks, and hoots, climaxed by a loud heronlike squawk or bark. Prior to and during copulation the female produces a call reminiscent of the begging calls of chicks and juveniles, the latter rapid, chattering *sher-richt* notes. The chicks also produce bill-snapping sounds when being handled or otherwise disturbed (Nero, 1980; Mikkola, 1983).

Habitats and Ecology

In North America the broad range of the great gray owl encompasses a variety of vegetational types, ranging from subalpine coniferous forests through dense boreal and montane coniferous forests to stunted forests transitional to arctic tundra. Nesting is commonly done in stands of mature poplars (*Populus* spp.) adjacent to muskegs. Islands of poplars or aspens amid stands of spruce or pine are common breeding locations, as are similar groves or marginal strips of often-stunted tamaracks (*Larix laricina*) in wetter sites (Nero, 1980). In the Sierra Nevadas of California the birds breed in mixed-conifer forests and red fir (*Abies magnifica*) forests (at about 900–1800 m and 1800–2700 m elevation, respectively), especially in dense forest stands bordering meadows. Nesting habitats might be limited by the number of available nest sites, typically old nests of other raptors; in Oregon all sampled forest types contained nests (Bull and Henjum, 1990; Bull and Duncan, 1993).

In Manitoba the birds favor tamarack during summer, apparently avoiding jack pine (*Pinus banksiana*), black spruce (*Picea mariana*), open treeless areas, and habitats with a dense shrub layer. Factors affecting habitat selection include relative availability of microtine prey, suitable perches, and shrub density (Servos, 1987). Most Saskatchewan breedings have been in tamarack–black spruce forested wetlands, with 25 of 27 suspected nestings within 500 meters of such habitats (Harris, 1984). Although within areas of tamarack forests, 14 nest sites in Minnesota were associated with black ash (*Fraxinus nigra*) and basswood (*Tilia americana*), the forks of which provide better nest sites for raptors than the surrounding scrub tamaracks (Spreyer, 1987). Preferred winter habitat in Alaska consists of the ecotone between grassland meadows and tall willows, balsam poplars (*Populus balsamea*), and white spruce (*Picea alba*) (Osborne, 1987).

In the western Palearctic the great gray owl is mainly associated with dense and mature lowland or sometimes montane coniferous forests that are dominated by pines, spruces, and firs, sometimes interspersed with birches (*Betula*) (Cramp, 1985). Most hunting is not done in such forests, but rather in adjacent open habitats, including marshes and cleared forests (Mikkola, 1983). Probably a combination of abundant small (up to about 100 g) rodents occurring in semi-open habitats such as meadows or muskegs where they can be readily captured, plus proximity to dense coniferous forests offering both roosting and nesting sites, are primary aspects of breeding habitats.

During late summer and fall the birds are prone to move higher into lodgepole pines (*Pinus contorta*) forests, but they also use lower-altitude ponderosa pine (*Pinus ponderosa*) forests during fall and winter (Verner and Boss, 1980; Winter, 1986). In winter the birds often move out of the forest to hunt in open fields having scattered trees, scrub patches, weedy areas, and fencerows (Brunton and Pittaway, 1971). In mountainous areas they tend to move to lower elevations, where the snow cover is thinner (Bull and Henjum, 1990).

Population density estimates for North America are few, but Bull and Henjum (1987) found 5 nesting pairs in one 290-hectare study area, and 7 in an area of 937 hectares. Spreyer (1987) noted that in Minnesota as many as 8 nests in a single year occurred within a 52-square-kilometer area. In Sweden variations in breeding density of from 7 pairs in 20 square kilometers to 9 pairs in 100 square kilometers (0.09–0.35 pairs per sq km) have been noted, and in one location 7 pairs occupied an area about 3 kilometers in diameter (Cramp, 1985). Nesting densities of 0.74 and 1.72 pairs per square kilometer have been documented in Oregon, 1.88 pairs per square kilometer in Manitoba, and 2.66 pairs per square kilometer in California (Bull and Henjum, 1990; J. Duncan, 1987; and Winter, 1986).

A nesting season home range of approximately 260 hectares, with a maximum diameter of about 2.3 kilometers, was estimated for great gray owls in the Grand Teton area of Wyoming by J. Craighead and Craighead (1956), based on sight records of unmarked birds. A winter home range of 45 hectares (maintained by 1 bird over an 11-day period) was estimated by Brunton and Pittaway (1971) in Quebec. Bull, Henjum, and Rohweder (1988a) estimated a much larger mean home range of 67.3 square kilometers for adults in Oregon, with some males hunting as far as 3.2 kilometers from their nests.

Movements

It is well known that great gray owls are irregularly irruptive or migratory, with periodic invasions into various northern states and southern Canadian provinces (Eckert, 1984; Nero, 1969). In the winter of 1983–1984 more than 400 birds were seen in southern Ontario alone, the numbers peaking in January (*American Birds*, 1984, 38:312). A more recent major invasion occurred in 1991–1992. Nero (1980) thought that these winter invasions might often be the result of a combination of years of good reproductive success followed by prey declines, or perhaps the birds are forced out of breeding areas because of deep snow accumulations or icy crusts that affect hunting success. Some winter birds drop to extremely low body weights of only about 30 percent normal mass, and yet may survive (Nero, 1980). There is some evidence that winter incursions may to a large degree be made up of immature birds; Nero and Copland (1981) noted that 20 of 24 birds banded during winter along the Trans-Canada

Highway in southern Manitoba were immatures. Nero (1980) also noted that 2 females that bred successfully one year were repeatedly seen the following winter within a mile or two of their nest sites. Radio-tracked individuals have traveled as far as 40 kilometers in 24 hours, and up to 650 kilometers in 3 months (Bull and Duncan, 1993).

Postfledging movements of juvenile birds are sometimes quite extensive, judging from European banding data. In Finland 11 juveniles moved up to 226 kilometers, and in Sweden 16 juveniles moved up to 490 kilometers from the nest. At least the Swedish movements were not correlated with rodent population levels, but instead the dispersal pattern was random. A few long-distance movements of adults, including 2 females that moved 110 and 430 kilometers over periods of 2–4 years, have also been reported (Cramp, 1985). One long-distance movement of an immature was mentioned by Nero (1980), the bird being a nestling banded near Winnipeg and recovered the following winter about 753 kilometers southeast in extreme southern Minnesota. In an Oregon study, 11 radio-tagged juveniles traveled 8.8–31.4 kilometers from their nests in one year, while 11 adults moved 3.1–42.9 kilometers during the same period, suggesting that little if any age difference in mobility occurred there (Bull and Henjum, 1985). Among a larger sample of 21 young Oregon birds, none went more than 50 kilometers from their nest, and 3 young followed for 2 years went no farther than 28 kilometers (Bull and Duncan, 1993).

Foods and Foraging Behavior

In spite of its large size, the great gray owl subsists almost entirely on relatively small rodents. Mikkola (1983) determined that of nearly 5200 prey items from the breeding season, 87.7 percent were of prey species averaging from 10 to 49.9 grams as adults, and only about 10 percent were of species averaging more than 100 grams. Studies at 61 nest sites in Finland and Scandinavia indicated that about 94 percent of the prey items were rodents, and *Microtus* species alone comprised nearly 75 percent, with *Clethrionomys* the second most important genus, adding about 10 percent. Birds contributed only about 1 percent. When breeding-season data are analyzed on a biomass basis, small microtine voles are responsible for 86.5 percent of the total; larger mammals (mostly of *Arvicola* voles), about 9 percent; and birds, about 2 percent. Outside the breeding season the biomass representation of small voles declined somewhat, the latter two prey categories totaling about 20 percent of the estimated biomass consumption (Cramp, 1985).

Although North American studies are far less extensive, a similar rodent-based dietary picture emerges. A rangewide survey of prey-analysis studies has been provided by Bull and Duncan (1993), using 4 regional studies, and involving a total of 7647 prey items. *Microtus* voles comprised 30.8–83.5 percent of the total items, *Thomomys* pocket gophers 0–57.9 percent, *Sorex* shrews 1.4–2.2 percent, *Peromyscus* mice 0–3.8 percent, *Clethrionomys* voles 0–13.9 percent, birds 0.1–2.2 percent, and other items 2.1–10.9 percent. Among the largest prey recorded are snowshoe hares and spruce grouse (*Falcipennis canadensis*). The greatest degree of dietary overlap with other owls involves medium-sized owls such as the long-eared owl rather than the great horned owl, which typically takes much larger prey. Juvenile great gray owls are sometimes killed by great horned owls, and even adults may occasionally be killed by them (Bull and Duncan, 1993).

Winter (1986, 1987) estimated the average weight of 662 prey items in California as about 75 grams, with pocket gophers (*Thomomys bottae*) contributing about 57 percent of the prey items and nearly 80 percent of the prey biomass. *Microtus* voles were of secondary importance, comprising 33 percent of the prey items and an estimated 17 percent of the total biomass. In Oregon, breeding-season prey consisted of about 58 percent *Microtus* voles and 34 percent pocket gophers (*Thomomys talpoides*) (Bull and Henjum, 1985). In the Grand Teton National Park area these 2 prey types likewise constituted 93 percent of the prey identified in one study (Franklin, 1985). The use of pocket gophers as summer prey has also been observed in Montana (Tryon, 1943). Limited observations in Quebec (Brunton and Pittaway, 1971) suggest that there the birds subsist almost exclusively on *Microtus* voles during winter, and *Microtus xanthognathus* comprised 66 percent of a sample of more than 200 pellets from Alaska, with other microtines contributing 28 percent, and miscellaneous mammals and birds the remainder. Oeming (1955) similarly reported a concentration of *Microtus* voles in Alberta. Both Bent (1938) and Nero (1980) suggested that other mammalian species such as squirrels, moles, rats, young rabbits and hares, and weasels are also taken, as well as birds, usually quite small but sometimes as large as ducks and grouse.

Great gray owls prefer to hunt in relatively open country where scattered trees or forest margins provide suitable vantage points for visual or auditory searching. Winter (1987) found that about 90 percent of monitored birds' time was spent within 124 meters of an open meadow. In the winter the birds hunt primarily in early morning and again from late afternoon to dusk, with little or no nocturnal activity, judging from Brunton and Pittaway's (1971) observations. Oeming (1955) also reported that, prior to the nesting season, most hunting is done in late afternoon, but while feeding young both daytime and nocturnal hunting may be done. Similar observations during winter in Finland suggest that the birds prefer to hunt at dusk, but modify their crepuscular tendencies to include daytime during midwinter, when the day is very short, and especially during dull, overcast days. On the other hand, during the short nights of summer at high latitudes the birds concentrate their foraging around midnight, although the great need for food during the nestling period may force the male to be active throughout the daylight hours (Mikkola, 1983).

There is good evidence that the great gray owl has remarkable visual acuity and is able to see small rodent prey running across the snow at distances of up to 200 meters. Additionally they are able to locate and capture live prey from deep beneath the snow by acoustic clues alone (Nero, 1980). This is done by dropping down from a perch or a nearly motionless hovering position above the invisible prey, reaching down with their legs, and crashing through the snow to depths of about 30 centimeters, rarely even to 45 centimeters. Snow cover strong enough on the surface to support an adult human can be penetrated in this way. Tryon (1943) saw an owl crash through the roof of a feeding runway of a pocket gopher's burrow to get at the animal below.

Social Behavior

As a seminomadic species, great gray owls would not be expected to have permanent pair bonds or strong nesting site tenacity, and this generally appears to be the case. If food is locally abundant over a period of years the females may return to nest at the same sites, with records of a nest used for as long as five years, but at other times they may move elsewhere. Similarly, some young birds return to breed near their natal areas, while others may breed as far as 100 kilometers away (Cramp, 1985; Mikkola, 1983). Judging from limited data, both hand-raised and wild females can sometimes breed at a year of age, but an age of at least two years might be expected for initial breeding, and first breeding at three years may be typical (Bull and Duncan, 1993). The pair bond is apparently monogamous but of unknown duration, and it is not maintained outside the breeding season (Glutz von Blotzheim and Bauer, 1980). However, at least along the West Coast, where winter wandering is probably not so common as in Canada, pairs are likely to remain together so long as both survive. Pair bonds with an earlier mate may also be reestablished the following year if prey populations remain high. Occasional polygyny has been suspected but not yet proven in great gray owls (Bull and Duncan, 1993).

When perched, the birds typically remain almost motionless while standing close to the main bole of the tree, where their barlike plumage pattern allows them to blend into their surroundings remarkably well. When aware of approach by humans, they may assume an upright, sleeked posture with the eyes remaining open and the breast rather than the wing directed toward the intruder (Figure 58, *right*). When the bird is about to attack an intruder, the bill is snapped, the head feathers are fluffed, and the wings are spread slightly and somewhat drooped before takeoff (Figure 58, *left*).

The two most evident aspects of courtship behavior in great gray owls are courtship feeding and mutual preening. Pair-forming behavior may begin as early as November or as late as two weeks before egg laying. Nero (1980) regarded mutual preening as one of the most significant aspects of pair-bonding behavior and found that it could be easily elicited from adults of both sexes as well as from subadults. Even badly injured owls would respond to his tilting the top of his head toward them by running their beaks through his hair, gently nibbling on the scalp, and often pulling on a few hairs. Similarly, Oeming (1955) observed mutual preening in captive birds. The birds would first stand with breasts touching and face-to-face as the male rubbed his beak over the female while uttering a humming sound; he would then circle her in a similar manner. Males

have also been observed "combing" the breast feathers of the female with their talons, and although males apparently initiate mutual grooming the female may actually groom her mate more (K. McKeever, quoted in Nero, 1980).

Courtship feeding begins in midwinter (lasting from January to mid-April in Manitoba), the female beginning to hoot softly and shifting her weight from leg to leg when she sees her mate carrying a prey animal. Stimulated by the female, the male flies to perch beside her, closes his eyes as he leans toward her, and holds out the prey for her to receive. The female seizes it with closed eyes and a slight mewing sound, thereby helping to form or reestablish the pair bond (Nero, 1980). J. Duncan (1987) reported seeing an immature male feeding a mated female at the nest, apparently representing the first record of possible nest-helping among owls, although the possibility of this has been suggested for long-eared owls.

Nero (1980) described one attempted copulation that occurred in late February. The male flew into a tree where he was shortly joined by the female, who perched on the same branch some 10 feet higher up. The male then flew and, cupping his wings, braked and dropped momentarily on the female's back. They then separated and flew away. In another incomplete observation the male was observed vigorously flapping his wings during copulation, while one or both birds uttered a peculiar rasping screech. Shortly after that the male flew away and the female resumed hunting.

Nest visits may begin as early as November in Manitoba, with the male uttering a nest-showing or advertisement call, while the female calls in response. When she visits the nest site she

Figure 58. Great gray owl behavior, including (*left*) preattack posture and (*right*) concealment posture. After drawings in Mikkola (1983).

often sits and makes scraping movements. The male may then fly off, followed by the female. He may thus show her several possible nesting sites, the final choice presumably being made by the female. Selection of a nest site may in part be influenced by the relative local prey population, and this factor may also affect the timing of initial egg laying (Cramp, 1985; Mikkola, 1983; Bull and Duncan, 1993).

Breeding Biology

Egg records in North America are rather limited, but 15 records from Alberta are from March 23 to May 15, with 8 occurring between April 9 and May 1. Three records from Alaska and arctic Canada are from May 15 to July 19 (Bent, 1938). In Alberta most nests have complete clutches by April 15, with the earliest record of a complete clutch being March 23 (Oeming, 1955). In Ontario eggs have been reported between April 29 and June 5 (Peck and James, 1983), and in the Sierra Nevadas of California breeding occurs from late February to mid-June, with a peak from mid-April to late May (Verner and Boss, 1980; Winter, 1986). Early April was reported as the earliest laying time by the Craigheads (1956) for the Grand Teton area, and Nero (1980) stated that laying may begin as early as mid-March, presumably referring to the area around Winnipeg. The mean date for laying is early April in Manitoba and early May in Wyoming and Idaho. Heavy snows may retard the start of laying. No second broods are produced, but there is one record of renesting after the loss of a brood of young nestlings (Bull and Duncan, 1993).

Of 185 nests found in Finland (Mikkola, 1983), about 83 percent were twig nests originally built by raptors or corvids; 13 percent were on stumps; and the remainder in miscellaneous locations. Of 106 nests, 45 percent were in "damp heath" coniferous forests; 35 percent were in spruce bogs; 11 percent in "dry heath" coniferous forests; and the remaining 9 percent in pine peat bogs or herb-rich forests. About half the nests had marsh areas located within 1000 meters, and nearly half had an area cleared by felling within 500 meters. The majority of the stick nests had originally been made by goshawks (*Accipiter gentilis*); those of buzzards (*Buteo buteo*) comprised the next most common category.

Franklin (1985) noted that 9 of 15 nest sites in the Grand Teton area were in broken-top snags, and almost 80 percent of the active nest sites were reused at least once. Of 52 nests in Oregon, half were in old raptor nests, 21 percent on artificial platforms, 19 percent on broken-top snags, and 10 percent in mistletoe clumps (Bull

and Henjum, 1985). All of 5 California nests were located on the tops of large snags (Winter, 1980). Of 32 Canadian (apparently mostly Manitoban) nests mentioned by Nero (1980), 16 were in man-made structures; 10 were in completely artificial nests; and 6 were in rebuilt natural nests. Oeming (1955) stated that in Alberta favored nesting areas are among poplar woods, which often are lightly mixed with conifers and usually are close to areas of muskeg that are used for hunting. Among 23 sites from Alberta, 15 were in aspens (*Populus tremuloides*), 3 in balsam poplars (*P. balsamifera*), 3 in black spruces (*Picea mariana*), and 2 in tamaracks (*Larix laricina*). They were typically in old, unmodified raptor or crow nests averaging 13 meters aboveground (Oeming, 1955). In spite of early statements to the contrary, there is no good evidence that the owls enlarge, line, or otherwise modify their nest sites in any way except to deepen the cup of the nest. There are a few records of ground nests in Canada (Bull and Duncan, 1993).

Females lay eggs at a rate of about 1 per day, although longer intervals may sometimes elapse, especially for the eggs laid later in a clutch. Among 241 European clutches the range of clutch size was 1–9 eggs, with an average of 4.4 (Mikkola, 1983). Twenty-three Alberta nests ranged from 2 to 5 eggs, with an average of 3.2 (Oeming, 1955). Evidently European clutch sizes increase from south to north, and they are also apparently influenced by local food conditions. Replacement clutches usually are laid 15–30 days after the loss of the first nest (Bull and Henjum, 1987). There are reports that in good vole years as many as three clutches may be laid, although only one brood per year is raised (Mikkola, 1983).

The female does all the incubation, which normally requires 28–29 days, while the male performs all the hunting duties, often in open areas only a few hundred meters from the nest. The female receives the prey from her mate with the bill and consumes it herself or, after the young have hatched, passes it on to them, after first tearing it to bits if the owlets are very small.

Hatching of the eggs typically occurs at intervals of from 1 to 3 days, with the young weighing about 37–38 grams at hatching. Within 5 days after hatching they will normally almost have doubled their hatching weights, and by two weeks old will have attained a weight of about 500 grams, which attests to the importance of an abundance of food at this time. There are cases of young increasing in weight from 40 to 225 grams in a single week. The owlets normally leave the nest at 20–29 days, when weighing 425–630 grams. By then they are surprisingly agile at climbing trees, even though they are incapable of flight. Actual fledging probably occurs before they are 55 days old, but even after this they are likely to remain near the nest. They stay within the nesting territory for some months, watched over by the female. They probably become independent and begin dispersing at about 4–5 months (Cramp, 1985; Mikkola, 1983). Great horned owls are apparently serious predators on young birds (Bull and Henjum, 1987). There is seemingly a high mortality rate of young birds; Nero and Copland (1981) noted that 88 percent of 50 great gray owls found dead one winter in Manitoba were young of the year. Among 193 owls found dead over a 15-year period, 157 were killed by collision with motor vehicles; 26 had been shot; and 10 died from miscellaneous causes (Nero, Copland, and Mezibroski, 1984).

Although adult great gray owls may consume about 150–200 grams of food per day on average, during a 50-day study period a young male and female averaged 76.4 and 80.6 grams of food, respectively. This provides some idea of the enormous weight and number of prey that must be provided by a pair of birds (and primarily the male) if they are to raise a brood successfully (Mikkola, 1983).

Among a sample of 42 Finnish nests whose clutch sizes were known, 80.5 percent of the eggs hatched, and 72.1 percent of the chicks left the nest, for an overall reproductive success rate of 58 percent. The average number of fledged young per successful nest was 2.4, with humans being responsible for the largest number of egg and chick losses (Mikkola, 1983). Among a sample of 69 nesting attempts in Oregon, 75 percent of first nestings were successful, with northern ravens a major cause of egg losses. The average mortality of radio-tagged juveniles was 46 percent during their first year, as compared with 8–29 percent for adults (range of 3 years) (Bull and Henjum, 1987). Franklin (1987, 1988) reported a 71 percent nesting success rate for 17 breeding attempts in the Grand Teton area, with an average of 2.5 fledged young per nest. Other nesting success rates (1 or more young fledged per nest) have ranged from 78 percent in Oregon to 81 percent in Minnesota and Manitoba (Bull and Duncan, 1993). In one study, the success rate for birds using artificial platform nests was higher (83 percent) than for natural nest sites (Bull and Henjum, 1990). The mean number of fledged young per nest in these several studies has ranged from 2.7 to 3.0. Franklin (1987) estimated that there is a 58 percent chance of an egg resulting in a fledged chick. Maximum known longevity under natural conditions is 12 years and 9 months (Patuxent Wildlife Research Center banding data).

Evolutionary Relationships and Conservation Status

The great gray owl is a quite distinct form and frequently has been given monotypic generic status by taxonomists. However, more recent classifications have placed it within the rather large genus *Strix,* albeit with no obvious close relatives. It seems possible that the Ural owl (*Strix uralensis*) and its southern counterpart the tawny owl are the nearest living relatives to the great gray owl; the great gray and Ural owls are widely sympatric in Eurasia.

The status of the great gray owl in North America is difficult to judge, but Nero (1980) made an educated guess that the total population may be in the neighborhood of 50,000 birds, most of which are certainly found in Canada. There have been reviews of the species' status in Manitoba (Nero, Copland, and Mezibroski, 1984) and Saskatchewan (Harris, 1984), as well as a California survey (Winter, 1980). Studies by Franklin (1985) have shown the species to be fairly common in northwestern Wyoming and adjacent Idaho, where he found evidence of 67 territories; in Oregon Bull and Henjum (1985) located more than 50 nests in three years. Breeding almost certainly occurs in Washington, but its occurrence in that state is virtually undocumented. There are several breeding records for Minnesota, at least one for northern Wisconsin, and several probable records from Michigan (Brewer, McPeek, and Adams, 1991). During the 1997–2000 Christmas Bird Counts the species was detected on an average of 25 counts, or about 1 percent of all U.S. and Canadian counts. In terms of all localities where the species was detected, the great gray owl was the 12th most widely reported of the 18 North American owl species then encountered. The maximum number ever reported on a single CBC was 28, at Sac-Zim, Manitoba, in 1996. The average total number of birds counted in all locations over the four-year period 1997–2000 was 71.5. In 2000 the total number reported was 206 birds, or well above the average of 27 for the previous three years, and the maximum number seen at a single location was 24 birds. That winter marked one of the largest recorded invasions of great gray owls into southern Canada and the northern United States in decades. During the 1990s the mean maximum total for a single location was 9.4 birds (for 9 years). During the 1980s the mean maximum number seen was 2.7 at any single location (for 8 years; these data were not summarized for 2 years). The Canadian population has recently been estimated at 10,000–50,000 pairs (Kirk and Hyslop, 1998) and is believed to be fluctuating but stable.

Long-eared Owl   *Asio otus* (Linnaeus) 1758

Other Vernacular Names:
American long-eared owl (*wilsonianus*); cat owl; western long-eared owl (*tuftsi*).

North American Range (Adapted from AOU, 1983.)

Breeds in North America from southern and eastern British Columbia, northern Yukon, southern Nunavut, northern Saskatchewan, central Manitoba, central Ontario, southern Quebec, New Brunswick, Prince Edward Island, and Nova Scotia south to northwestern Baja California, southern Arizona, southern New Mexico, northern Texas, central Oklahoma, Arkansas, Missouri, central Illinois, western and northern Indiana, northern Ohio, Pennsylvania, New York, and New England. Winters in North America from southern Canada south to southern Mexico. Also widely distributed in Eurasia and in northern Africa. (See Figure 59.)

North American Subspecies (Adapted from AOU, 1957.)

A. o. wilsonianus (Lesson). Resident from southern Manitoba east to Nova Scotia, and south to northern Oklahoma and Virginia.

A. o. tuftsi Godfrey. Resident from southern Nunavut, southern Yukon, and southern British Columbia east to Saskatchewan, and south to northwestern Baja California and western Texas.

Measurements

Wing, males 284–302 mm (ave. of 14, 292), females 288–303 mm (ave. of 11, 293.9); tail, males 121.5–157.5 mm (ave. of 14, 147.6), females 143.5–160 mm (ave. of 11, 151) (Ridgway, 1914). The eggs average 40 × 32.5 mm (Bent, 1938).

Weights

Earhart and Johnson (1970) reported that 38 males averaged 245 g (range 178–314), and that 28 females averaged 279 g (range 210–342). Breeding-season males from Montana, 223–304 g (ave. of 55, 261 g), females 289–409 g (ave. of 49, 337 g) (Marks, Evans, and Holt, 1994). The estimated male:female mass ratio is 1:1.14–1.29. Mikkola (1983) reported that 22 males and 20 females of the Eurasian population averaged 288 and 327 g, respectively. The estimated egg weight is 22 g. The estimated egg-to-female proportional mass ratio is 6.5–7.9 percent.

Identification

In the field. These birds spend the daylight hours perched near tree trunks in rather dense foliage, making them nearly invisible. If disturbed, the relatively thin body posture and erected ear tufts are quite distinctive. In flight, the ear tufts are nearly invisible, and the birds are very similar to the short-eared owl, but are generally darker throughout. The upper dark wrist markings are less contrasting with the rest of the wing than is apparent in the short-eared owl; the pale area at the base of the primaries is more orange toned. The trailing edge of the wing lacks the definite pale band typical of short-eared owls, and the wing tips are usually more distinctly barred with black and buff than in that species. The tail banding is also narrower and less evident. The usual territorial "song" of the male is an indefinite series of widely spaced *hoo* notes, with as many as eight seconds sometimes separating individual notes. Nonrepetitive wing clapping in flight is also commonly performed by territorial birds.

In the hand. The distinctive long ear tufts, the reddish brown facial disk with narrow black rim markings around the orange-colored eyes, and a contrasting lighter "moustache" and "eyebrow" area around the bill make for ready recognition. The ear openings are very large and asymmetrically located, and only the outermost one or two primaries are emarginated on the inner webs. Sex identification of adults may be verified by the presence of a brood patch in breeding females, and attempted but not always attained by a combination of linear and weight measurements, and the tendency for females to be richer and darker colored throughout. Males have generally paler and less heavily streaked plumages, and silvery, whitish, or pale buff wing coverts (Pyle, 1997b). The natal down is white or ochre, and the juvenal plumage is generally russet-tinted and broadly barred with blackish brown and grayish white, the latter predominating anteriorly. The juvenal remiges and rectrices are uniform in color and amount of wear in first-year birds; the rectrices are narrower than in adults and the pale cross-bars on both the remiges and rectrices are narrower and more numerous. The iris is initially brownish in juveniles, but becomes yellow to orange yellow by the third year (Pyle, 1997b).

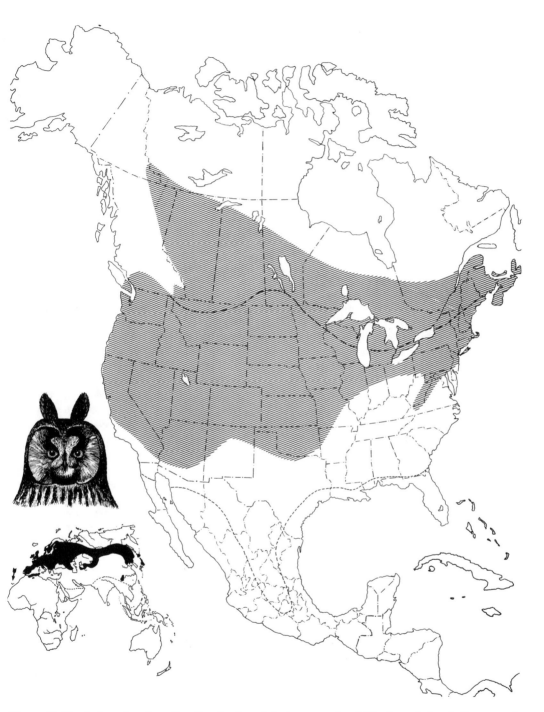

Figure 59. North American breeding distribution of the long-eared owl. The long dashes indicate usual northern wintering limits; the short dashes indicate usual southern wintering limits. Extralimital distribution shown in inset.

Vocalizations

The male's advertisement song consists of an extended series of low-pitched (ca. 400 Hz) *hoo* notes (such as those made by blowing over the top of a narrow-necked bottle) that are usually evenly spaced at intervals of about 2.5 seconds, though the intervals may be varied from about 2 to 8 seconds. Typically about 30 such hoots are uttered prior to a display flight, but from 10 to 200 may be uttered. The song is usually uttered by perched birds but may also be produced in flight or, uncommonly, from the ground. The song may also be uttered as part of a duet, with the female uttering her nest call, a variable nasal and buzzing note something like that produced by blowing through a paper-covered comb, or reminiscent of a sheep's or lamb's call (Mikkola, 1983; Cramp, 1985).

Males normally begin advertisement singing about dusk and may continue until about midnight, with a second and shorter period of vocal activity before sunrise. Their display flights also begin at dusk and end at dawn; these consist of erratic flight among the trees, with slow wingbeats interspersed with glides and occasional wing clapping below the body. Unlike the repeated wing claps of the short-eared owl, these are performed irregularly and nonrepetitively (Cramp, 1985). Complete song periods in Britain may last up to 5 hours, especially between January and March, when singing may occur nearly every evening. The song may be heard for up to a kilometer's distance; the longest song sequences are given prior to a display flight or when "marking" territories (Scott, 1997).

The female's nest call is most commonly uttered just prior to nest selection, and from then until egg laying occurs she calls regularly at intervals of about 2–8 seconds, often from the nest tree, apparently to attract the male or in response to his wing clapping. Female song may be heard at distances of up to about 200 meters; it lasts 1–2 seconds, or rarely longer, sounding like *shooo-oogh*. Unlike the male's lower-pitched hoots, it has an audible sighing element, especially toward the end, and overall has a distinctly nasal or burr-like timbre (Scott, 1997).

Males and more rarely females utter a cat-like hissing during or immediately after prey exchange, and both sexes utter barking and repeated *ooak* or *oo-ack* sounds when intruders approach the nest. A descending wheeze is uttered as a contact-alarm note, which in males sounds like a catlike hiss and in females is more querulous and higher in pitch. Typical defensive hissing notes similar to those of other owls are also produced. Warning calls, which immediately silence the young, are uttered as long, repeated *psii* notes or as short, loud *chawit* calls (Cramp, 1985).

Twittering notes accompany mutual preening, and during copulation a series of clear notes is produced by one or both of the participants, perhaps only the female. A variety of other notes have also been described but are rather difficult to classify, including sounds made during or after food transfer, while feeding young or toward begging young perhaps to quiet them, during attacks on predators, and in other situations. These total perhaps at least 12 adult vocalizations, plus several additional calls typical of nestlings. Sonograms representing many of these have been published (Cramp, 1985; Glutz von Blotzheim and Bauer, 1980).

Habitats and Ecology

At least during the breeding season long-eared owls are closely associated with fairly open woodlands, forest edges or patches, or similar partially wooded habitats of coniferous, deciduous, or mixed composition. In Britain, more than 200 nesting areas were found to be most often (33 percent) associated with small tree plantations, copses, or scattered trees on moorlands, heath, or mosses, followed by blocks of deciduous, coniferous, or mixed woods (24.5 percent), small plantations, shelter belts, or hedgerows in various agricultural settings (24 percent), and scrub or wooded clumps on coasts or wetlands (15 percent) (Glue, 1977a). In Finland all of 91 nests were no more than 500 meters from cultivated land of some type, and in areas of larger woods and forests the birds occupy only their very margins (Mikkola, 1983). In Ontario the birds breed most often in dense coniferous woods and reforestation groves that are more often wet than dry, and progressively less often in mixed and deciduous woods (Peck and James, 1983). Optimum habitat for the species may perhaps be characterized as including open spaces with abundant prey and short vegetation suitable for hunting, plus nearby wooded cover providing both roosting and nesting opportunities (Glutz von Blotzheim and Bauer, 1980).

The Craigheads (1956) judged that the presence of coniferous woods seemed to be a very important part of winter habitat needs for roosting cover, with the birds fanning out from such places to hunt in open fields. Bent (1938) similarly noted an association between long-eared owls and coniferous forests in eastern North America. However, he also noted that in the West and Southwest they seemed equally adapted to deciduous timber associated with lakes and streams, especially where heavy growths of climbing vines (providing dense roosting cover) also occur. R. Stewart (1975) characterized their habitat in North Dakota as fairly dense thickets or groves of small trees and the brushy margins of more extensive forested tracts. In the Sierra Nevadas of California the species extends from blue oak (*Quercus douglasii*) savanna habitats upward into ponderosa pine and black oak (*Q. kelloggii*) types, providing there is nearby riparian habitat (Verner and Boss, 1980).

Normally the breeding density of long-eared owls is fairly low, from 10 to 50 pairs per 100 square kilometers, with associated territory sizes in Finland typically between 50 and 100 hectares (Mikkola, 1983). However, density estimates are sometimes complicated by the presence of non-breeders; thus in Scotland about 17 percent of the 9–18 pairs per 10 square kilometers were non-breeders (Village, 1981). In central Europe a typical density might be in the range of 10–12 pairs per 100 square kilometers (Cramp, 1985; Glutz von Blotzheim and Bauer, 1980). Estimated densities in Michigan and Wyoming varied from about 10 to 100 pairs per 100 square kilometers, and the estimated home range of birds in Wyoming riparian habitat was judged to be about 55 hectares (J. Craighead and Craighead, 1956). Knight and Erickson (1977) estimated the breeding density along the Columbia River to be about a pair per 12 linear kilometers of riverine habitat. Along the Snake River in Idaho an average of 0.28–0.42 nesting pairs per square kilometer was estimated, as compared with other estimates of from 0.64 to 1.55 pairs per square kilometer elsewhere in southern Idaho (Marks, 1986).

Although usually well dispersed and territorial during the breeding season and perhaps sometimes locally territorial throughout the year (probably depending upon food supplies), breeding birds often forage outside their defended areas, which centers on the immediate vicinity of the nest. Additionally during winter aggregations of birds will often cluster in favored roosting areas, from which they forage outwardly over sometimes rather large home ranges that are evidently uncontested areas (Cramp, 1985). In various parts of Britain and Europe the numbers of owls thus aggregated have ranged from about 6 to 50 birds, the number doubtless varying in response to the distribution and abundance of food supplies (Mikkola, 1983). Similarly, roosts of from 7 to 50 birds have been observed in various parts of North America at least as far south as Arizona (Bent, 1938). Favorite roosting sites are sometimes used year after year by similar numbers of owls, suggesting a high level of roost fidelity. One such roost site consisted of densely foliaged conifers ranging from 4 to 8 meters high, with branches almost reaching the ground. The birds roosted low in these trees in cold weather, and higher up under warmer conditions (D. Smith, 1981). In another four-year study, the number of owls present in a New Jersey roost varied inversely with mean winter temperatures, but there was a strong fidelity to specific roost trees each winter. These were conifers that had dense foliage concealing most or all of the main trunk. Communal roosting was preferred by the owls to solitary roosting, perhaps because of reduced predation risks (Bosakowski, 1984). Several other hypotheses to explain communal roosting in owls have been proposed but remain unproven.

Movements

There can be no doubt that some degree of actual migratory or nomadic activity occurs among North American long-eared owls. Bent (1938) mentioned regular movements in Florida during the winter, the presence of winter roosting groups south of the breeding range in Arizona, and two cases of birds having been banded and recovered at great distances. Houston (1997) reported that a long-eared owl banded in Saskatchewan was later recovered (perhaps in the same year) in Oaxaca, Mexico, some 3890 kilometers away. Another Saskatchewan bird was recovered five years after banding about 2370 kilometers away, in Mississippi. One bird banded during summer in Alberta was recovered two winters later in Utah, and another, banded during spring in California, was recovered the following fall in Ontario, a most unusual displacement. One remarkable bird, banded as a nestling in Scotland, was recovered dead in Iceland, 13 years later and1230 kilometers distant, where breeding by the species is unknown. Another bird, banded as an adult in Great Britain, was recovered dead a month later in the former USSR, 3279 kilometers away (Scott, 1997).

Banding data in central and western Europe indicate that there is a general random dispersal during the first fall and winter (rarely to distances of 1200 or even 2300 km), but older birds either tend to overwinter on breeding areas (where weather permits and rodents are abundant) or congregate elsewhere as necessary to find abundant food. These winter aggregations may primarily involve first-winter birds, although this is still unproven. In Great Britain individual birds will sometimes show breeding site fidelity in successive years or overwinter in the same patch of woods. However, in Finland as many as 80 percent of the birds are nomadic, breeding only when and where food supplies permit. Thus there are marked year-to-year oscillations in breeding densities and clutch sizes as *Microtus* populations vary annually (Mikkola, 1983; Cramp, 1985).

Foods and Foraging Behavior

An extensive review of long-eared owls' prey selection both in Europe and North America has been provided by Marti (1976), who noted that voles (*Microtus* spp.) are the most common prey in most (31 of 42) studies, comprising 30–84 percent of prey items, while *Peromyscus* mice were most common in 5 North American studies and *Perognathus* mice in two others from western North America. Mammals accounted for 98 percent of almost 24,000 North American prey items and 89 percent of more than 37,000 European prey. At least 45 species of mammals (up to the size of hares) have been reported as prey in North America, and at least 23 species in Europe. Birds were found in higher frequencies (10.9 percent vs. 1.7 percent) in Europe, with house sparrows (*Passer domesticus*) the most commonly reported avian prey in both areas. All told, 35 species of birds (up to the size of grouse) have been identified among prey items in North America, as compared with 55 from Europe. Other types of prey occurred in insignificant quantities. The mean weight of all the prey items from North America was 37 grams, and 32.2 grams for all the European prey. Daily consumption by adult wild birds has generally been estimated as about 40–60 grams, or the equivalent of 1–3 average-sized prey. Throughout its range the species appears to be a rather specialized (stenophagous) feeder, being dependent on a relatively few species of small mammals regardless of habitat or location. Marks (1984) reported that, judging from 4208 prey items from Idaho, the owls fed opportunistically on a variety of small mice, with considerable food variation between sites but not between years, and with prey size apparently being the most important single factor in food selection.

A broad survey of food studies from North America was provided by Marks, Evans, and Holt (1994), summarizing data from 10 states and from 16 studies. The total number of prey items in this summary approached 35,000, and in 8 of the studies *Microtus* was the most common prey, followed by *Peromyscus* in 5, *Perognathus* in 2, and *Dipodomys* and *Thomomys* in 1 each. The incidence of mammals ranged from 93.5 to 99.9 percent, with birds comprising the rest. Locally or occasionally lizards, snakes, and mammalian orders other than rodents are taken, including bats, rabbits, moles, shrews, and even small weasels. Bird prey are usually no larger than small passerines,

but a ruffed grouse (*Bonasa umbellus*) has also been reported as prey. Compared with some sympatric owls, the long-eared seems to be more of a generalist in its prey selection (Marti, Steenhof, et al., 1993). Extensive data are available on the foods of the long-eared owl in Europe and Britain, as summarized by Glutz von Blotzheim and Bauer (1980), Mikkola (1983), and Cramp (1985). Samples from central Europe are dominated by common voles (*Microtus arvalis*), which form about two-thirds of more than 120 prey items, with other small mammals comprising about 26 percent, birds about 8 percent, and other materials in trace amounts. Data from Britain and northern Europe show a slightly lower proportion (47–65 percent) of common voles, but microtid voles as a group nevertheless dominate the samples, with murid mice and rats of secondary importance, while shrews and birds are minor supplemental components. On a relative biomass basis the percentage of the diet made up of murine or microtine rodents varied from about 67 to 98 percent of the total in 12 different European and British areas, with larger mammals, mostly comprised of *Arvicola* voles and rats (*Rattus*), ranging up to 20 percent of the total, birds up to 17.4 percent, and shrews rarely more than 2 percent. Apparently larger mammals and birds are the major alternative food whenever or wherever small rodents are unavailable, and shrews are generally avoided (Cramp, 1985).

Hunting is done by quartering open grounds at heights of about 50–150 centimeters. Prey is captured by the bird suddenly stalling, then dropping down with talons spread, thereby pinning the animal to the ground as it absorbs the shock of the bird's weight. Sometimes the bird may hover briefly over vegetation before plunging down on prey (Mikkola, 1983).

Observations by Getz (1961) in Michigan indicated that the birds prefer hunting in open, grassy situations such as old fields, rather than in timbered areas that might be closer to their roosts and offer a larger potential food supply or in mammal-rich marshy areas having extensive ground cover that helps to shield the prey from view. The birds are normally nocturnal hunters, typically leaving their roosts less than an hour after sunset and terminating their first phase of hunting before midnight. A second phase begins after midnight and ends less than an hour before sunrise. However, in high latitudes the short summer nights may force the birds to hunt in bright sunlight, and daytime hunting during summer has occasionally been seen in lower latitudes (Mikkola, 1983), perhaps in conjunction with foraging pressures associated with the feeding of dependent young.

Social Behavior

In areas where these birds are sedentary and remain on their territories throughout the year it is likely that they are essentially permanently monogamous, with the pair bond being renewed annually. Marks, Dickinson, and Haydock (1999) confirmed monogamous mating by testing 59 nestlings in 12 nests and found no evidence of extra-pair fertilizations in Montana. However, there is a single report of a male paired with two females on a territory in the Netherlands, and Scott (1997) also reported a case of apparent bigyny in England, the females' two nests located only 20 meters apart. Both nests were successfully hatched, each producing and fledging two chicks. Additionally, second broods in a single season have been reported several times in Britain (Scott, 1997) but apparently not yet in North America.

It is common for the same nesting area to be used year after year and occasionally for the same nest site to be used in successive years, although a nearby site may also be utilized (Cramp, 1985). Observations in Idaho (Marks, 1986) indicate that nest sites are most likely to be reoccupied in the following year if they were successful the previous season (48 percent of successful nests being reused), strongly suggesting reuse by the same birds. Marks (1986) found that four males nested only 0.5–1.5 kilometers from their natal sites, and he judged that loose nesting colonies may develop from such philopatric tendencies. He further suggested that cooperative nest defense, and possibly even the feeding of young other than their own by adults, might result from this trait.

Sexual maturity occurs in the first year of life, and courtship begins with the male starting to utter his territorial advertisement calls. In Britain resident birds may begin calling as early as late October or early November, but become especially active after the first of the year. Other males may occupy and begin advertising territories in March and April (Cramp, 1985). Territories of paired birds are probably first occupied by the male, followed a few days later by the female.

During the period of territorial defense and associated advertisement calling the male also performs somewhat circular or zigzag display flights above his territory, occasionally calling in flight. Such flights begin at dusk and end at dawn, the bird flying at near treetop level, with undulations caused by alternate flapping and gliding, and producing wing-clapping noises with every third or fourth wingbeat. Up to 20 or so such claps below the body may be performed in a single flight, but since these are done singly rather than in rapid succession little if any altitude is lost. Females also clap their wings, but much less frequently than males, and sometimes the two will fly side by side in upward ascent, followed by duetting wing-claps on their downward arcs (Scott, 1997). At peak intensity, males perform their display flights while circling over prospective nests, in which the females sit and respond with their nest calls (Cramp, 1985). Or, the female may utter "sighing" calls as the male wing-claps above and around her, or hovers directly in front of her. Finally, he may land on her back and attempt to copulate (Scott, 1997). From the time that territories are occupied the pair begins to roost close together, the female sometimes sharing the male's roost site. Later, when the nest site is selected, the female is likely to roost there, with the male roosting some distance away.

Nests are selected by the female as soon as she joins the male on his territory, by running and maneuvering through the tree branches as the male circles nearby. Once having selected the site she makes circular flights around it, sometimes wing clapping. She also flies repeatedly to the nest and sits there for long periods (Cramp, 1985). Copulations often occur near the nest. The male precedes copulation with calls and display flights, sometimes followed by strong wing-waving signals and tilting body movements performed while perching near the female or on the nest (Figure 60). The female lies flat across her perch, usually a branch, her wings spread and slightly drooped, and the male quickly mounts, extending his wings to maintain balance during treading (Mikkola, 1983). Treading usually lasts only a few seconds, but may take up to several minutes (Scott, 1997). Copulation has also been observed occurring on the ground, when it was preceded by a precopulatory duet (advertising call by male, nest call by female). The male then displayed aerially and glided to the ground, followed shortly by the female, after which copulation occurred (Cramp, 1985). Copulations are fairly regular from the egg-laying period through early chick-rearing and may even occur during the fledging period of the young (Scot, 1997). This would suggest a possible role for copulation in maintaining pair bonds.

Mutual nibbling or preening is frequent both before, after, and sometimes even during copulation, and self-preening is common afterward. Mutual preening occurs most frequently just before egg laying but is common during nest site selection (Scott, 1997). Preening of the female by the male was observed while she was tending nestlings (Cramp, 1985). Courtship feeding of the female by the male is also a basic part of social behavior, beginning before the female starts incubation and continuing until the young are well grown.

207

Figure 60. Long-eared owl behavior, including (*A–B*) precopulatory male calling postures, (*C*) copulation, and (*D*) concealment posture. After drawings in Mikkola (1983).

Like most owls, the long-eared has an impressive defensive posture when trapped or defending young, involving tail and wing spreading and producing a nearly circular outline when viewed from the front. When attempting to remain inconspicuous on a roost the bird adopts a sleek upright posture, with the "ears" fully extended (the black streaks formed by these feathers extending down through the eyes almost uninterruptedly, terminating in the black facial disk patterning). The nearer folded wing is brought forward and upward to cover the flanks, and the entire appearance of the bird is transformed so as to closely resemble a bark-covered vertical extension of the tree branch (Figure 60).

Although the male long-eared owls demarcate their territorial boundaries with calls and display flights, they are surprisingly tolerant of conspecific neighbors, and boundary fights are unreported. Because of this high tolerance level, nests of long-eared owls are sometimes fairly close together, but the species is by no means a colonial nester (Scott, 1997).

Breeding Biology

The breeding season of long-eared owls in North America is fairly prolonged, with 42 egg records from New England, New York, New Jersey, and Pennsylvania extending from March 14 to May 30. Eight egg records from Indiana, Illinois, and Iowa are from March 20 to April 28; 21 from southern Canada are from April 12 to June 5; and 58 from California are from March 1 to May 23. The peak for most of these regions appears to be between mid-March and mid-May (Bent, 1938). A sample of 43 egg dates from Ontario are from March 19 to May 24, with half between April 15 and May 5 (Peck and James, 1983). There are indications in North America that renesting will sometimes occur within about 20 days following the loss of a clutch (Bent, 1938), and in Britain and Europe there are some reports that two broods may even occasionally be raised successfully (Cramp, 1985; Mikkola, 1983).

Clutch sizes in North America collectively average about 4.5 eggs, with a tendency for the clutch size to increase northward as well as westward (Murray, 1976). This is similar to the average reported for central Europe but greater than that determined for Britain and smaller than that typical of Scandinavia (Mikkola, 1983, Cramp, 1985).

Studies on nest site selection in Britain (Glue, 1977a) and Finland (Mikkola, 1983) indicate that old nests of carrion (hooded) crows (*Corvus corone*) and magpies (*Pica pica*) are by far the most frequently utilized substrate, comprising 84 percent of 239 and 95 nests in the two locations, respectively. Other large birds' nests, plus dreys of squirrels, made up nearly all the rest, with a few in clusters of natural tree growths, natural tree cavities, or on the ground in heavy cover. Nest boxes or nesting baskets constructed for various other birds have at times also been used. Conifers provided sites for 74 percent of 194 British tree nests and 66 percent of 101 Finnish nests, with nests situated an average of 6.7 and 8.2 meters aboveground, respectively.

Nest site data for North America are not so extensive, but of 112 nesting attempts in Idaho all the sites were old corvid stick nests in trees, with nest heights (averaging 3.2 m) and nest diameters (averaging 22.3 cm) apparently being important criteria for determining suitability as nesting sites for long-eared owls (Marks, 1986). Of 48 Ontario nests, the majority (40) were in conifers, which were nearly always living and were most often in pines (*Pinus*) or cedars (*Juniperus*). The heights of 57 sites ranged from 2.5 to 18.5 meters, but usually were between 5.5 and 9.0 meters, and most often were old crow nests (Peck and James, 1983).

Egg laying is done irregularly, with intervals of 1–5 days between eggs, so that a clutch of 7 eggs might be laid in 10 or 11 days. Incubation usually begins immediately with the laying of the first egg. Although it is normally done by the female alone, the male may rarely sit on the eggs for short periods. It usually lasts for 25–26 days, but sometimes up to 30 days. Early in the incubation period the female may leave each evening to feed for a time, but later on in incubation she is much more reluctant to leave the nest. Hatching occurs over an extended period, which may be as much as 11 or 12 days in a nest with 6 owlets. While the nestlings are still quite young the male may continue to call during early morning and evening hours. Injury feigning by adult owls, apparently of unknown sex, has also been observed (W. Armstrong, 1958).

The hatchlings, whose eyes open after 5 days, are covered initially by white down. By a week after hatching black feathers begin to appear around the base of the bill, and bill snapping may accompany disturbance. After another 5 days the entire face is a black mask, and belly striping is evident. When the owlets are 20–26 days old they typically leave the nest and begin "branching," although actual fledging does not occur until they are 30–40 days old and their remiges and rectrices have become well developed. They gradually become independent of their parents when about 2 months old, but may continue to make food-begging calls until about 50 days of age (W. Armstrong, 1958; Mikkola, 1983).

Nesting success of more than 112 nests studied by Marks (1986) averaged 46 percent in 2 successive years, with a minimum of 3.7 young fledged per successful nest and a possible maximum of 4.15 fledged young. Most nest losses were the result of predation, with relative access by raccoons (*Procyon lotor*) apparently an important factor in nesting success. Nesting success in 4 North American studies summarized by Marks, Evans, and Holt (1994) ranged from 70 percent to 100 percent, and the mean number of young fledged per successful nest from 3.0 to 4.5. Houston (1997) noted that 139 broods in Saskatchewan averaged 3.6 young at banding during a half-century of study, but in 2 years of high vole populations they averaged 4.1 and 4.5 young per successful nest.

Of 78 nesting efforts in Britain, 59 percent failed completely, usually by losses of clutches (Glue, 1977a). In another study of 58 pairs studied over a 4-year period, 83 percent laid, 63 percent hatched 1 or more young, and 57 percent were able to fledge young, with an average of 3.2 young fledged per successful nest (Village, 1981). The birds bred at higher densities, earlier, and more successfully during years when voles were abundant than when they were scarce (Village, 1985). Scott (1997) reported that the average number of young fledged per nest in Britain increases with clutch size up to 5 eggs (3.9 young) but declines with 6-egg clutches to 3.5 young. The overall mean success for 68 nests and 222 hatched eggs was 3.02 fledged young per nest (Scott, 1997). Nesting success levels seem to be generally higher on the continent, at least in Germany and Spain, than in Britain or Finland, perhaps because the species is better adapted to environmental conditions in central Europe (Mikkola, 1983). Of 193 British long-eared owls banded and recovered through deaths, 56 percent were dead within a year, 36 percent were recovered at 1–5 years, 6.7 percent at 5–10 years, and only 1 percent at more than 10 years. Maximum known longevity in North America under natural conditions is 11 years and 1 month (Patuxent Wildlife Research Center banding data). There are somewhat questionable records of a 15-year-old bird in Britain (Scott, 1997) and an amazing 27-year, 9-month record of longevity for Europe (Cramp, 1985).

Evolutionary Relationships and Conservation Status

Probably the nearest living relative of this species is the stygian owl. It is substantially darker in color than the long-eared owl and generally is associated with more shaded and more humid environments.

In Europe this species appears to fluctuate markedly with rodent populations but has decreased in Britain and possibly elsewhere (Cramp, 1985). Probably there has been a comparable decrease in North America, mainly as a result of forest cutting and destruction of grovelands and riparian habitats, especially in western states such as California (Verner and Boss, 1980). It is state-listed as endangered in Illinois, threatened in Iowa, and a species of special concern in California, Montana, North Dakota, Wisconsin, Missouri, and throughout all of New England except for Maine (Marks, Evans, and Holt, 1994). During the 1999 and 2000 Christmas Bird Counts the species was detected on an average of 158 counts, or about 9 percent of all U.S. and Canadian counts. In terms of all localities where the species was detected, the long-eared owl was the seventh most widely reported of the 18 North American owl species then encountered. The average total number counted among the 100 counts having the highest species totals was 381 birds, or about 4 percent of all owls reported in the tabulated summaries of counts for the United States and Canada during 1999–2000. The maximum number reported on a single CBC was 65, at Hamilton, Ontario, in 1961. The Canadian population has recently been estimated at 10,000–20,000 pairs (Kirk and Hyslop, 1998) and is believed to be stable or declining.

Stygian Owl *Asio stygius* Wagler, 1832

Other Vernacular Names:
Búho-cornudo oscuro; Lechuza estiga (Spanish).

Range

Ranges residentially in Mexico along the Pacific slope from Durango south locally to Guerrero and Chiapas, and along the Atlantic slope from central Veracruz south locally to Guatemala; also known from Cozumel Island and Belize, where its residential status is uncertain. From there it extends south through Nicaragua and Honduras discontinuously to northern South America and patchily east of the Andes south to northeastern Argentina. It also occurs in Cuba and on Hispaniola. (See Figure 61.)

North American Subspecies

A. s. robustus Kelso. Mexico south to Belize and Nicaragua.

Measurements

Wing, male 305 mm, 2 females 340–347 mm (ave. 343.5); tail, male 157 mm, 2 females 169.5–171 mm (ave.170.2) (Ridgway, 1914). No mensural information is available on the eggs.

Weights

One female from Colombia weighed 675 g (A. Miller, 1952). One male from Belize, 391 g (S. Russell, 1964). These few numbers suggest that a substantial mass difference between the sexes may exist, but much more data from birds of the same region are needed.

Identification

In the field. This impressive owl might be easily confused with a long-eared owl, although it is larger and much darker overall, especially on the underparts. It also has a dark charcoal-black, rather than brown, facial disk. The forehead has a distinctive contrasting whitish diamond-shaped or triangular patch extending up from the eyes and narrowing to a point between the long "ears." A very narrow whitish border also outlines the facial disk, and white spotting becomes more prominent on the otherwise chocolate brown breast and sides. The upperparts are slightly flecked with white, and the primaries and tail are barred with grayish and white, respectively. The eyes are dark yellow. See also "Vocalizations."

Figure 61. Distribution of the stygian owl. Mexican distribution after Howell and Webb (1995). The inset map shows the South American distribution.

211

In the hand. This is a medium-sized (wing 300–350 mm), long-tufted owl, with an unusually dark overall plumage, and the only Mexican owl with dark ear tufts and an equally dark facial disk. It has feathered tarsi but naked and relatively small toes, and a rather weak and narrow bill. Like other *Asio* species it has very large external ears that are asymmetrically developed and have large opercular flaps anteriorly. Natal and later developmental plumage stages are undescribed, as are sexing criteria for adults.

Vocalizations

Like the long-eared and striped owls, which have extended series of well-spaced notes as territorial songs, males of this species utter usually loud *wuof* notes, each with a downward inflection, at intervals of several seconds (6–10 sec reported in northern Mexico; shorter intervals reported from Belize). The call is somewhat like that of a large dog's slightly threatening "woof." Females utter more catlike notes in response, and both sexes utter extended series of scratchy notes when excited. Tape recordings of the species' song from Belize (race *robustus*) and several South American sites were provided by J. Hardy, Coffey, and Reynard (1999). Examples of the song recorded from Belize indicate intervals of 3–4 seconds between the individual *wuof* notes, which are acoustically quite similar to those of the striped owl and long-eared owl.

Habitats and Ecology

Stygian owls are montane owls in Mexico, with a preference for pine and pine-oak forests, so their breeding range closely corresponds with that of these forests. They have also been described as a cloud forest species. Their usual elevational limits in Mexico are from about 1000 to 3000 meters (but sometimes they occur much lower, perhaps as nonbreeders). Their altitudinal range is considerably lower in Belize, where the highest mountains only reach about 1000 meters. In South America they range widely, over elevations from about 100 to 3000 meters.

Movements

Apparently residential and sedentary, although it is possible that some wandering by immatures or nonbreeders might occur, perhaps accounting for some U.S. and Cozumel Island occurrences.

Foods and Foraging Behavior

Small mammals, including bats, and birds are common prey for this nocturnal hunter, as are reptiles, crustaceans, and larger insects. In Haiti a ground dove (*Columbagallina*) was reportedly eaten. However, no detailed analyses of food intake are available for this elusive species.

Social Behavior

Almost nothing is known of the social behavior of these birds.

Breeding Biology

There is no information on breeding in Mexico. In Cuba, one nest was found on the ground, among shredded palm leaves, while another was in a tree. Stick nests of other species are sometimes used, and the typical clutch seems to be of two eggs. A breeding season of April and May has been suggested for Hispaniola.

Evolutionary Relationships and Conservation Status

Part of the large genus *Strix,* the stygian owl seemingly is essentially a very dark version of the long-eared owl, and thus must be a close relative.

The stygian owl is closely associated with pine and pine-oak forests, and although generally believed to still be fairly common, its fate is conjoined with that of these valuable and relatively accessible (to loggers) forest types in Mexico. The Cuban and Hispaniolan populations are now extremely rare and perhaps endangered, mainly owing to deforestation but also to persecution by superstitious natives, who regard these ominous-sounding owls as evil omens. Like other nocturnal owls, this species' effective concealment during the day makes the actual status of this species almost impossible to judge with any degree of certainty. Stotz et al. (1996) listed it as occurring in at least 13 Latin American countries and patchily distributed to rare. It is considered endangered by the Mexican government (Enriquez-Rocha, Rangel-Salazar, and Holt, 1993).

Short-eared Owl *Asio flammeus* (Pontoppidan) 1763

Other Vernacular Names:
grass owl; marsh owl; northern short-eared owl (*flammeus*); prairie owl.

North American Range (Adapted from AOU, 1983.)

Resident or variably migratory in North America from northern Alaska, northern Yukon, Nunavut, northern Quebec, northern Labrador, and Newfoundland south to the eastern Aleutian Islands, southern Alaska, central California, northern Nevada, Utah, northeastern Colorado, Kansas, Missouri, southern Illinois, northern Indiana, northern Ohio, Pennsylvania, New Jersey, and northern Virginia. Additionally resident on Hispaniola and adjacent islands; also breeds in Cuba, in South America and in Eurasia. The northernmost populations are migratory during winter, with individuals of unknown origin sometimes wandering to the southern United States and central Mexico, rarely to Guatemala. (See Figure 62.)

North American Subspecies

A. f. flammeus Pontoppidan. Distributed in North America as described above and also widely distributed in the Old World.

Measurements

Wing (of *flammeus*), males 298–330 mm (ave. of 23, 312.9), females 300–326 mm (ave. of 16, 312); tail, males 136.5–161.5 mm (ave. of 23, 148.3), females 142–158.5 mm (ave. of 16, 152) (Ridgway, 1914). The eggs average 39 × 31 mm (Bent, 1938).

Weights

Earhart and Johnson (1970) reported that 20 males averaged 315 g (range 206–368), and that 27 females averaged 378 g (range 284–475). Mikkola (1983) reported that 10 males and 4 females of the Eurasian population averaged 350 and 411 g, respectively. Twenty-eight incubating and brooding females from Montana averaged 411.8 g (range 333–553 g) (Holt and Leasure, 1993). The estimated male:female mass ratio is 1:1.17–1.2. The estimated egg weight is 19.5 g. The estimated egg-to-female proportional mass ratio is 4.7–5.1 percent.

Identification

In the field. Usually found in much more open country than the long-eared owl, the only other North American owl of similar size and appearance. This species is often seen in low flight over marshes or prairies, where its dark wrist markings on the upper and lower wing surfaces are apparent (also present but usually less conspicuous in long-eared owls), and the pale area at the base of the primaries is more buffy or tawny than rusty. A mothlike flight is characteristic, with a more wavering movement than is typical of the long-eared owl, the wingbeats being shallower and slightly more rapid. Additionally, the birds appear paler overall in color, especially ventrally, and have a conspicuous pale buffy trailing edge on the secondaries, together with more strongly contrasting blackish wrists and wing tips (the primaries usually forming small buffy "mirror" areas beyond the blackish wrists rather than having definite cross-banding basally). Although the birds often roost singly or communally in trees during winter, at other times they more often land amid tall grassy vegetation where they are usually impossible to see. When perched, especially on the ground, the birds assume a more slanting and less upright posture than other similar-sized owls.

In the hand. This medium-sized owl is recognizable by its short, inconspicuous (usually invisible) ears, and a rounded facial disk in which the yellow eyes are surrounded by fairly thick blackish rims but a whitish outer facial disk, as compared with thinner blackish rims and a more rusty outer disk in the long-eared owl. The dark breast stripes lack the long-eared's anchorlike lateral extensions, and the upperparts are generally tawnier (ochraceous) than those of the long-eared owl, which tend more toward grayish. As with the long-eared owl, the wing is relatively long and narrow (the ninth primary the longest), with only the one or two outermost primaries emarginated, and the ear openings are asymmetrically situated. The presence of a brood patch in breeding females may facilitate sex identification. Sex identification of adults may also be attempted by a combination of linear and weight measurements, and the tendency for females to be richer and darker throughout, with heavy brown streaking on the underparts and large brown spots on the creamy to buff underwing coverts. Males have generally paler and grayer plumages, sparse underpart streaking, and more whitish and less spotted underwing coverts (Pyle, 1997b). The natal down is light buff dorsally, whiter below, and darker on the sides of the mantle (Mikkola, 1983). The juvenal plumage is dark sooty brown dorsally, the feathers broadly tipped with ochraceous-buff

213

Figure 62. North American breeding distribution of the short-eared owl. The long dashes indicate usual northern wintering limits; the short dashes indicate usual southern wintering limits. Extralimital distribution shown in inset.

or russet tones. The face is mostly uniform brownish black, with a distinctive white "moustache" and "beard." The underparts are wholly plain pale dull ochraceous or buffy, tinged anteriorly with dark grayish. Little reliable ageing information is available for this species; but in juveniles the iris color is brownish to brownish yellow, and the remiges and rectrices are probably uniform in color and wear. Other age criteria such as those mentioned for the long-eared owl probably also apply (Pyle, 1997b).

Vocalizations

This is a relatively silent owl, a characteristic perhaps at least partly associated with its relatively diurnal activity pattern and its open-country rather than forested habitats. The male's territorial advertisement song is a pulsing series of numerous brief notes that sound like *voo-hoo-hoo* . . . or *boo-boo-boo.* . . . It often lasts about 4 seconds and resembles the noise made by an old steam engine. Its fundamental frequency is less than 500 Hz, with overtones up to about 1000 Hz, thus it carries well in open country in spite of its seemingly low volume. The notes are uttered at a rate of up to 4 per second, occasionally in series of up to about 20 notes, and the series may be repeated 5–6 times an hour. Song is most often performed during a display flight, often when the male is flying into the wind. It may also be performed by perching birds, with the head pumping slightly and the gular area swelling and deflating in rhythm with the calling. Besides being used as a territorial signal it may also be uttered during precopulatory display and by males perched near incubating females (Clark, 1975; Mikkola, 1983). Females typically respond to the male's advertisement song with a barking *keee-ow* call, a call that is also uttered by males. In both sexes it is the commonest call, serving a variety of functions, such as during threatening situations or in alarm, when it may be uttered 3–4 times in rapid succession. In the male this call tends to be more disyllabic than in the females and lower in pitch (Clark, 1975; Mikkola, 1983).

When adults are threatened at the nest, they produce a variety of shrill notes, barking sounds, and raucous squawking noises. They also utter hissing sounds and perform bill snapping under these conditions. When young, nestling owls utter chittering *psseee* sounds as low-intensity threats, and *psssssss-sip* notes for food begging. On the whole, however, they are relatively silent, at least as compared with woodland owls. Loud wing-clapping noises during display flights are common; however, and unlike those of the long-eared owl, which are less frequent, several are produced in rapid succession (to about 10 per sec), or about as rapidly as one can slap both hands on the thighs alternately (Holt and Leasure, 1993). Wing-clapping is also used as an aggressive display to drive away intruders from the male's territory.

Habitats and Ecology

The ecology of this species is largely associated with open habitats, including in the winter such areas as old fields, grain stubble fields, hay meadows, pastures, and inland or coastal marshes. In summer its North American breeding habitats include prairies, grassy plains, and tundra (Clark, 1975). Mikkola (1983) describes the breeding habitats as including moorlands, marshlands, bogs, and dunes, and sometimes previously forested areas that have been cleared. In Britain the species favors open areas such as grassy moorland heaths, newly afforested hillsides with long grasses, extensive areas of rough grazing lands, marshes, bogs, long heather areas, sand dunes, and inshore islands. Substantial areas of suitable resting and nesting cover and productive nearby hunting areas with an abundance or superabundance of small mammals, are probably dominant factors in selecting breeding habitats (Cramp, 1985). Winter roosts typically are characterized by shelter from the weather, close proximity to hunting areas, and relative freedom from human disturbance (Clark, 1975). They often consist of fairly dense coniferous vegetation, with similar characteristics to those used by long-eared owls, and rarely may even be shared with that species. It is likely that the presence of snow on the ground (when the birds become much more conspicuous) is a prime stimulus for the owls to abandon ground roosting and begin roosting in dense tree vegetation (Bosakowski, 1986). An exhaustive review of the habitat needs of this species and the effects of management practices on it was recently provided by Dechant et al. (1999b).

Territories on short-eared owls vary greatly in size, perhaps in relationship to relative prey abundance. In Alaska, where lemmings are often abundant, they may be as small as 20 hectares (Pitelka, Tomich, and Trichel, 1955a), but territories in areas farther south have ranged from 55 to 82 hectares (Holt and Melvin, 1986; Clark, 1975; Holt, 1992). Population densities probably likewise vary greatly geographically and according to densities of prey populations. Thus in Manitoba an area of 15.5 square kilometers supported up to 9 breeding territories, or an approximate density of about 0.6 pairs per square kilometer. In various European study areas breeding densities ranged from about 1 to 6 pairs per square kilometer

(Glutz von Blotzheim and Bauer, 1980), and Lockie (1955) reported finding 30–40 pairs occupying an area in Scotland of 14.2 square kilometers after a rodent plague. Breeding territories in 4 European studies were found to vary from averages of 15 hectares (in Germany) to 200 hectares (in Finland) (Mikkola, 1983). In Scotland territorial sizes increased from averages of about 18 hectares to 137 hectares as food supplies diminished (Lockie, 1955). Small hunting territories (of about 6 ha) may be defended in winter, and these are sometimes maintained and enlarged for later breeding purposes. Territories are seldom violated by other birds, although occasionally undefended areas may be used by various birds (Clark, 1975).

On the breeding grounds the birds appear to have relatively small home ranges. Available information would suggest that defended breeding territories and home range boundaries are essentially the same in this species. The large aggregations of wintering birds in some areas, of up to as many as 100 or rarely even 200 birds, certainly reflect a local superabundance of prey, and probably these birds simply range freely around their communal roosts in search of prey (Cramp, 1985). At such times birds may roost within a meter of one another and even on breeding areas nests have been found as close as 55 meters apart (Holt and Leasure, 1993).

Movements

This is a relatively migratory species, at least near the northern edge of its range, but the exact pattern and magnitude of migration in North America is rather uncertain. Certainly there are some fairly long-distance north-south movements evident from banding returns, but additionally some movements are nearly east-west in orientation. Displacements of more than 1000 and up to nearly 2000 kilometers have been documented for banded birds; of 8 of these, 3 were cases of probably juvenile dispersal, but at least one was an adult moving southward during fall (1730 km south-southeast in about 50 days). On the other hand, a juvenile banded in Massachusetts in October was found dead in almost exactly the same location the following June (Clark, 1975). There are still only relatively few recoveries of banded birds in North America (47 through 1992) and not enough to separate simple nomadism from true migratory behavior (Holt and Leasure, 1993). Houston (1997) obtained no recoveries from 246 banded birds, but judged that because of the influx of owls during years of high vole populations the birds must be somewhat nomadic.

Banding data from Europe are more complete, and those from Finland indicate that in addition to a distinct north-south migratory tendency there is also evidence of nomadism, with some Finnish birds wandering east to the former USSR and some west to Scotland. Similarly some Dutch and West German birds, banded as juveniles, were eventually recovered as far north as northern Scandinavia and as far east as Sverdlovsk (now Ekaterinburg), Russia (Cramp, 1985).

Foods and Foraging Behavior

Clark (1975) summarized essentially all the information available on the year-round foods of short-eared owls in North America, which at that time totaled 25 studies that collectively analyzed nearly 10,000 pellets. Of this total, 94.8 percent of the identified prey were mammalian, and 5.1 percent were birds. Of the mammals, about 61 percent were *Microtus* voles, which are common in grassy habitats. Many of the bird species that have been identified among pellet remains in North America are similarly open-country or marsh species, including various sandpipers (*Calidris* spp.), killdeer (*Charadrius vociferus*), western meadowlark (*Sturnella neglecta*), horned lark (*Eremophila alpestris*), red-winged blackbird (*Agelaius phoeniceus*), and Virginia rail (*Rallus limicola*). More recent summaries have been provided by Holt (1993) and Holt and Leasure (1993), which include pellet analysis information from 11 studies and involve more than 20,000 prey items. In these, the incidence of mammalian items ranges from 79.4 to 99.8 percent, and birds from 0.2 to 15.1 percent, with 3 studies having some crustacean remains and 1 having insect remains present. Of the mammals, *Microtus* voles comprised more than 78 percent of all items. These data confirm the relatively narrow trophic range of this species as compared, for example, with prey generalists such as the great horned owl.

Clark (1975) suggested that the high incidence of voles in the species' diet is not so much a result of the owls preferring them and seeking them out as it is a matter of the species' affinity for open-country habitat and their tendency, as opportunistic hunters, to take whatever prey species happens to be most vulnerable to them. However, Colvin and Spaulding (1983) found that more *Microtus* voles were taken in their study than chance based on relative availability would predict, suggesting to them that the birds prefer larger voles than smaller *Peromyscus* prey for reasons of energy efficiency, and tend to concentrate their hunting times during major periods of vole activity.

Food analysis studies in Europe have been summarized by Mikkola (1983) and Cramp

(1985). Breeding-season studies in four European countries, involving more than 4000 prey items, indicate a generally high incidence of *Microtus* voles in the remains in three of the four areas, with shrews replacing voles as the most common prey items in the fourth area. Autumn foods in Finland also showed a predominance of *Microtus* voles in pellets, while winter-season studies from five west Palearctic areas likewise indicated that voles were of primary importance in all but one area (Glue, 1977b). (In Ireland, the one exception, few voles are present and wood mice [*Apodemus* spp.] and brown rats [*Rattus norvegicus*] predominated.) When calculated on a relative prey biomass basis, results from five west Palearctic areas totaling more than 23,000 prey items showed that small murine and microtine rodents comprised from 55 to nearly 100 percent of the biomass, with the exception again of Ireland, where wood mice and brown rats contributed nearly 80 percent to the biomass (Cramp, 1985).

Short-eared owls primarily hunt by prolonged coursing flights, usually less than 2 meters above the vegetation, typically flying into the wind. During such hunting sorties they may hover momentarily almost motionless in the air and then quickly descend vertically on the prey. Extended ternlike hovering over potential prey locations is sometimes also used, for periods of as long as 30 seconds. Occasionally they will watch from low vantage points such as fence posts until they see a prey, and then fly out to pounce upon it. Of a total of more than 600 attempted pounces observed in one study, at least 20 percent were successful (Clark, 1975). Limited observations by Marr and McWhirter (1982) suggest that adult owls were more successful in their hunting (19 of 33 strikes successful) than immatures (2 of 13 successful). In Europe the birds have been reported to hunt at all times of the day and night and may be more diurnal than any other European owl, but apparently they favor late afternoon hours and early evening (66 percent of observed hunts in Finland occurred between 3:00 and 9:00 P.M.) (Mikkola, 1983). Clark's (1975) observations generally agreed with these findings, with apparent peaks in winter hunting beginning about 4:30 P.M., the owls evidently hunting until from one to three mice had been captured. He believed that daytime hunting was largely limited to those times when they were not able to obtain enough preferred foods during the night.

Social Behavior

Short-eared owls become sexually mature during their first year of life and form monogamous pair bonds that probably last only for a single breeding season. Breeding by year-old females has been reported both in North America and in Europe. Polygyny has been suspected in Europe and North America, but there is only limited evidence to support this view (Holt and Leasure, 1993).There is no good information on nest site fidelity, although in one nest the remains of an earlier, well-rotted nest that apparently dated from the year before could be seen (Bent, 1938). In Clark's (1975) study the breeding territories of several Manitoba pairs remained fairly similar during two successive years. However, over much of the species' northern range it is relatively nomadic, which would militate against repeated matings with the same individual and nesting in the same location in subsequent years. However, Holt and Leasure (1993) had no recaptures of females during three years of banding.

Courtship displays begin in late winter. Clark (1975) first observed wing clapping in mid-February, shortly after which the birds began to disperse from communal winter roosts and resume their territories. As the season progressed, wing clapping became more frequent, apparently serving not only to provide an advertisement of the male's territory to prospective mates and to expel potential rivals but perhaps also to synchronize reproductive cycles in the established pairs.

After taking off on a display flight, the male climbs rapidly with a rhythmic wingbeat reminiscent of rowing strokes, the wings momentarily hesitating at the top of the upstroke and quickly "bouncing" up again at the end of the downstroke. While climbing, the male may perform repeated below-the-body wing claps, as well as sometimes soaring or hovering into the wind while uttering his courtship song. The hovers are interspersed with shallow, descending glides that end with a wing clap, followed again by climbing. The whole sequence may be ended by a spectacular rocking descent called by Clark (1975) the "sashay flight," in which the wings are held in a steep dihedral and the bird rapidly loses altitude while it oscillates from side to side (Figure 63). The descent may also be interrupted by one or two leveling-offs, followed by more glides, wing claps, and further descents. During aerial flights two other display activities sometimes also occur. These include the "underwing" display, in which the lower surface of the wings are presented to the view of another owl, perhaps as a means of achieving sex recognition (males are whiter below than females) or as a challenge, and the "skirmish," during which two owls hover before one another, sometimes entangling their talons and possibly even spinning or cartwheeling downward for some distance. Clark (1975) believed this latter behavior is at least partly aggressive in moti-

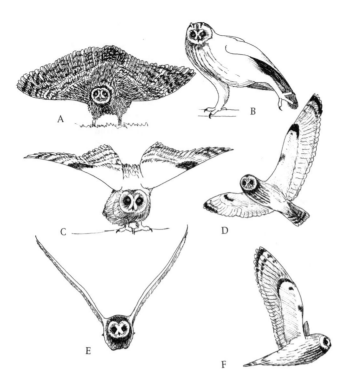

Figure 63. Short-eared owl behavior, including (*A*) defensive wing-spreading posture, (*B*) leg and wing stretching, (*C*) double wing stretching, (*D*) "sashay flight," and (*E–F*) "underwing display." After drawings in Clark (1975).

vation rather than pure "courtship." Females have been observed to join males in their courtship flight on a few occasions (Holt and Leasure, 1993).

Wing clapping is often done from a fairly high altitude, while the bird is circling its territory. It suddenly claps its wings several times below the body, while quickly losing altitude. A typical display sequence is terminated after 15–20 claps have been produced (Mikkola, 1983; Clark, 1975). During the courtship period the birds usually display regularly, beginning in late afternoon and continuing until after midnight, but on cloudy days display may occur at almost any hour of the day. Aerial display flights have occasionally been reported to occur during autumn and spring migration as well, when they clearly do not help to advertise breeding territories.

Although mutual preening has not yet been observed, courtship feeding of the female by her mate precedes copulation. In this species copulation takes place on the ground. After capturing a prey item, the male lands in his territory within sight of the female, who calls in return to his hooting as he approaches to land. Typically the female flies to the male, who holds the recently captured prey in his beak and presents it to her as he partially opens his wings. Then he turns into position alongside the female and mounts her, spreading his wings to maintain balance. Treading lasts about four seconds, after which the female flies to her nest scrape and settles down on

it, as if eggs were already present (Clark, 1975; Mikkola, 1983).

The face of a short-eared owl is remarkably "expressive," depending on the degree to which the "ears" are raised and the feathers between and above the eyes are flattened. When in defensive situations, as when a female is approached on the nest, the birds adopt a distinctive, rather cat-like, facial appearance, with the eyes partially closed and the pupils dilated, almost hiding the yellow iris (Figure 64). Except when the owls are agitated or curious, the ear tufts are hidden among the other forehead feathers. Under intense duress, a nesting bird may perform injury feigning on the ground, as well as aerial dives with wing clapping (Bent, 1938).

There is no information on nest site selection and nest building, which is presumably, but not certainly, done by the female. The short-eared owl is one of the very few owls that actually constructs its own nest, although few observations have been reported on this morph of reproduction. Apparently the bird makes a scrape in the substrate prior to lining the nest with stalks of stubble or other vegetation. At least some of the latter may at times have to be transported to the nest from some distance (Clark, 1975). There is also at least one case known in which the eggs were found to be laid on bare ground, and two others in which a new nest was built over a previous year's nest and unhatched eggs (Holt and Leasure, 1993).

Nest sites are selected in various open-

Figure 64. Normal (*left*) and defensive (*right*) appearances of the short-eared owl. After photographs.

country but usually well-vegetated habitats. Of 63 sites tabulated by Clark (1975), more than half were in grasslands, about a quarter in grain stubble, and the rest were located in haylands or low perennial vegetation. Tall and rank vegetation, such as cord grass (*Spartina*) and alfalfa (*Medicago*), appear to be preferred cover plants. Of 13 Ontario nests, 5 were in abandoned grassy fields, 3 in heath bogs, 2 each among short grasses of airport fields and in cattail (*Typha*) marshes, and 1 on tundra. Some were slight ground depressions; some were merely cups of dried weeds, flattened grasses, or both; and one had a canopy of tall grasses above it (Peck and James, 1983). Of 28 Montana nests, surrounding vegetation was 85 percent grasses and the rest herbs or mixed herbs and grasses, with 90 percent of all surrounding vegetation less than half a meter in height (Holt and Leasure, 1993).

Breeding Biology

Once the nest site has been readied, egg laying begins. Six egg records from Alaska and arctic Canada are from June 10 to 30, and 9 from Alberta and Manitoba are from May 5 to June 20. Seventeen records from the Dakotas and Minnesota are from March 20 to June 12, and 7 from Nebraska, Kansas, and Illinois range from April 8 to May 17 (Bent, 1938). R. Stewart (1975) reported that 26 North Dakota egg dates range from April 4 to August 1. Ten active Ontario nest records are from April 14 to August 1, with half in the period May 6–19 (Peck and James, 1983).

Murray (1976) reported that the average clutch size of 186 North American nests was 5.61 eggs, with a clinal increase toward the north and a small but statistically insignificant increase toward the west. Holt and Leasure (1993) summarized data from 8 North American regions, where mean clutch sizes ranged from 5.7 eggs (Massachusetts) to 8.8 eggs (Manitoba). Similarly in Europe the clutch sizes average higher near the northern end of the breeding range, but averages vary from about 6.9 in Germany to 7.4 in Finland, similar to those typical of North America. Clutch

sizes are known to average larger during years when food is abundant, with incredibly large clutches of as many as 13–16 eggs being reported during vole plague years (perhaps the result of two females laying in the same nest). In a few cases two broods have reportedly been raised successfully in a single breeding season, which may help to account for the generally long breeding periods that have sometimes been reported. Additionally, replacement clutches after loss of the first one are also apparently fairly common (Mikkola, 1983; Cramp, 1985).

Eggs are laid at intervals of 1–2 days, and incubation, starting with the first egg, requires 24–29 days, rarely longer. This is probably done largely or entirely by the female, although some unproven reports of both sexes incubating have been published. Additionally, two females have been observed incubating the same nest, which might also further confuse the interpretation of sex roles in incubation (Mikkola, 1983).

At hatching the young average about 15–17 grams, and after 3–4 days they can support themselves upright and begin to beg for food by wing flapping and uttering calls. Their eyes are fully open by about the eighth or ninth day, and by 10 days they often weigh more than 10 times their original hatching weight. This remarkably rapid rate of development may be related to their relatively exposed nesting sites and high vulnerability to predation. They may begin to leave the nest when only 14–18 days old, or even as young as 12 days old. Thus they do a considerable amount of walking and running (sometimes wandering nearly 200 meters from the nest) before they actually fledge at 24–36 days (Clark, 1975; Mikkola, 1983; Holt and Leasure, 1993).

Reproductive success of the short-eared owl is often fairly low, especially where agricultural practices make the nests vulnerable to destruction by crows, hawks, foxes, farm equipment, or purposely set fires (Mikkola, 1983). The best North American data on breeding success come from 29 Montana nests, where 235 eggs hatched 174 chicks, or 74 percent hatching success, and 159 prefledged chicks dispersed from the nests

(Holt and Leasure, 1993). Houston (1997) reported that 78 broods averaged 3.1 young at the time of banding, but by then some of the oldest chicks had probably already left the nest. In a peak vole year 6 nests held 43 chicks, and the most productive nest ever recorded contained 4 eggs and 7 chicks. Of 121 eggs in 17 West German nests, 44 young hatched (36 percent) and 33 fledged (27 percent), averaging only 1.9 fledged young per active nest. Besides deaths caused by siblings, crows and foxes may be important causes of loss of eggs and young in Europe (Cramp, 1985). Bad weather, changes in prey density, and habitat changes have all contributed to variations in productivity in North America (Holt and Leasure, 1993).

Evolutionary Relationships and Conservation Status

Probably the nearest living relative to this species is the African marsh owl (*Asio capensis*), which is ecologically very similar to the short-eared owl but is more nocturnal and has dark brown rather than yellow iris coloration. It generally replaces the short-eared owl in Africa, especially south of the equator where the short-eared owl is absent (Mikkola, 1983).

The status of this widespread and rather nomadic species is difficult to assess, but in Europe its breeding range has contracted in some areas and expanded slightly in others. In North America, there was a statistically nonsignificant annual population trend of −14.3 percent, based on annual Breeding Bird Surveys done between 1966 and 1993 (Price, Droege, and Price, 1995). A more recent Internet summary of these data suggest a 3.4 percent annual population decline from 1966 to 2000, which is statistically significant (http://www.mbr-pwrc.usgs.gov/bbs/bbs.html). Declines have been especially severe in the Northeast, where at least seven states have listed the species as endangered, threatened, or of special concern. It is classified as endangered in several eastern states, including Pennsylvania and Michigan. Maximum known longevity under natural conditions in North America is 4 years and 5 months (Patuxent Wildlife Research Center banding data). There is a European survival record of 12 years and 9 months (Cramp, 1985).

During the 1999 and 2000 Christmas Bird Counts the species was detected on an average of 277 counts, or about 15 percent of all U.S. and Canadian counts. In terms of all localities where the species was detected, the short-eared owl was the fourth most widely reported of the 18 North American owl species then encountered. The average total number counted among the 100 counts having the highest species totals was 794 birds, or about 8 percent of all owls reported in the tabulated summaries of counts for the United States and Canada during 1999–2000. The maximum number ever reported on a single CBC was 145, at Delaware Reservoir, Ohio, in 1990. The Canadian population has recently been estimated at 20,000–40,000 pairs (Kirk and Hyslop, 1998) and is believed to be declining.

Striped Owl   *Pseudoscops clamator* Vieillot, 1808

Other Vernacular Names:
striped horned owl, Búho-cornudo cariblanco, Búho listado, Tecolote griton (Spanish).

Range

Ranges residentially in Mexico along the Atlantic slope from southern Veracruz to central Chiapas and Tabasco, also in Belize and eastern Guatemala, and southward along the Pacific slope from Chiapas through all of Central America (except for the driest areas and heavy forests) to Panama; in South America from the eastern slope of the Andes across most of eastern South America (except the lower Amazonian basin) to central Argentina. (See Figure 65.)

North American Subspecies

P. c. forbesi (Lowery and Dalquest). Southern Mexico south to Costa Rica and Panama. The validity of this race has been questioned, and it has sometimes been merged with the nominate form.

Measurements

Wing, males 240–244 mm (ave. of 3, 241.3), females 240–273 mm (ave. of 6, 255.8); tail, males 127–129.5 mm (ave. of 3, 128.7), females 132.5–150 mm (ave. of 6, 139.8) (Ridgway, 1914). The claw-lengths of this species are unusually long, suggesting it is an effective predator of larger vertebrates (Voous, 1988). Two Panamanian eggs measured 43.7 × 37.4–37.8 mm (Wetmore, 1968). Some eggs (sample size unstated) from Suriname were 43.6–46.4 × 35.7–36.8 mm (Haverschmidt, 1968), and some laid by a captive in Paraguay averaged 44.7 × 36.5 mm (Krahe, 1981). Schönwetter (1967–1984) reported 42.4 × 34.0 mm and an associated estimated weight of 26.8 g for 16 Guyana eggs.

Weights

Weight of males 335–347 g, females 400–502 g (unspecified sample size) (König, Weick, and Becking, 1999). Haverschmidt (1968) reported male weights (unspecified sample size) as 347–408 g, and females as 480–546 g, indicating a male:female mass ratio of about 1:1.28. Stiles and Skutch (1989) reported a sex-unspecified weight of 440 g, and Voous (1988) gave a range of 235–305 g. Some eggs (unspecified sample size) from Suriname weighed 31–32.5 g (Haverschmidt, 1968), which represents about 6 percent of the adult female's mass.

Figure 65. Distribution of the striped owl. Mexican distribution after Howell and Webb (1995). The inset map shows the South American distribution.

221

Identification

In the field. This is the only Mexican owl with long ear tufts and a nearly white facial disk that is narrowly rimmed with black. Its underparts are buffy, with vertical brown streaks, resembling a short-eared owl, and in flight as well as in typical habitats it also resembles a short-eared owl (which occurs in Mexico only as a winter visitor and mainly to the north of the striped owl's range). See also "Vocalizations."

In the hand. This is a medium-sized (wing 240–275 mm) and rather pale-colored owl that has long ear tufts, brown eyes (not yellow, as sometimes portrayed) that are surrounded by a whitish facial disk, feathered toes, a fairly stout bill, and large, asymmetrical external ears with well-developed opercular flaps. Voous (1988) reported that an owl he measured exhibited meatus openings of 37.5 and 42 millimeters similar to measurements in an earlier report, but with the left opening larger than the right. The postaural ear flaps were 4.0 and 5.5 millimeters wide, and the preaural flaps were 9 millimeters wide on both sides. The ear condition closely approaches that of the long-eared owl, and argues for their congeneric affinities. As adults the sexes are apparently identical, but females average substantially larger. Natal and later developmental plumage stages are undescribed.

Vocalizations

The usual call of this species is a single, prolonged and nasal hoot, or *ahooooo,* rising and then falling in pitch and volume, and lasting about a second. It also utters this same note as a series of well-spaced and similar-sounding hoots that are spaced at intervals of several seconds, presumably as a territorial advertisement. A different description is of a series of rather rapid and nasal *ah* notes given at the rate of about five per second. Tape recordings of the species' song from Venezuela were provided by J. Hardy, Coffey, and Reynard (1999). One sequence from Venezuela consists of three spaced and abrupt *ahooooo* notes uttered in 30 seconds, followed by a harsher squealing note. The female's call is similar to the male's but higher in pitch. Barking notes are produced by both sexes during duetting, and single but similar notes are produced by the males when they are disturbed. Young birds may utter rather high-pitched and whistled screams with downward inflections.

Habitats and Ecology

This is primarily a lowland species, ranging in Mexico from sea level to 900 meters, and inhabiting open habitats such as savannas, marshes, gallery forests in more open lands, forest clearings, agricultural lands, and occasional plantations. It has been collected from three Mexican states (Enriquez-Rocha, Rangel-Salazar, and Holt, 1993). Similar habitats are used farther south in Central America and South America, where it sometimes occurs to elevations in excess of 1000 meters.

Movements

Apparently residential.

Foods and Foraging Behavior

This is mainly a nocturnal hunter, with hunting beginning near dark. The birds often quarter low over open country, in the manner of short-eared owls, dropping down quickly on any prey they flush out. Many of these are small grassland rodents, such as spiny pocket mice (*Liomys*), vesper mice (*Calomys*), gerbils (*Reithrodon*), introduced Old World rats (*Rattus*), and marsh rats (*Holochilus*). Their prey also includes a few ground-foraging birds such as eared doves (*Zenaida*) and quail doves (*Geotrygon*), reptiles such as lizards, and large insects, especially orthopterans (Hoyo, Elliott, and Sargatal, 1999). In Suriname they have been found to prey on rice rats (*Oryzomys*), spiny mice (*Proechimys*), cane mice (*Zygodontomys*), and rat-tailed opossums (*Metachiris*) (Haverschmidt, 1968). In southeastern Brazil they apparently concentrated on rodents (frequency of occurrence in pellets 53 percent, estimated biomass 65.6 percent), with arboreal mice (*Oryzomys*) important prey items. Birds, small arboreal marsupials (*Marmosa*), insects, and bats comprised, in descending order, smaller prey quantities (Motta-Junior and Talamoni, 1994). Pellets and prey remains from birds in Argentina indicate a majority of rodents (55.4 percent of all 56 prey items, 77 percent of estimated biomass total), especially *Rattus* rats, in the diet. Bird remains made up 23 percent of the total items, but only 3.3 percent of the estimated prey biomass. A captive pair regularly ate newly hatched domestic chickens but never tried to attack wild birds that sometimes entered their cage.

Striped owls are not so long-winged as short-eared owls but have a similar mothlike appearance in flight. Like the other *Asio* species, their highly developed external ear structure probably allows for hunting under completely dark conditions, and their dark brown eyes might suggest that they are more nearly nocturnal than crepuscular birds.

Social Behavior

Little is known of social behavior in the wild. The pair of birds kept by Goodman and Fisk (1973) mated frequently, and performed much conversational hooting by day as well as by night. The female would cluck when she wanted to be fed, and would become agitated and scream if he delayed in tending to her. It was also the female that would shriek in alarm when possible danger approached. When the male was agitated he would utter warning sounds similar to heavy breathing. Social roosting of up to about 15 individuals has been observed in Venezuela. Such roosts are usually in low thickets, but ground-roosting has also been reported.

Breeding Biology

Little information exists for Mexico, but a January–February breeding period has been mentioned for El Salvador. In Costa Rica the birds similarly breed between December and March, and may place their eggs on the ground or on a mat of epiphytic plants growing on a palm tree, a few meters aboveground. In Panama there are breeding records for December and January, including 2 nests with 2 young each in January, and a nest with 2 eggs in December. All 3 of these nests were found during different years at the same site, an elevated area of ground in a citrus grove, where the surrounding grass was more than a meter high (Wetmore, 1968). In South America its breeding occurs over a broad range of months, until as late in the year as September and October, especially toward the southern part of its range.

Krahe (1981) bred these birds in captivity for several years, obtaining 83 eggs in total. Another captive breeding was reported by Goodman and Fisk (1973). In the latter case the male made several shallow scrapes prior to each nesting. One was accepted by the female and made deeper, but only a few bits of leaves and litter were pulled into the scrape by the female. She produced 10 clutches over a period of several years, 3 with 4 eggs each, and 7 with 3. The eggs were laid at approximate 60-hour intervals, and incubation by the female began with the laying of the first egg. Incubation lasted 33 days, with the chicks hatching 1–5 days apart. For the first few weeks the female left the nest only to defecate. On several occasions the nest was moved somewhat while eggs or chicks were present, up to as much as 5 feet. In each successful breeding only a single chick was reared, the other hatched young disappeared within 5 days of hatching, and the male fed the incubating and brooding female. For the first two weeks the chick was closely brooded, eventually under the female's wings when it became too large to get beneath her. She would also shelter it from the sun by raising her wing over it, until it was almost fully grown. The female would not allow the mate to feed their young until she had done so, and she sometimes prevented the male from feeding the chick while she was trying to teach it to feed for itself. The first failed effort at flight by a chick was seen at 37 days, but by 46 days the youngster was flying for short distances. At 66 days a chick was starting to feed for itself, and at 132 days it was expelled by the parents from their territory.

Evolutionary Relationships and Conservation Status

Recent molecular studies indicate that this species should probably be included within *Asio* (Hoyo, Elliott, and Sargatal, 1999). Such a merger has been done by Howell and Webb (1995) and König, Weick, and Becking (1999). However, S. Olson (1995) believed on morphological grounds that a continued generic distinction is valid. Remarkably, a captive mixed mating between a striped owl and a barn owl produced fertile eggs that survived about two weeks of incubation (Fleig, 1971), which is rather remarkable in view of the current degree of taxonomic separation of these two groups.

The striped owl occupies many of the same open grassland habitats as does the short-eared owl in North America, and as such is not dependent on forests for nesting or hunting. It is thought to be generally uncommon in Mexico. It actually avoids dense forests and is attracted to agricultural fields, so clearing of tropical forests for agricultural purposes probably favors this species. Stotz et al. (1996) listed it as occurring in at least 21 Latin American countries, but patchily distributed and uncommon. It is considered endangered by the Mexican government (Enriquez-Rocha, Rangel-Salazar, and Holt, 1993).

Boreal Owl *Aegolius funereus* (Linnaeus) 1758

Other Vernacular Names:
arctic saw-whet owl; funereal owl; Richardson's owl (*richardsoni*); Tengmalm's owl (English
vernacular name used in Europe for *A. f. funereus*).

North American Range (Adapted from AOU, 1983.)

Breeds in North America from northern tree line in central Alaska, central Yukon, southern Nunavut, northern Saskatchewan, northern Manitoba, northern Ontario, central Quebec, and Labrador south to southern British Columbia, central Alberta, central Saskatchewan, southern Manitoba, northeastern Minnesota, western and central Ontario, southern Quebec, and New Brunswick; also breeds locally in the higher mountains of Washington, Idaho, Montana, Wyoming, Colorado, and northern New Mexico. Winters generally in the breeding range, but in North America wanders south in the plains irregularly to North Dakota, southern Minnesota, central Wisconsin, southern Michigan, southern Ontario, New York, and New England, casually to southern Oregon, Illinois, Pennsylvania, and New Jersey. Also distributed widely in northern Eurasia. (See Figure 66.)

North American Subspecies (Adapted from AOU, 1957.)

A. f. richardsoni (Bonaparte). Resident in North America as described above.

A. f. magnus (Buturlin). Accidental in Alaska; resident in northeastern Siberia.

Measurements

Wing (of *richardsoni*), males 163–171.5 mm (ave. of 7, 168.4), females 171.5–182.5 mm (ave. of 5, 178.3); tails, males 96–106 mm (ave. of 7, 98.6), females 95.5–107 mm (ave. of 5, 104) (Ridgway, 1914). The eggs of *richardsoni* average 32.3 × 26.9 mm (Bent, 1938). Hayward and Hayward (1991) reported male and female wing measurements of 163–179 mm (ave. of 50, 172.5) and 174–198 mm (ave. of 53, 183.6), respectively. In both cases some measurements were repeats of the same birds obtained on separate occasions.

Weights

Hayward and Hayward (1991, 1993) reported adult male weights of 93–139 g (ave. of 50, including some repeat measurements, 117.3) and female weights of 132–215 g (ave. of 53, including some repeats, 166.8), for birds of the Idaho population. This is one of the highest male:female mass dimorphism ratios (1:1.42) reported for any North American owls. Korpimäki (1987b) similarly found that females of the Finnish population average 43 percent heavier than males at the start of the breeding season. Glutz von Blotzheim and Bauer (1980) reported that 74 males of the general Eurasian population (Tengmalm's owl) averaged 101 g (range 90–113) and that 96 females averaged 167 g (range 126–194). Mikkola (1983) noted that 89 males and 100 females of the same population averaged 123 (range 116–133) and 168 (range 150–197) g, respectively. The estimated egg weight is 12 g; 13 freshly laid eggs averaged 11.7 g (Hayward and Hayward, 1993). The estimated egg-to-female proportional mass ratio is 7.2 percent.

Identification

In the field. This inconspicuous owl is usually seen perched amid the branches of conifers, often standing motionless in an upright position near the trunk in the manner of a long-eared owl. It closely resembles the northern saw-whet owl, but besides being larger it has grayish brown rather than chestnut streaking on the breast, and the general dorsal tone is brownish gray rather than reddish brown. Juvenile birds are mostly sooty black, except for whitish "eyebrow" markings and a few white wing spots. It tends to be silent and has a wavering mode of flight. The male's primary territorial song is a rapidly trilled series of *hoo-poo-poo* notes, with about 16 pulsed notes typical (range 11–23) uttered in less than three seconds. At times a softer but more prolonged version, lasting up to a minute and with 9–10 notes per second, is uttered (Hayward and Hayward, 1993).

In the hand. The rather large head and facial disk of this otherwise small owl are distinctive, although the eyes are relatively small and the ear tufts are rudimentary. The crown is spotted with white, and contrasting, somewhat pearl-like white spotting also occurs widely on the upperparts, while the underparts are marked with grayish brown in indefinite streaks. As with the related northern saw-whet owl, the ear openings are asymmetrical, and only the two outermost primaries are emarginated. No reliable plumage criteria exist for sexing adults, but breeding females exhibit brood patches, and females generally have tawnier faces than males. Hayward and Hayward

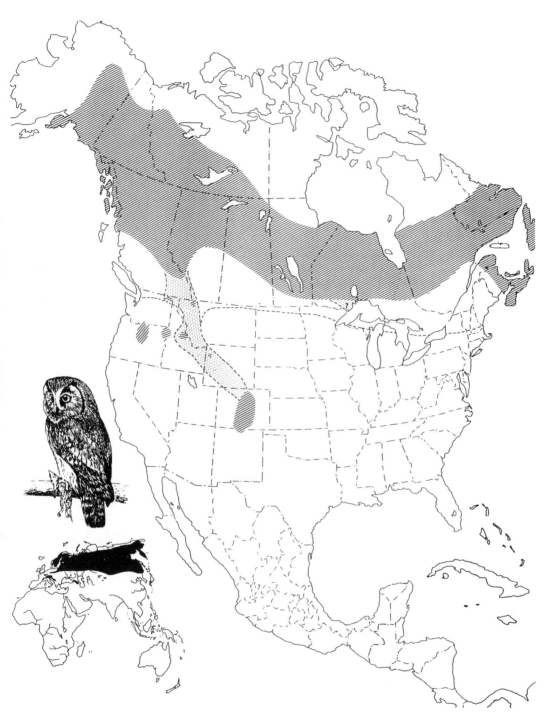

Figure 66. North American breeding distribution of the boreal owl. The dashed line indicates usual limits of wintering vagrants or migrants, and the stippled area indicates regions of probable local breeding. Eurasian distribution of Tengmalm's owl shown in inset.

(1991) provided a discriminant function analysis that provides for complete sexual separation of mated pairs. Nestlings are initially partly covered with short, velvety down that is buffy white above and white below, with bare pink skin between. The juvenal plumage is unusually dark chocolate brown, except for strongly contrasting white "eyebrows" and cheeks. Adult contour feathers begin to develop around the facial disk at about 50 days, and are grown within about a month (Bondrup-Nielsen, 1978). The remiges of first-year birds are uniform in color and wear, with the tips of the outer primaries gradually becoming more pointed through wear. The central rectrices are relatively narrow, and the white cross-bars of the rectrices tend to be chevron-shaped; the tertials also average narrower and are more pointed than in adults (Pyle, 1997b).

Vocalizations

Although a considerable literature exists on the biology and associated vocalizations of the Eurasian Tengmalm's owl, relatively little had been provided for the boreal owl. Meehan's (1980) and Bondrup-Nielsen's (1978, 1984) studies have rectified this situation, and these studies provide a convenient means of describing and summarizing the form's vocalizations. Hayward and Hayward (1993) have also provided a good summary of the American form's vocalizations.

The male boreal owl's primary advertisement vocalization is the "staccato song," which is uttered during nighttime hours (mainly from dusk until midnight) from late winter through early spring. Unlike early descriptions that describe it as bell-like or similar to the sound of water dripping (both of which descriptions better apply to and may perhaps refer to saw-whet owl calls), it instead resembles the winnowing noise of the common snipe (*Gallinago gallinago*), being a trill of essentially constant pitch (averaging 740 Hz) and lasting 1.3–2.3 seconds (averaging 0.8–1.8 sec in the 2 above-mentioned studies), with about 10 uniformly pulsed notes per second. It gradually gains in amplitude before stopping abruptly. This is the only really loud vocalization of the species, carrying up to more than 3 kilometers on clear, cold nights, and easily audible at 1.5 kilometers. The song is uttered persistently, often for 20 minutes or longer, night after night, while the male is unpaired. Early in the singing season the song is produced by males perched near potential nest sites for brief periods in early evening, but the intensity gradually increases until singing lasts most of the night. A subdued version of this song may be produced under various conditions, such as for a minute or two early in the evening, after a long pause in singing, after being disturbed during the day, or when the male is approaching the nest with food.

A prolonged version of the primary song, lasting continuously up to a minute (rarely to 15 minutes with brief pauses) differs in that it is not divided into phrases. This "prolonged song" is softer than the primary song and is produced from the start of courtship until the onset of incubation. It is uttered when a female enters the male's territory, whereupon he flies to his nest hole, enters it, and sits in the entrance while singing. Generally, prolonged singing seems to follow the staccato song in the reproductive cycle and apparently replaces it (Bondrup-Nielsen, 1978). It seems to function in pair-forming and pair-bond maintenance, and thus serves as a true song (Hayward and Hayward, 1993).

Meehan (1980) did not recognize the prolonged and subdued staccato songs as distinct from the primary song, but stated instead that the paired male utters a repeated single note with an indeterminate number of units, which he called the "murmured song." It is uttered by males from the time of the female's arrival to the territory until about the first week of incubation. After the initial mate attraction by the staccato song, the male's song in the "murmured" version apparently serves to announce the location of the nest, the presence of an intruder, and territoriality of the male. The greater variability of this version, in length and volume, apparently is more suitable for these varied functions than is the more stereotyped primary song. During the incubation period the males also utter a brief trill while transferring prey to the female or dependent young.

The *moo-a* (Bondrup-Nielsen, 1978) or *hooh-up* (Meehan, 1980) note is a varied call of males, ranging from the sound of a child crying to a tree creaking, and consists of a prolonged note that gradually increases in pitch and volume before dropping terminally. In these studies it was typically uttered to announce arrival of prey at the nest. It was also was produced in response to playbacks of the staccato song, and thus presumably is partly agonistic in function. Another short warning or aggressive call that is used by both sexes is the *skiew* or screech call, which is sometimes also uttered in response to playbacks of the staccato song but tends to drop rather than rise in frequency. It typically ends abruptly, with the birds' snapping their mandibles. Hissing is also used by females when defending the nest. The *chuck* call is a harsh, brief, frequency-modulated call that is uttered by the female when she is on a male's territory, often in response to hearing the subdued version of the staccato song. The female also produces a "peeping" (Bondrup-Nielsen) or "cheeping" (Meehan) call, likewise in response to the subdued staccato song, when in flight over the territory, or when on the nest. It is the major vocalization of the female, used throughout the breeding season, primarily as a begging call but also as a contact call. It is also used later in the breeding season when bringing prey to the young.

Young boreal owls produce harsh chirping or peeping calls that seem to stimulate the female to feed them. After fledging this call is not so harsh but may be uttered as the birds fly from tree to tree, ceasing after they have been fed. The nestlings also produce a chatter call that on one occasion was heard when the young were trying to get under the female to be brooded (Bondrup-Nielsen, 1984).

Habitats and Ecology

General habitat affinities of the boreal owl in North America can be characterized as including forests ranging from pure deciduous to mixed and pure coniferous composition (Bent, 1938). Additionally the species is cavity-dependent for nest sites, typically using old woodpecker nests, especially those of northern flickers (*Colaptes auritus*). Its prey base consists of small forest-adapted rodents, especially microtine rodents, which are primarily captured nocturnally. Similar habitat needs seem typical of the Eurasian Tengmalm's owl, but it is primarily associated with the dense coniferous forests of the taiga zone, with a special preference for spruce but also ranging into mixed forests, subalpine coniferous forests, and pine forests of lower montane slopes. There, the availability of suitable nesting holes may be a more important ecological factor than relative small microtine rodent abundance, since the birds seem to be able to shift readily to mice and shrews (Mikkola, 1983).

More specific habitat preferences of the boreal owl have been studied by Bondrup-Nielsen (1978) in Ontario and by Palmer (1986) in Colorado. Palmer determined habitat preference by analyzing habitat composition within which 36 radio-tagged owls maintained home ranges during his study. These occurred at elevations of about 2800–3200 meters, with highest densities above 3000 meters in areas where mature spruce-fir forests were interspersed with numerous subalpine meadows. The tagged owls avoided large and unbroken stands of pines, many of which were very dense, and also avoided stands of quaking aspens, a habitat type favored by saw-whet owls of the same area. Additionally, boreal owls used sites with somewhat larger trees present than apparently preferred by saw-whets. Areas

serving as boreal owl territories were often ones with *Vaccinium* and *Arnica* ground cover, along with other forbs, but with fewer grasses or sedges than typical of saw-whet territories. Further, the favored boreal owl sites had higher densities of red-backed voles (*Clethrionomys gapperi*) and lower densities of deer mice (*Peromyscus maniculatus*), which are the respective most important food species of these two owls in that region. Bondrup-Nielsen (1978) found that virgin forests were preferentially utilized, and mixed conifer-deciduous habitats were preferred over purely coniferous ones, probably at least in part because of the presence of available nesting cavities.

In western Montana the birds are apparently limited to old-growth spruce-fir forests above 1525 meters (Holt and Hillis, 1987), while more widely in the northern Rockies of Idaho and Montana they have generally been found in subalpine fir (*Abies lasiocarpa*) or western hemlock (*Tsuga heterophylla*) forests above 1585 meters (Hayward, Hayward, and Garton, 1987). In northeastern Washington they have also been found in spruce-fir forests above 1500 meters in the Selkirk and Kettle ranges (O'Connell, 1987). In Colorado they occur in climax spruce-fir forests, mainly above 2900 meters (Ryder, Palmer, and Rawinski, 1987). In the northern Rocky Mountains of Wyoming, Montana, and Idaho boreal owls mainly occur in the subalpine spruce-fir zone, where 90 percent of all territories were located by Hayward, Hayward, and Garton (1993). Nesting sites there were limited to old-growth spruce-fir stands with complex vegetational structure, where prey densities were 2–10 times greater than elsewhere and where the open forest structure facilitated hunting.

Roost site characteristics of boreal owls were determined by Palmer (1986) and by Hayward, Hayward, and Garton (1993). In the latter study, done in Idaho, it was determined that all 13 roosting trees were coniferous and that the vegetation within the estimated home range of the birds was less than 2 percent deciduous. Tree densities immediately around the roosts were slightly (but not statistically) higher for boreal owl roosts than for those of saw-whet or western screech-owls, and the roosting trees had smaller average diameters and slightly lower average heights than those used by the other species. However, the boreal owl roosts averaged slightly higher (6.9 m) in the trees, and typically the birds perched immediately beside the tree bole, rather than well out on the branch as typical of saw-whets. Palmer (1986) examined 74 roosts in Colorado and found that all were in rather dense conifers, usually Engelmann spruces (*Picea engelmanni*), and on steep slopes. This species of conifer, as well as the others used, provides a greater amount of vegetational cover above the owl than beneath, simultaneously offering protection from overhead attack and unobstructed visibility for seeing prey below. Roosting trees averaged about 14 meters high and had average breast-height diameters of about 34 centimeters. The birds typically roosted about 5 meters up and frequently perched closely adjacent to the bole. There were no apparent seasonal variations in roost site preferences, and only among afternoon roosts was the compass orientation of the roost nonrandom, when the roosts tended to be oriented toward the eastern and southeastern sides of the tree. Bondrup-Nielsen (1978) found that the birds preferred to roost in trees with their lower foliage confined to the outer half of the branch, as in the balsam fir (*Abies balsamea*). The birds usually roosted on naked branches, sometimes in exposed sites, and on average about 6 meters aboveground.

Few population density estimates are available for this species, at least in North America. Data for the Tengmalm's owl from Europe (summarized by Glutz von Blotzheim and Bauer, 1980, and by Cramp, 1985) indicate rather low general densities, although in some favorable locations or years of good small rodent populations the territories may be in easy sound contact, sometimes separated by a few hundred meters or so. Locally in Sweden, nests have been found as close as 250–300 meters apart, or as many as 5 in an area of 0.25 square kilometers, but in low-density areas nests averaged about 3.5 kilometers apart. During peak years in Sweden and Germany the density has ranged from 0.48 to 1.5 pairs or nests per square kilometer. Over a fairly large study area (nearly 25 sq km) of Finland the density varied from 0.08 to 0.33 pairs per square kilometer during various years (Korpimäki, 1981). Bondrup-Nielsen (1978) estimated a density of about 1 bird per 11 square kilometers in 2 different Canadian study areas and years. Distances between territories and nests varies with prey abundance; in North America nests have been found as close as 500 meters, and the home ranges of each of 13 monitored owls was found to overlap that of another monitored owl by at least 50 percent (Hayward, Hayward, and Garton, 1993). In this area large home ranges were typical, those of 13 averaged 1451 hectares in winter, and 15 had home ranges averaging 1182 hectares in summer.

Apparently true territorial defense in boreal owls is limited to the immediate vicinity of the nest hole, which is defended by the male. Palmer (1986) estimated the home ranges of 2 male boreal owls by various methods, estimating average breeding season home ranges to be 296 hectares, as compared with ranges averaging 1132 hectares during the postbreeding period. The ranges of the 2 tagged owls overlapped partially (about 25 percent of the total combined area) during the breeding season and almost completely (about 98 percent) during the postbreeding period.

Using estimates based on roost locations, Hayward (1983) estimated a 32-hectare home range for a single female. He later estimated from radio tracking that the yearly home range of adults averaged about 1500 hectares, with winter home ranges larger than summer ones. The winter home ranges also averaged lower in altitude and about 2.3 kilometers away from the summer range. Substantial range overlap occurred with adjacent owls during both seasons (Hayward, Hayward, and Garton, 1987). Using estimates based on roost locations, a home range of about 1000 hectares was estimated for male Tengmalm's owls by Holmberg (1982). Bondrup-Nielsen (1978) found that 12 males had territories averaging about 185 meters wide (based on maximum observed distances between singing trees), with an average radius of 85 meters (from main singing trees). The total hunting area was considerably larger than the courtship territories; 3 such areas were estimated at 1–5 square kilometers.

Besides being the occasional prey of larger species of owls, the boreal owl occurs sympatrically with the northern saw-whet owl over a large part of North America. Palmer (1986) investigated possible resource competition between these two species, and judged that it would most likely occur in habitat selection, nest site selection, or overlapping prey selection. Differences in macrohabitat selection possibly reflected habitat preferences of primary prey, which are voles (*Microtus* and *Clethrionomys*) for boreal owls and somewhat smaller mice (mainly *Peromyscus*) and shrews (*Sorex*) for the saw-whets. Microhabitats used by the two appeared to overlap considerably in his study, differing in singing territory characteristics mainly in amounts of deciduous cover present (more in saw-whets), the average tree height (higher in boreals), and amounts of grass and shrubs present in the understory (more grass and fewer shrubs in boreals). Hayward (1983) found some differences in vegetational canopy densities at various tree levels between the two. Both species utilize similar-sized tree cavities; probably saw-whets can exploit slightly smaller-diameter cavities than boreals, but the degree of overlap must be quite substantial. Interestingly, Palmer noted that saw-whets exhibited antagonistic behavior when boreal songs were broadcast, but not vice versa, suggesting that perhaps boreal owls may be able to displace saw-whets from contested nesting sites that are big enough to accommodate them.

Movements

The best available information on migrations in this species is from Europe, where the periodic population movements seem to correspond to population cycles in small mammals, at intervals of about 3–5 years. Until recently these movements were thought to involve both sexes and all age groups, the birds moving rather randomly and opportunistically into new foraging and nesting areas according to environmental conditions. However, evidence has accumulated to suggest that adult males tend to remain sedentary, while females and young birds are relatively more migratory or nomadic (Mikkola, 1983; Hayward and Hayward, 1993).

Less is known of possible large-scale migrations in boreal owls in North America, although periodic irruptions do occur from time to time, bringing large numbers of birds into southern Canada and the northern and northeastern states (Anweiler, 1960; Catling, 1972c). Irruptions studied by Catling (1972c) suggest that these typically peak in late winter and involve both sexes. However, the birds may remain south of their breeding ranges until as late as April or May, suggesting that nonbreeding birds are largely involved in these irruptions. Most movements apparently occur within the limits of the boreal forest. However, sometimes large-scale southward movements are evident, frequently in synchrony with similar movements of great gray owls and northern hawk-owls, suggesting that migrations may occur in response to common prey-base scarcity. In Idaho, two males left home ranges they had occupied for more than a year when the local snow cover became crusted-over, and three females left their home ranges shortly after brooding of their young had terminated. One of these moved at least 17 kilometers, the others beyond radio-tracking distances (Hayward and Hayward, 1993).

Day-to-day movements of territorial boreal owls were studied by Palmer (1986), using three radio-tagged birds. The interseasonal average daily movements among 113 roosts was 708 meters for the three birds, with movements least during the courtship period, increasing during the summer (to an average of 1.5 kilometers in August), and declining in the fall after the first snowfall. In the Colorado mountains, although at least some males remain at high elevations throughout the year, others may wander extensively; these are presumably nonterritorial birds.

Foods and Foraging Behavior

Summaries of foods consumed by the Eurasian Tengmalm's owl (Mikkola, 1983) indicate that during the breeding season small mammals comprised from about 92 to 98 percent of the prey individuals taken in four different areas of Europe, with bank voles (*Clethrionomys glareolus*) the most important species in Finland, short-tailed voles (*Microtus agrestis*) the most important in Sweden, and various mice, including dormice (*Muscardinus avellanarius*), most important in central Europe. Outside the breeding season they feed largely or almost entirely on small mammals, with microtine voles again being the most important food source. When analyzed in terms of prey biomass, winter and breeding-season foods included 62–83 percent small microtine or small murine rodents, with birds comprising 11–29 percent and shrews 3–11 percent, while the biomass composition of summer foods from Sweden and eastern Germany averaged only a few birds and relatively few shrews, but 86–89 percent small murines and microtines (Korpimäki, 1981; Cramp, 1985). Females are on average about 43 percent heavier than males and tend to take a larger proportion of the relatively heavier but slow voles, while the smaller males tend to take a larger proportion of agile prey such as birds (Korpimäki, 1987b).

Studies in North America are generally quite similar to those just cited. Voles of the genera *Microtus* and *Clethrionomys* are utilized primarily, with smaller prey including *Sorex* shrews and *Peromyscus* mice when locally available (Bondrup-Nielsen, 1978; Catling, 1972c; Eckert, 1979; Hayward, 1983). Palmer (1986) found that *Clethrionomys* voles comprised 54 percent of 72 prey items identified in Colorado, with *Microtus* voles another 25 percent and birds totaling 7 percent. Catling (1972c) noted that 62 of 72 prey items (86 percent by number, 91 percent by biomass) that he recorded from 75 pellets obtained in southern Ontario were of *Microtus* voles, with the remainder made up of moles, shrews, and *Peromyscus* mice. Prey as large as flying squirrels (*Glaucomys*) are sometimes taken (Eckert, 1979). Hayward, Hayward, and Garton (1993) examined 914 prey items from the northern Rocky Mountains, of which red-backed voles (*Clethrionomys gapperi*) comprised 35 percent of summer and 49 percent of winter foods. Adults of this major prey species average 26.5 grams, and the largest prey item found was the northern flying squirrel (*Glaucomys sabrinus*) averaging 140 grams. Marti, Korpimäki, and Jaksic (1993) calculated a geometric mean weight of 19.2 grams for prey in the Rocky Mountains, and 22.2 grams in Alaska. Food niche

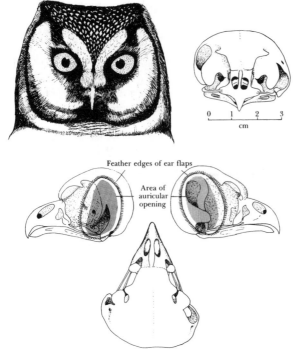

Figure 67. Head and skull characteristics of the boreal owl, showing skull asymmetry associated with external ear specialization. After Shuffeldt (1900) and Norberg (1978).

breadth tends to expand northward in North America.

Although in Europe they are regarded as essentially nocturnal hunters (Mikkola, 1983), boreal owls observed by Palmer (1986) often hunted during the day (observed 27 times by him) as well as at night. The species' small size and maneuverability allows it to hunt in forested areas, using low perches from which it makes generally short flights to strike ground-dwelling prey, at times probably using primarily acoustical clues to localize them (Norberg, 1970). Palmer observed that the birds often captured voles under moderate shrub cover by plunging through the shrubs to reach the prey. Moving prey were taken more frequently than stationary animals, suggesting that auditory clues may be important in locating them. The remarkable asymmetry of the external ear structure in this species (Figure 67) is probably important in prey localization (see Chapter 3). Food caching, with subsequent thawing of frozen rodents by performing brooding-like behavior over them, has also been reported in this species, as well as in the northern saw-whet owl (Bondrup-Nielsen, 1977).

Social Behavior

This bird has an unusual slimmed concealing posture, with the facial disk erected into shallow "horns" and the near-side wing brought forward and upward almost to the level of the beak. This posture is quite different from the normal resting and distinctly fluffed attentive postures. Young birds sometimes assume a "choking" posture when food begging, and disturbed birds have also been observed performing an apparent displacement-sleeping posture prior to fleeing. All these positions are illustrated in Figure 68.

Sexual maturity and breeding occur within a year of hatching in this species (Glutz von Blotzheim and Bauer, 1980). Although several cases of nest site tenacity (use of the same nesting site in successive years) have been reported, at least two cases are known of birds breeding a few hundred meters away from a previous nesting site, and in one case an owl bred at least 510 kilometers away from the previous year's site (Wallin and Andersson, 1981). Relatively little is known of pair bonding in this species, although in common with other nomadic or irruptive species, it would seem unlikely that lifelong pair bonding is typical. There is at least one record of the same male and female breeding with one another during two consecutive years (Korpimäki, 1981), as well as a case of double brooding by a year-old female (Kellomaki, Heinonen, and Tiainen, 1977; Mikkola, 1983). Double brooding in Europe is thought to be fairly frequent in good vole years, but has not yet been established for North America. Occasionally a pair bond persists through two seasons, but this would seem to be exceptional.

There are also numerous published accounts of polygyny (successive bigyny) by this species, with the male taking on a second mate during the same breeding season, and several known cases of successive biandry, with the same female attempting to raise two broods in the same breeding season that were fathered by different males (Solheim, 1983; Hayward and Hayward, 1993). In the biandrous cases the average clutch sizes have not differed between the broods, but the young of the second nesting have sometimes exhibited higher mortality rates. In the cases of the bigynous nestings the periods between the laying of the first and second clutches have ranged from 18 to 30 days, with the nests 500–2100 meters apart, while in the biandrous nestings the clutches have been laid from 50 to 63 days apart and the nests separated by distances of 500–10,000 meters. There are also known cases of trigyny (a male mated to three females) known from Europe, and a male may advertise up to as many as five potential nesting cavities within its home range (Hayward and Hayward, 1993). It is believed that relative nest-hole abundance, food availability, and nest predation levels might all affect mating strategies in the species, and thus considerable variation in nesting biology might occur over different parts of the nesting range.

Males do not defend large territories, but rather confine their territorial activities to the nest sites themselves, which apparently are often in limited supply and usually consist of old woodpecker holes. In Europe these are mostly made by black woodpeckers (*Dryocopus martius*), and their entrances typically measure 70–120 × 90–175 millimeters, although sometimes those of smaller woodpeckers are used that have entrance diameters of as little as 54–75 millimeters (Sonerud, 1985; Lindhe, 1966). Those measured in Idaho had entrances averaging 102 millimeters high and 95 millimeters wide, and throughout North America the cavities used are mostly those of pileated woodpeckers (*Dryocopus pileatus*) and the smaller ones made by northern flickers (*Colaptes auratus*). In regions where natural cavities are scarce artificial nest boxes are often used. Such boxes are only used two years successively if they are cleaned between breeding seasons, and they are only rarely used by the same individuals (Hayward and Hayward, 1993).

Males advertise their presence and their control of potential nesting sites by prolonged singing behavior, which in the Colorado Rockies may begin as early as mid-February and last until the latter half of June, with a peak in late April that corresponds to day length periods of about 14 hours (Palmer, 1986). This same general relationship between photoperiod and breeding in the species has been observed in Europe for the Tengmalm's owl, suggesting possible photoperiodic control of breeding. The courtship period may end at about the time that minimum tempera-

Figure 68. Boreal owl behavior, including (*A*) normal and (*B*) attentive appearance of adults, (*C*) begging posture of nestling, (*D*) erect concealment posture, and (*E*) pseudo-sleeping posture. After Glutz von Blotzheim and Bauer (1980).

231

tures remain above freezing and snow disappears from the ground (Bondrup-Nielsen, 1978). However, Palmer (1986) found no such close correlation with minimum nightly temperatures and did not believe that snow depth played a role in timing of breeding in the Colorado Rockies. He observed relatively long courtship periods (male singing durations) of 31–119 days in various years, with individual owls singing for about 19–49 days, although one apparently unsuccessful male sang for 102 days. Meehan (1980) reported individual male courtship periods of 6–51 days, with unsuccessful males having somewhat shorter periods than successful ones, which was in contrast to Palmer's findings. Bondrup-Nielsen (1978) found overall courtship periods of 28–55 days, with individual males singing for periods of 8–10 days. Singing generally begins shortly after sunset and may extend through the night, although often with diminishing intensity after midnight. Wind and precipitation have negative effects on singing, while the combination of a clear, calm night with a bright moon and only slightly subfreezing temperatures appears to favor singing (Palmer, 1986).

Breeding Biology

More is known of nest site selection for the Eurasian Tengmalm's owl than for the boreal owl. Of 148 nest sites analyzed for that race, most were in open forest habitats, the birds avoiding closed forests and showing a preference for nesting in boxes on isolated trees in clear-cut stands. The scarcity of suitable natural cavities in Finland and Scandinavia, brought on by drastic reductions in populations of the black woodpecker, has caused the Tengmalm's owl to accept nesting boxes in increasing numbers, sometimes even those placed on cow sheds or similar locations close to human habitation. The breeding season in Finland may start as early as the end of February in good vole years, and at such times the clutch size averages larger than during poor vole years. Thus in northern Finland clutches may average as high as 6.2 eggs in peak years or as low as 4.5 eggs in poor years. Clutch sizes also tend to decrease southward in Europe, so that in Germany they may average 5.7 in good years or only 2.7 in poor vole years (Mikkola, 1983). Finally, temporal variations in clutch sizes may occur within a single breeding season (Korpimäki, 1981, 1988a). The initial onset of breeding in females usually occurs among yearlings, whereas males typically begin breeding when two years old (65 percent, compared with 16 percent of yearling males). In both sexes a few birds do not begin breeding until three years old.

In North America fewer nests have been found, but clutch sizes typically range from 3 to 7 eggs, usually numbering 4–6 (Bent, 1938). The mean of 11 Idaho clutches was 3.25 eggs (range 2–4); a separate sample of 31 nests averaged 3.57 eggs (range 2–5) (Hayward and Hayward, 1993). Of 3 nests (one of which was used in consecutive years) found by Palmer (1986), all were in conifer snags. These averaged 7.3 meters aboveground and had entrance diameters of 78, 80, and 100 millimeters. The largest was a natural cavity, while the others were in excavations that were probably made by northern flickers. Five presumptive (based on male behavior) nest sites found by Bondrup-Nielsen (1976) were at heights of 11–17 meters and had entrance diameters of 6–7 × 6–17 centimeters, with cavities 20–25 centimeters wide and 5–35 centimeters deep. Cavities in Idaho averaged about 13 meters aboveground, or just above the tree's mid-point, and their inside dimensions averaged 31 centimeters deep and 19 centimeters wide (Hayward and Hayward, 1993).

Four nest-initiation (initial egg laying) dates determined by Palmer (1986) in Colorado were from April 17 to June 1. Idaho laying dates range from April 12 to May 24; Minnesota laying dates are from March 30 to April 12; and Alaska records extend from March 27 to May 5 (Hayward and Hayward, 1993). A sample of 8 egg dates from southern Canada range from April 11 to June 9 (Bent, 1938). A nest found in Ontario hatched between May 22 and May 27 (Bondrup-Nielsen, 1976). One found in northern Minnesota hatched after June 25 (Eckert, 1979), while another began hatching on May 25 (Matthiae, 1981).

Based on data from the Tengmalm's owl, eggs are probably laid at approximate 2-day intervals, with females laying 4-egg clutches averaging slightly shorter intervals than those with 3-egg clutches (Norberg, 1964; Korpimäki, 1981). The birds are normally single-brooded, but replacement clutches are sometimes laid after clutch loss, and there is one record of a female raising 2 broods in Finland during a single season (Mikkola, 1983). Females begin incubation with the laying of the first or sometimes second egg, resulting in asynchronous hatching. The incubation period lasts 28.5 days on average (range 25–32), and the young are more or less continuously brooded for another 15–23 days. The fledging period is 28–36 days, averaging about 32 days, with the last-hatched young having a slightly shorter fledging period than the others. They become independent of their parents at 5–6 weeks of age and are sexually mature in less than a year (Korpimäki, 1981; Cramp, 1985).

In a brood studied by Bondrup-Nielsen (1978), the eyes of the young opened at 10–12 days after hatching, and by that time the young were largely covered with emerging brown juvenal feathers. Their remiges began to appear at 13–14 days, when defensive bill snapping was also first noted. At 17 days the distinctive white band appeared above the eyes. Based on observations of the Tengmalm's owl, fledging probably occurs at 30–31 days. The postjuvenal molt began at 48–52 days, and by 3 months the birds were in their definitive adultlike plumage.

Breeding success data for North America are limited, compared with those of the Tengmalm's owl. However, 10 of 16 Idaho nests produced no young, but the 6 successful nests had an average of 2.3 young leaving their nests (Hayward, 1989; Hayward and Hayward, 1993). The Tengmalm's owl in Finland was found to have an 85.4 percent hatching success rate (for 701 eggs), and a fledging success rate of 68.2 percent, or an overall reproductive success of 53.6 percent (Korpimäki, 1981). Of 210 hatched nests in Sweden, 60 percent produced at least 1 fledged young, and an average of 4.6 young were raised per successful pair (T. Norberg, cited in Cramp, 1985). Primary causes of losses among eggs and young seem to be desertion and predation, and the risk of nest predation apparently has a strong influence on the female's choice of nest sites and her incubation behavior in this species (Sonerud, 1985). The annual survival of adults in Idaho has been estimated as 46 percent, or substantially less than various estimates for Europe, where survival of first-year males has been estimated at 50 percent, and that of older males as 67 percent (Hayward, 1989; Hayward and Hayward, 1993). Too few birds have been banded to provide significant longevity information.

Evolutionary Relationships and Conservation Status

There can be no doubt that the boreal owl and northern saw-whet owl are very close relatives, and a possible pattern of speciation from an ancestral forest type to a more northern, larger, and conifer-associated boreal owl form and a more southerly, smaller, and deciduous-associated saw-whet owl form can be readily visualized, perhaps during the corresponding boreal and temperate differentiation of the Arcto-Tertiary flora during middle to late Cenozoic times.

The population status of the boreal owl is another matter, and it must remain largely conjectural, owing to the bird's elusive, mostly nocturnal nature and its primary association with relatively inaccessible areas of coniferous forests. Thus although the species has been known to occur in Colorado for more than 80 years, and probably has been a resident since Pleistocene times, it has only fairly recently been proven to breed there (Palmer and Ryder, 1984; Ryder, Palmer, and Rawinski, 1987). The known Colorado breeding range now extends to the New Mexico border (Kingery, 1998) and somewhat beyond into northern New Mexico (Stahlecker and Rawinski, 1990). In recent decades it has also been documented as a breeding bird in Ontario (Bondrup-Nielsen, 1976), Minnesota (Eckert, 1979), Washington (Batey, Batey, and Buss, 1980), and Idaho (Hayward and Garton, 1983). It also breeds locally in northwestern Wyoming and western Montana (Hayward and Hayward, 1993). Territoriality and probable nesting have been found in the Selkirk Mountains and Kettle Range of northeastern Washington (O'Connell, 1987). It may also breed in Utah and northern California, and perhaps even in northern New England (Hayward and Hayward, 1993). All the evidence thus suggests that a continuous or nearly continuous breeding population extends down the Rocky Mountains from Alberta to northern New Mexico, and from Washington south locally into Oregon, in the Cascades, Blue, and Wallowa mountains (Gilligan et al., 1994). During the 1997–2000 Christmas Bird Counts the species was detected on an average of 8 counts. In terms of all localities where the species was detected, the boreal owl was the 15th most widely reported of the 18 North American owl species then encountered. The average total number counted during these 4 years was 15.25. The maximum number ever reported on a single CBC was 7, at Dillingham, Alaska, in 2000. The total number seen in 2000 was 21 birds, or well above the average of 10 for the previous 3 years. The Canadian population has recently been estimated at 20,000–100,000 pairs (Kirk and Hyslop, 1998) and is believed to be fluctuating but possibly stable.

Northern Saw-whet Owl *Aegolius acadicus* (Gmelin) 1788

Other Vernacular Names:
Acadian owl (*acadicus*); Queen Charlotte owl (*brooksi*).

Range (Adapted from AOU, 1983.)

Breeds from southern Alaska, central British Columbia including the Queen Charlotte Islands, central Alberta, central Saskatchewan, central Manitoba, central Ontario, southern Quebec, northern New Brunswick, Prince Edward Island, and Nova Scotia south to the mountains of southern California, locally in the Mexican highlands to Oaxaca, and to southern New Mexico (Magellan and Sacramento mountains), southeastern Arizona (Chiricahua Mts.), western South Dakota, central Minnesota, northern Illinois, southern Michigan, central Ohio, West Virginia, western Maryland, and New York; also breeds locally in the mountains of eastern Tennessee and western North Carolina. Winters generally in the breeding range, but part of the population (mostly immatures) migrates south regularly to the east-central and southeastern United States and casually to southern California, southern Arizona, the Gulf coast, and central Florida. (See Figure 69.)

Subspecies

A. a. acadicus Range as indicated above except for the Queen Charlotte Islands.

A. a. brooksi. Endemic to and resident in the Queen Charlotte Islands, British Columbia.

Measurements

Wing, males 133.5–139 mm (ave. of 8, 136.3), females 135–146 mm (ave. of 9, 141.7); tail, males 65–70 mm (ave. of 8, 67.4), females 69–73 mm (ave. of 9, 71.3) (Ridgway, 1914). The eggs of *acadica* average 29.9 × 25 mm (Bent, 1938).

Weights

Earhart and Johnson (1970) reported that 27 males averaged 74.9 g (range 54–96), and that 18 females averaged 90.8 g (range 65–124). Fall mixed-age birds (2 independent samples) from New Jersey included male means of 78.8 g (65 birds) and 78.4 g (80 birds), and female means of 94.6 g (626 birds) and 95.0 g (629 birds) (Duffy and Matheny, 1997). The associated male:female mass ratios range from 1:1.2 to 1.21. Twenty-two breeding males of *acadicus* averaged 77.4 g and 36 females 131.1 g. (Cannings, 1993). Their associated mass ratio of 1:1.7 is the highest reported for any North American owl and is probably typical

only of breeding birds, when females are at maximum weight. A mean male : female mass ratio based on museum specimens (39 males, 30 females) that were also reported by Cannings is 1:1.25, or similar to that of the fall samples noted above. The estimated egg weight is 9.7 g. The estimated egg-to-female proportional mass ratio is 7.4–9.9 percent.

Identification

In the field. This tiny owl, with no ear tufts but a relatively large head, is usually seen perched motionless in conifer trees during the day, when it can often be approached closely. The distinct chestnut streaking of the underparts and the whitish streaking (rather than spotting) on the crown set it apart from the larger boreal owl. The territorial song of the male is a measured series of mellow single notes that may be continued almost indefinitely. Some representative postures of the saw-whet owl are shown in Figure 70.

In the hand. This owl is similar to the boreal owl, but besides being smaller (wing under 150 mm, instead of more than 160 mm; tail under 85 mm, instead of more than 100 mm) it is more rusty brown throughout, and the crown is narrowly streaked with white. The sexes are essentially identical in plumages as adults, although females may have slightly tawnier faces. Wing-length differences have provided a means of sexing fall migrants in Ontario, but regional mensural variation makes it unlikely that those data can be universally applied (Pyle, 1997b). A more complex and effective method of sexing fall birds has been provided by Brinker et al. (1997), using a combination of wing chord and mass data. Chicks are initially covered with white down that persists about two weeks. Juveniles have the superciliary region and anterior forehead white, in strong contrast with the uniformly blackish brown or lighter brown of the auricular region. The rest of the crown and remainder of upperparts are plain deep brown; chin and sides of throat are dull white; throat, chest, and breast are plain brown, lighter than color of upperparts. The rest of underparts are plain tawny buff or cinnamon buff. According to Evans and Rosenfield (1987), juveniles can be identified through their first year by the presence of a single generation of remiges, as compared with the presence of two generations of remiges, as is typical of virtually all adults. The central rectrices of first-year birds are narrower

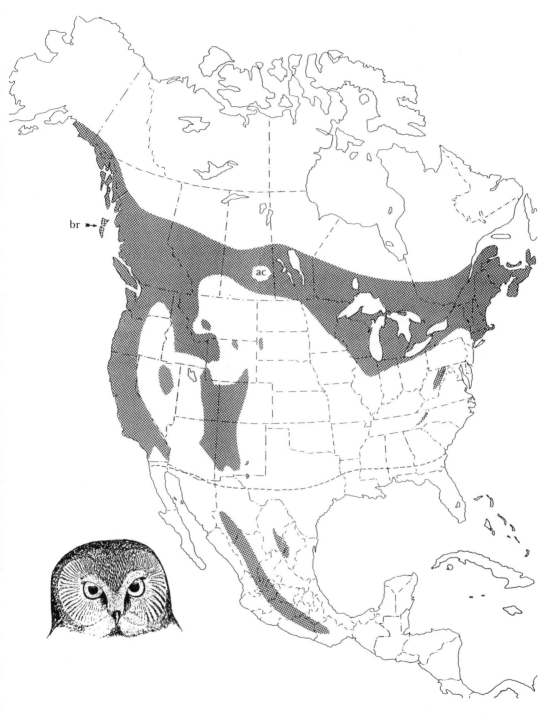

Figure 69. Breeding distribution of the northern saw-whet owl, including races *acadicus* (ac) and *brooksi* (br). The dashed line indicates usual southern limits of wintering vagrants or migrants.

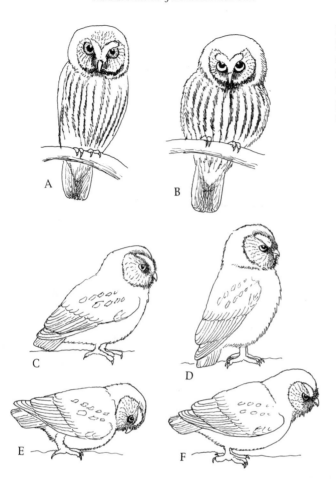

Figure 70. Northern saw-whet owl behavior, including adults in (*A*) erect or so-called concealing posture, and (*B*) fluffed or "fright" posture (after Sealy and Hobson, 1987). Frontal threat display sequence (*C–F*) is also shown (after sketches by C. Collins, 1993).

and more pointed than in adults, and the same is true of the tertials (Pyle, 1997b).

Vocalizations

These birds are vocally active for only about two or three months, mainly from March to May, which corresponds to their breeding period. Otherwise they remain almost mute through the fall and winter periods when on migration and on wintering areas; when captured at that time they are likely to merely snap their beaks defensively (Walker, 1974).

During the breeding season a much wider variety of acoustic signals is doubtless used, although a complete survey of their repertoires remains to be done. Bent (1938) summarized the available early literature on vocalizations, which seem to include at least two major male vocalizations. The first of these is the male advertisement song. This song is heard primarily early in the spring. In Maine it reportedly starts during February, reaches a peak in March, and nearly always ends by early May, rarely extending into the first week of June. The song is uttered most strongly just before daybreak but also occurs during night

(especially moonlit nights) and sometimes during daytime hours under cloudy conditions. It was described by Bent (1938) as a series of repeated monosyllabic *whoop* or *kwook* notes that are uttered at the rate of about 3 notes per 2 seconds with clocklike regularity. Simpson (1972) described the song as a series of resonant, bell-like cooing notes repeated at the rate of 1 or 2 per second, often lasting for several hours without a break. This singing mainly occurred between early April and mid-June, with vocalizations outside of this period erratic and unpredictable. Swengel and Swengel (1987) noted that what they termed the "series song" is the male's advertisement signal, usually an extended series of short, clear notes, often uttered in response to playbacks. The intervals between notes are rather variable, as is the total duration of the call sequence. Cannings (1993) stated that the notes are uttered at a mean frequency of 1.1 kHz, at the rate of about 2 notes per second. It is audible for up to about 300 meters in forests, and up to a kilometer over water. Females respond to the male's song with a softer, lower-pitched, and more rapidly repeated version. Some descriptions of this song might fit a saw-whetting sound, but

Cannings (1993) believes that the so-called saw-whetting call is the species' *ksew* call, which is uttered by both sexes and is not a territorial advertisement signal.

In the mountains of northern Colorado little or no singing occurs in fall, but it begins in late winter (from as early as late January to as late as early April) and is essentially over by the end of April. Similar seasonal schedules have been reported from Wisconsin (Swengel and Swengel, 1987). Individual saw-whets have been heard singing for periods of 70–93 days, averaging 81.5 days. Singing activity in Colorado is essentially nocturnal, usually starting within an hour after sunset and continuing until sunrise. Low temperatures seem to have little effect on singing activity, but singing was never heard during periods of heavy snowfall. Cloud cover also apparently does not affect singing rates, and moderate winds may actually increase the rate of saw-whet singing, although high winds have a negative effect. The degree of moonlight may also have a slight positive effect on singing and may stimulate the birds to begin initial singing for the season (Palmer, 1986). In British Columbia singing begins in February and peaks in late February or March; it may continue on into June, with a minor resumption in August and September (Cannings, 1993).

In addition to the male advertisement song, and the *ksew* call, several additional vocalizations have been described. They are summarized by Cannings (1993), who provided sonograms of four of these. One is a nasal whine or wail, lasting up to 3 seconds, of unstated possible function. Another is a short series of whistled notes uttered by the male when he is approaching the nest with food. The female utters a high-pitched *tssst* call in response to the male's advertising song, or when he is visiting the nest.

Several other miscellaneous notes have also been described. Johns, Ebel, and Johns (1978) noted that nesting males have two distinct calls, including a loud, harsh territorial call, uttered from different trees within about 10 meters of the nest, and a second softer call uttered when the owl is approaching or leaving the nesting tree. Webb (1982a) stated that after uttering preflight calls at dusk, the male will fly to the nest tree and approach the female. Depending on the stage of the nesting cycle, the male will either enter the nest cavity (early in the cycle) or perch near the hole (eggs or owlets present). The flight call uttered during this approach is a very rapid staccato burst of calls, to which the female responds with soft *swee* notes of rising inflection. When the two birds are closest to one another there is a crescendo of chattering *chuck* notes.

Very young owlets produce peeping or chirruping notes of 6–8 elements, having dominant frequencies at about 2.5 kHz and repeated at the rate of about one per second (Cannings, 1993). Defensive beak snapping appears after about 7–10 days. The high-pitched squeaking notes produced by young birds are replaced around or after fledging with coarser begging calls. These are rasping and sibilant or hissing notes similar to that made by steam jets escaping from a nozzle. These are similar to the *tssst* notes produced by females when soliciting copulation (Cannings, 1993).

Habitats and Ecology

In the eastern and northeastern parts of the United States the northern saw-whet is variably common, and there it occupies an altitudinal range appreciably lower than in the West, though it still generally breeds above 1500 meters in the southern Appalachians, typically in spruce-fir forests. Wintering occurs at lower elevations and in varied habitats ranging from conifers to hardwoods and open country. The owl typically roosts in rather dense, often young conifers at the edge of extensive woodlands, and it forages in both woods and open fields. Probably swampy areas among deep coniferous forests are preferred over dry, deciduous woods for breeding. In any case breeding habitats are typically fairly mature forests, with a mixture of large trees, both living and dead, and with medium-sized woodpecker cavities present.

In Colorado northern saw-whet owls are vertically distributed during the breeding season from about 1900 meters in riparian habitats to more than 3000 meters in spruce-fir forests. In his study of ecological distribution and habitat use by singing saw-whet owls in northern Colorado, Palmer (1986) found saw-whets to be distributed from 1770 to 3170 meters in elevation, with densest concentrations in riparian areas having associated deciduous trees, primarily quaking aspen (*Populus tremuloides*). On 15 saw-whet territories this tree covered an average of about 16 percent of the surface area. Like the boreal owl, the saw-whet avoided large and unbroken stands of pines but otherwise tended to select habitat types in proportion to what was locally available. Thus a considerable amount of lodgepole pine (*Pinus contorta*) and Douglas fir (*Pseudotsuga menziesii*) was usually present on their territories, as well as mountain shrub, meadows, ponderosa pine (*Pinus ponderosa*), and miscellaneous substrates in diminishing abundance. The ground cover was diverse as compared with that of boreal owl territories; it was the typical ground cover of mesic environments, with substantial amounts of grass.

Territories occupied by saw-whets had some areas of south-facing slopes that were snow-free, probably facilitating hunting for those individuals that might remain on territories through the winter.

Surveys in North Carolina by Simpson (1972) indicated that advertising birds there were always associated with spruce-fir forests, although 17 of 49 locality records were from areas where hardwoods comprised more than 90 percent of the forest canopy, with conifers existing only as scattered, solitary trees. He judged that the apparent abundance of the owl in mixed forest habitats, which support a large population of small rodents, may in part reflect the relatively abundant prey species in these sites.

Regardless of tree species present, saw-whets are dependent upon woodpecker cavities, mainly those of northern flickers (*Colaptes auritus*) in Colorado, for their nest sites (Webb, 1982a). These cavities, averaging about 7.5 centimeters in diameter, are normally in dead trees having a minimum diameter of 30.5 centimeters at breast height and are at least 4.6 meters above ground level (Thomas, 1979). All but 2 of 15 territorial saw-whets in central Colorado were found in ponderosa pine (*Pinus ponderosa*) or mixed pine–Douglas fir (*Pseudotsuga menziesii*) forests between 1700 and 2600 meters elevation; the other two were in higher subalpine forests. Dense forests are generally avoided (Cannings, 1993).

In Idaho, saw-whets have also been found to exhibit a habitat preference for deciduous riparian areas, and when in coniferous habitats they select perching and roosting sites with a relatively high density of midelevation (2–4 m) coniferous canopy vegetation. The birds often roost about 4 meters aboveground, in trees about 22 meters high, and in densely foliated areas of the outer half of the branch. Roosting trees are usually in areas where tree density is relatively thick, and roosts are chosen that seemingly provide both thermal cover and hiding advantages (Hayward, 1983; Hayward, Hayward, and Garton, 1993).

Observations of saw-whets roosting on spring migration in southern Ontario suggested that the birds preferred roosting in the thickest hemlock (*Tsuga*) trees available, usually between 1.5 and 3 meters aboveground and with excellent concealment from above (Catling, 1971). During winter they seem to prefer to roost in rather small trees (about 6 m tall) that offer a combination of concealment and a relatively open zone below the canopy that allows for easy approach or escape below the tops of the trees (Mumford and Zusi, 1958). The small roosting trees may be under a canopy of old-growth conifers (Boula, 1982) or sometimes along the margins of upland forest groves (Randle and Austin, 1952). When disturbed during the day, roosting saw-whet owls typically assume a concealing posture in which the body plumage, especially the breast and upper back region, is compressed; the wing nearest the intruder is raised to the level of the bill and its upper surface directed toward the intruder; the crown feathers are raised and the feathers above and between the eyes fanned out; and the eyes are widely opened (Catling, 1972a). This results in a posture essentially identical to that typical of alert boreal owls (see Figure 68).

Home ranges in saw-whet owls are still only rather poorly studied, but Forbes and Warner (1974) estimated the home range of a single radio-tagged bird as 114 hectares during winter, of which 31 hectares were infrequently used. Studies in Michigan during the winter months indicate that wintering birds may be moderately sedentary during that season, perhaps ranging over an area of about 40 hectares and utilizing a number of different roosting sites (Mumford and Zusi, 1958). Palmer (1986) estimated the home range of one bird as 78 hectares, and Cannings (1987) estimated that of another as 142 hectares.

There are also few estimates of population densities, but Simpson (1972) estimated a density equivalent to 1.1 advertising males per linear kilometer of census transects in spruce-fir habitats of North Carolina, and he cited other observations of similar densities in the area. Hardin and Evans (1977) indicated that a maximum breeding density might be a pair per 40 hectares (2.5 pairs per sq km). Swengel and Swengel (1987) estimated a density of 5.3 males per square kilometer, based on estimated numbers and locations of calling males responding to playbacks, but they did not have any banded or radio-marked birds in their population. A later estimate (Swengel, 1990) indicated an even higher maximum breeding density of up to 12.9 singing males per square kilometer. However, most breeding habitats probably support no more than a pair per square kilometer (Cannings, 1993).

Movements

This is one of the most migratory of the noninsectivorous owls in North America. Its migrations probably range all the way from minor vertical movements to lower areas during winter in mountainous regions of western North America to fairly extensive horizontal movements in eastern and northern portions of the species' range. Some vertical migration may also occur in the southern Appalachians, although this is still poorly documented (Simpson, 1972). Judging from banding records (Holroyd and Woods, 1975), in the eastern parts of North America the spring migration is from March 1 to May 31, ending by mid-April in New Jersey but starting in late April in Michigan. The fall migration extends from September 1 to November 30, starting in early October in Michigan and Wisconsin and ending about mid-November in Ohio and Pennsylvania. There are apparently two main migratory corridors in eastern North America. One reaches down through the Ohio River Valley; the other extends along the Atlantic coastal lowlands from Maine to North Carolina. At least in some parts of their range northern saw-whet owls may be more nomadic than truly migratory, temporarily settling in regions of high food abundance that they had encountered during their movements in the non-breeding season.

Fall migratory movements of these owls are sometimes quite substantial. This is likely to be especially the case when rodent populations may be likely to become low; early reports of large movements of owls in southern Ontario occurred prior to the worst parts of the winter, judging from accounts summarized by Bent (1938). A summary of fall migration records at Prince Edward Point, Ontario (Weir et al., 1980), indicates that the yearly variations in migration amplitude are a reflection of the relative numbers of yearling birds in the samples captured. In one of three years studied, the juveniles migrated significantly earlier during fall than did the adults, but in the other two years females of both age classes migrated earlier than did males. In the total sample young birds made up a majority (58 percent) of the captured sample, which seems likely to be a much higher proportion than would be found in the population as a whole and indicates the probable greater tendencies for young to be more mobile than adults. Most of the owls were captured during October, with half the total being caught during the second and third weeks of that month. The size of the nighttime captures was related positively to the presence of northwest winds and clear skies, a pattern similar to that found for nocturnally migrating passerine birds as well as diurnal birds of prey. Using data from birds captured and from recaptures at various migration point, Brinker et al. (1997) calculated mean nightly movements of 14.1–23.1 kilometers, and one short-term estimate of 5.6 kilometers of movement per hour of darkness. There was a significant decline in the percentage of males captured with declining latitude, suggesting that females are more likely to migrate farther south than males, and immatures more likely to migrate greater distances than adults. Owls moving south of Pennsylvania are mainly females, and the

southeastern coniferous forests appear to be a major wintering area for these birds. Autumn movements of banded birds to and from a banding station in northeastern Wisconsin were studied by Erdman et al. (1997). Four juveniles documented for periods of from 12 to 41 nights moved an average of 21.8–33.3 kilometers per night, distances comparable to the average (29.6 km) of four adults similarly estimated by banding and recapture data.

A similar study of spring migration patterns was performed at Toronto, Ontario, by Catling (1971). There, as well as in western New York, the spring migration extends from mid-March until the end of April, with major influxes coinciding with clear nights having light winds. Based on a rather small sample size, the sexes probably migrate simultaneously, and the migration coincides with that of various small passerines that at least in part are used by the owls as a source of food.

Along the western shore of Lake Michigan in Wisconsin, the fall migration of saw-whet owls extends from late September to late November, with a peak about the third week of October. There the migration intensity appears to be positively related to westerly winds and the passage of a cold front, and no obvious temporal differences in migration by age classes is apparent. Of 168 birds captured by Mueller and Berger (1967), 59 percent were first-year birds, again suggesting a higher tendency for migration in young birds than in older age classes.

Foods and Foraging Behavior

Bent (1938) summarized the early literature on saw-whet foods, which indicates a strong dependence on mice, especially woodland mice. Randle and Austin (1952) examined 173 pellets from southwestern Ohio and found that all but 60 of these contained a single rodent or shrew skull each, the remaining ones made up entirely of mammalian fur. Of the skulls present, *Peromyscus* mice comprised about 48 percent, *Microtus* voles 23 percent, and *Blarina* shrews and *Synaptomys* bog lemmings together another 10 percent. Errington (1932a) found in southern Wisconsin that most (70 percent) of 72 pellets contained skulls of *Peromyscus* mice, with voles (*Microtus*) secondarily represented, plus the remains of a single dark-eyed junco (*Junco hyemalis*). Likewise, of 77 pellets from Oregon that were examined by Boula (1982), *Peromyscus* remains comprised nearly 80 percent of the total; all the remainder of the prey items were other small mammals except for 4 percent comprised of passerine birds. *Peromyscus* mice were also a major component of the pellets examined by Catling (1972b) in Ontario and by Palmer (1986) in Colorado. Of about 400 pellets examined by Swengel and Swengel (1987), about 80 percent were of *Peromyscus* and the remainder comprised of voles and shrews.

Graber (1962) examined more than 350 pellets obtained in central Illinois between November and April and found that deer mice (*Peromyscus leucopus*) comprised 70 percent of all the prey individuals identified, with white-footed mice (*P. maniculatus*) of secondary importance and various shrews, voles, and house mice (*Mus musculus*) of small occurrence. Bird remains were found in some pellets, especially during spring, and consisted of various small birds (swallows, chickadees, kinglets, sparrows), with the largest species represented being a northern cardinal (*Cardinalis cardinalis*). Graber (1962) estimated that the usual amount of food consumed per foraging period was 13.3–18.3 grams, or about what was required to keep adult captives alive without weight loss during 24-hour periods. C. Collins (1963) suggested that these estimates were too low, and that with two feedings per day, in early evening and again at dawn, the food intake might be substantially greater than this.

Cannings (1993) provided a current analysis of saw-whet foods, based on studies collectively representing 3583 prey items from eastern regions and 2924 items from western ones. In the East, *Peromyscus* species dominated the prey base (68.7 percent of the total items), with *Microtus* forms secondary (13.6 percent). In the West, *Peromyscus* types represented 32.0 percent, and *Microtus* 28.8 percent. Other mice and voles (mainly *Reithrodontomys*, *Clethrionomys*, and *Mus*) made up most of the rest, with birds contributing 1.8 (West) to 3.2 percent (East), and insects barely registering. The largest prey taken are juveniles of pocket-gophers, chipmunks, and squirrels among mammals, and rock doves (*Columba livia*) among birds. Most prey taken weigh less than 40 grams, but prey approaching that size might be preferred.

Few observations on hunting have been made in this seminocturnal species, but apparently most prey is captured on the ground by pouncing on it from above, mostly after short flights from elevated perches. The high degree of ear development would suggest that the birds probably can capture prey in total or near-total darkness. Palmer (1986) suggested that the species' relatively light wing loading as compared with boreal owls provides for higher maneuverability and allows them to hunt in relatively heavy shrub-dominated cover, a trait that has also been reported by Hayward (1983) as well as by Forbes and Warner (1974).

Social Behavior

Although normally monogamous, males are sometimes bigynous when prey is abundant, and there is even a record of trigyny. Successive biandry by females has also been suspected, in which the female abandons her first brood in the care of her original mate, to begin a second nesting with another male. It is believed that some males or pairs may maintain territories throughout the year, which would increase chances of the pair bond being continued for a second year (Cannings, 1993). However, the migratory tendencies of at least some populations would diminish this possibility, as would the seemingly high mortality rates of 40–50 percent.

Observations by Palmer (1986) in Colorado indicate that singing activity there peaks in April, or about the same time reported elsewhere in the species' range, corresponding roughly to the peak period for egg records in the northeastern states. Karalus and Eckert (1974) have described courtship as beginning with the male flying in a circle around a perched female 15–20 times before landing. He then utters a variety of calls and engages in a series of bobbing, shuffling, and foot maneuvers that gradually bring him closer to her. He occasionally carries prey during these maneuvers, dropping it a few inches away from her. Once she has swallowed it he breaks out into a series of tooting calls, at which time the female takes off, followed by the male. Copulation takes place in a tree at midheight and may be repeated several times a night over a period of several nights, according to these authors. Mutual preening has not been specifically noted but almost certainly occurs in the species.

Breeding Biology

Nesting in this species is almost always in woodpecker cavities, although a few nests have also been reported from natural tree cavities and more rarely other sites. Flickers (*Colaptes*) are the usual sources of such cavities, although hairy woodpecker (*Dendrocopus villosus*) cavities have also reportedly been used, and probably any woodpecker hole that is at least 7 centimeters in diameter will serve. Palmer (1986) found one nest in a woodpecker hole with an entrance diameter of 7.2 centimeters. One nest observed by Peck and James (1983) had an entrance diameter of 9 centimeters, and two had inside cavity diameters of 7.5 and 9 centimeters.

Nesting records are not numerous for this species, but in New England and New York there are 12 records between March 19 and July 3, with half from April 10 to 20. Thirteen active Ontario nests were found between April 1 and July 27, with 7 of these between April 10 and May 17 (Peck and James, 1983).

Clutch sizes are most often of 5 or 6 eggs, but ranges of 4–7 are common (Bent, 1938). Cannings (1993) reported an average of 5.67 eggs for 36 British Columbia nests. Pacific-coast clutches may average slightly smaller (3.79 vs. 4.96) than those from eastern North America (Cannings, 1993). A sample of 14 nests observed by Peck and James (1983) had from 1 to 7 eggs, with 6 being the modal clutch size and 4.85 eggs the average. Of the nests they observed, 9 were in old woodpecker cavities; another 5 were in unspecified cavities; and 1 was in a wood duck (*Aix sponsa*) nesting box. Of the tree nests, 10 were in deciduous trees and 2 in conifers; in 11 of 12 cases the trees were dead. The height of 13 nests ranged from 2.5 to 13.5 meters, with most from 3.5 to 6 meters aboveground.

The eggs are laid at intervals of 1–3 days, and usually 48–72 hours apart. Cannings (1987) found an average laying interval of 2.0 days. Only the female incubates, probably beginning with the laying of the second egg. Although there are some early estimates of 21-day incubation periods, other estimates (Terrill, 1931; Peck and James, 1983) are for periods of at least 26–28 days. Cannings (1987) reported an average incubation period of 27.3 days (9 nests). The young begin to open their eyes at 8–9 days of age, gradually continuing through the first 3 weeks of life. By the age of 26–28 days one observed owlet was able to fly about 15 feet from a log, but could not rise from the ground. By 3 or 4 weeks of age the wings are growing rapidly and the down is wearing off the tips of the juvenal plumage. In one case the young left the nest when the oldest was 27–34 days old, and they probably could fly fairly well by then (Terrill, 1931). Cannings (1987) found an average fledging period of 33.4 days, and an average fledging interval of 1.4 days, or slightly more abbreviated than the average laying interval.

Too few nests have been followed to provide much indication of typical nesting success. However, Cannings (1987) reported that 30 of 40 eggs in 9 clutches hatched. He also found (1993) that an average of 2.67 young fledged from 22 total nests, or 3.47 per nest from 17 successful nests. One bigamist male fledged at least 6 young from his 2 nests, and another up to 11 young. A trigamist male fledged a total of 10 young, with 2 of his 3 mates successfully producing fledglings. Banding return data are limited but suggest annual survival rates of about 40 percent for first-year birds, and 50 percent for older birds. Of 74 banded first-year birds, only 6 survived until their third birthday (Cannings, 1993). Maximum known longevity under natural conditions is 10 years and 4 months (Patuxent Wildlife Research Center banding data).

Evolutionary Relationships and Conservation Status

Certainly the saw-whet and boreal owls are close relatives, but in addition there are two other species of *Aegolius* occurring in the Western Hemisphere. One of these is the unspotted (or Central American) saw-whet owl (*A. ridgwayi*), occurring from southern Mexico to western Panama; the other is the buff-fronted owl (*A. harrisii*), occurring discontinuously in South America. Of these, the former is certainly a very close relative of *acadicus*, resembling the immature state of the northern saw-whet but lacking light markings on the wings and tail (Wetmore, 1968). A single juvenile saw-whet specimen from Oaxaca (south of the previously known breeding range of *acadicus*) was described by Briggs (1954) as a new race (*brodkorpi*) of the northern saw-whet rather than of *ridgwayi*, which is known from below the Isthmus in adjacent Chiapas. These two forms possibly overlap in southern Mexico and perhaps even hybridize, but too little is known of this to be certain of the situation (König, Weick, and Becking, 1999). The AOU *Check-list* (1998) currently considers the two as specifically distinct but comprising a superspecies. Richard Cannings (pers. comm.) has noted that the song of the unspotted saw-whet owl is seemingly identical to that of the northern saw-whet, which would support the view that the two should be considered conspecific. Recent classifications have retained their separate species status.

No good information is available on the status or possible population trends of this species. During the 1997–2000 Christmas Bird Counts the species was detected on an average of 159 counts, or about 9 percent of all U.S. and Canadian counts. In terms of all localities where the species was detected, the northern saw-whet owl was the sixth most widely reported of the 18 North American owl species then encountered. The average total number counted among the 100 counts having the highest species totals was 158 birds, or about 1.6 percent of all owls reported in the United States and Canada during the past few years. The maximum number ever reported on a single CBC was 61, at Quabbin, Massachusetts, in 1995. The Canadian population has recently been estimated at 50,000–150,000 pairs (Kirk and Hyslop, 1998) and is believed to be fluctuating but possibly stable. Cannings (1993) estimated a total national population of 100,000–300,000 birds that is perhaps slowly declining.

Appendix 1

Key to Genera and Species of North American Owls (Exclusive of Strictly Mexican Species)

A. Facial disk heart-shaped; claw of middle toe comblike; outer rectrices (tail-feathers) the longest ....
.... family **Tytonidae,** genus **Tyto;** barn owl (Figure 71C)

AA. Facial disk not heart-shaped; claw of middle toe not comblike; central rectrices the longest .... family **Strigidae**

    B. With distinct ear tufts present (if inconspicuous, then the eye rimmed broadly with black or the facial disk tinged with rusty brown)

        C. Larger (wing over 275 mm)

            D. Flanks and underparts barred; bill over 2 cm from nostrils to tip; adult weight over 1300 g ....
.... genus **Bubo;** great horned owl (Figure 71A)

            DD. Flanks and underparts striped; bill under 2 cm from nostrils to tip; adult weight under 500 g ....
.... genus **Asio** (2 spp.)

                E. With long (over 4 cm) ear tufts, plumage mottled or finely cross-marked below and more grayish above . . .
.... long-eared owl (Figure 71E)

                EE. With short (under 2 cm) ear tufts; plumage more heavily striped below and more buffy throughout.....
.... short-eared owl (Figure 71D)

        CC. Smaller (wing under 200 mm) .... genus **Otus** (4 spp.)

            D. Iris brown; toes naked; culmen from cere 8.5–10 mm; outer primary longer than longest secondaries, wing to 145 mm; weight to 70 g........ flammulated owl (Figure 72G)

            DD. Iris yellow; toes not naked, middle toe over 10 mm; culmen from cere over 10 mm, outer primary shorter than secondaries; wing to 190 mm, adult weight over 70 g

                E. Inner web of outermost primary with light bands or blotching; middle toe (excluding claw) no more than 14 mm; 4 primaries indented on inner web; wing 130–150 mm, weight 71–127 g ....
.... whiskered screech-owl (Figure 72F)

                EE. No white on inner web of outermost primary; middle toe over 14 mm; 5–6 primaries indented on inner web; wing 135–190 mm; weight 113–227 g

                    F. Plumage with dorsal linear black streaks and ventral thin crossbars; bill usually black; rufous phase rare; Rocky Mountains westward ....
.... western screech-owl

                    FF. Plumage with dorsal transverse streaks and ventral anchor-shaped crossbars; bill never black, often yellowish green; rufous plumage phase fre-

quent; east of Rockies . . . .
. . . . eastern screech-owl (Figure 72A)
BB. Ear tufts lacking or rudimentary
  C. Larger (wing over 200 mm)
    D. Mostly white, under tail coverts reaching tip of tail . . . .
    . . . . genus **Nyctea;** snowy owl
    DD. Not mostly white; under tail coverts not reaching tip of tail
      E. Tail graduated, about three-fourths as long as wing . .
      . . . . . . genus **Surnia;** northern hawk-owl (Figure 71F)
      EE. Tail rounded, about two-thirds as long as wing . . . .
      . . . . genus **Strix** (3 spp.)
        F. Wing over 400 mm; iris yellow; feathering of toes
          hiding base of claws . . . .
          . . . . great gray owl (Figure 71G)
        FF. Wing under 375 mm; iris brown; feathering of toes
          not hiding base of claws
          G. With spotted flanks and breast . . spotted owl
          GG. With barred breast and striped flanks . . . . . .
          . . . . barred owl (Figure 71B)
  CC. Smaller (wing under 200 mm)
    D. Tarsus more than twice the length of middle toe (exclud-
    ing claw). . . . . . genus **Athene;** burrowing owl (Figure 72B)
    DD. Tarsus about equal to middle toe plus claw
      E. Tail less than half length of wing (under 50 mm), of
      10 rectrices; tarsus with scant feathering; weight to
      50 g . . . . genus **Micrathene;** elf owl (Figure 72E)
      EE. Tail at least half the length of wing (over 50 mm), of
      12 rectrices; tarsus heavily feathered; weight over 50 g
        F. Wing over 130 mm; tail about half the length of
        wing, no "eye spots" present on nape. . . . .
        . . . . genus **Aegolius** (2 spp.)
          G. Wing at least 165 mm; spotted with white on
          crown . . . . boreal owl
          GG. Wing under 150 mm; streaked with white on
          crown . . northern saw-whet owl (Figure 72C)
        FF. Wing under 130 mm; tail about two-thirds length
        of wing; 2 "eye spots" on nape . . . .
        . . . . genus **Glaucidium** (2 spp.)
          G. Crown plain to lightly white-spotted; tail with
          6–7 whitish bars (3–4 visible from below). . .
          . . . . northern pygmy-owl (Figure 72D)
          GG. Crown heavily streaked with white; tail with
          7–8 light brown bars (3–5 visible from below)
          . . . . ferruginous pygmy-owl

Key to Genera and Species of Indigenous Mexican Owls*

A. Head tufted with earlike feathers on the crown
  B. Ear tufts long; larger species (wing over 200 mm, total length 35–45 cm)
    C. Tufts almost entirely dark, underparts streaked or coarsely
    barred. . . . . genus **Asio** (2 spp.)
      D. Facial disk white; heavily streaked below . . . .
      . . . . striped owl (Figures 73D, 74D)

---

*Excludes those Mexican species occurring only south of the Isthmus, and most U.S. species that extend variably south into Mexico and are included in earlier key (great horned owl, long-eared owl, short-eared owl, burrowing owl, flammulated owl, northern saw-whet owl, elf owl, and eastern, western, and whiskered screech-owls). The two U.S. pygmy-owls are included for ready comparison.

DD. Facial disk fuscous, heavily barred below. . . . .
. . . . stygian owl (Figure 74F)

CC. Tufts mostly buff or white; finely barred on underparts, facial
disk dark rufous . . . .
. . . . genus **Lophostrix;** crested owl (Figures 73B, 74E)

BB. Ear tufts short, sometimes almost invisible. . . . .
. . . . genus **Otus** (4 spp.)

C. Larger (wing over 165 mm; tail length 23–27 cm), underparts
distinctly streaked, with many crossbars; primary song of pulsed
"bouncing ball" phrase, with up to 15 gruff notes; only gray
plumage morphs known (superspecies *kennicottii*)

D. Eyes brown to golden brown; southwestern interior Mexico
. . . . Balsas screech-owl (Figure 74G)

DD. Eyes yellow; southwestern coastal Mexico . . . .
. . . . Pacific screech-owl

CC. Smaller (wing under 165 mm, total length 20–23 cm), song a
long trill; gray and red morphs known; eyes yellowish to brownish
yellow (superspecies *guatemalae,* with 2 semispecies or sibling
species)

D. Underparts more streaked, with scattered cross-bars; upperparts lighter and with whiter spotting; eyebrows flecked
with whitish, eyes golden yellow; song a prolonged quavering trill (6–15 seconds), rising to a crescendo and ending
abruptly . . . .
. . . . Central American (Vermiculated) screech-owl
(typical *guatemalae,* Mexico to northern Costa Rica)
(Figure 74H)

DD. Underparts more mottled and vermiculated, upperparts
darker, more yellowish throughout; eyebrows uniform
brown, eyes more brownish; song a short trill (3–5 seconds), guttural and froglike, fading toward end . . . .
. . . . vermiculated screech-owl
(typical *vermiculatus,* from Costa Rica south)
(Figure 74I)

AA. Head lacking distinct ear tufts
B. Larger (wing over 200 mm, total length 33–48 cm)
C. Chest broadly banded, lower breast unmarked or with a few nar-
row bars . . . .
. . . . genus **Pulsatrix;** spectacled owl (Figures 73A, 74A)

CC. Entire undersurface barred or streaked . . . . genus **Strix**
D. Sides and underparts buffy, heavily streaked with brown. . .
. . . . Mottled owl (Figures 73C and 74C)

DD. Lower breast white, heavily barred with black . . . .
. . . . black-and-white owl (Figure 74B)

BB. Smaller (wing under 200 mm, total length 14–20 cm) . . . .
. . . . genus **Glaucidium** (5 spp.)

C. Tail longer (> 53 mm); scapulars conspicuously spotted with white;
rufous, olive brown and grayish brown plumage morphs all exist

D. Crown spotted; tail bars white (3–5 visible beyond
coverts); montane habitats; double hoot-notes typical. . . . .
. . . . northern pygmy-owl

DD. Forehead streaked; tail bars brownish (4–5 visible beyond
coverts), lowland habitats; up to 60 single hoots in series
typical. . . . . ferruginous pygmy-owl

CC. Tail shorter (usually < 53 mm); scapular spotting inconspicuous,
rufous Dorsal plumage typical, but head often distinctly grayish;
single hoots typical

D. Head grayish; southeastern Mexico; 2–18 short ringing

hoots, equally spaced with short pauses between; 4–5 total tail bars (2–4 visible beyond coverts)................... .... Central American pygmy-owl (Figure 74L)

DD. Head brownish or brownish gray; western or northeastern Mexico; hoots with long pauses between, 6–7 tail bars (4–5 visible beyond coverts)

    E. Western Mexico; 2–24 short hoots with long pauses typical, females' plumage like males.............. .... Colima pygmy-owl (Figure 74K)

    EE. Northeastern Mexico; 1–3 longer, well-spaced hoots typical; females more rufous than males........... .... Tamaulipas pygmy-owl (Figure 74J)

Key to Structural Variations in External Ears of North American Owl Genera (Exclusive of Mexico)

A. Facial disk heart-shaped; ear opening small, with a small preaural ear flap (family Tytonidae) .... **Tyto** (Figure 71C)

AA. Facial disk not heart-shaped (family Strigidae)

  B. Dermal flaps lacking around auricular opening (subfamily Buboninae)

    C. Facial disks poorly developed; auricular opening smaller than diameter of eye .... **Surnia, Glaucidium, Micrathene, Athene,** and **Nyctea** (Figure 72B, D, E)

    CC. Facial disks well developed; auricular opening slightly wider than diameter of eye

      D. Auricular opening crescent-shaped and of same size on both sides .... **Otus** (Figure 72A)

      DD. Auricular opening oval and larger on right side .... **Bubo** (Figure 71A)

  BB. Dermal flaps present around auricular opening, which is much wider than diameter of eye, forming a dermal conch; external ear asymmetrically developed (subfamily Striginae)

    C. Cranium symmetrical; the preaural flap distinctly enlarged

      D. Right and left dermal flaps of equal size but the meatus vertically displaced on the 2 sides; the dermal conch crescent-shaped, and with a ligamentous bridge; the stapedial footplate unspecialized............... **Asio** (Figure 71D, E)

      DD. Dermal flaps and meatus larger on right side, the dermal conch kidney-shaped and without a ligamentous bridge; the stapedial footplate specialized .... **Strix varia** and **S. occidentalis** (Figure 71B)

    CC. Cranium and external ears distinctly asymmetrically developed and vertically displaced on the 2 sides

      D. Cranium slightly asymmetrical; preaural flap much larger than postaural flap .... **Strix nebulosa**

      DD. Cranium markedly asymmetrical; preaural flap slightly larger than postaural flap........ **Aegolius** (Figure 72C)

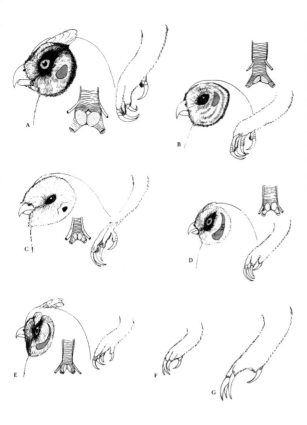

Figure 71. Head and tracheal structures, foot structures, or both of (*A*) great horned owl, (*B*) barred owl, (*C*) barn owl, (*D*) short-eared owl, (*E*) long-eared owl, (*F*) northern hawk-owl, and (*G*) great gray owl. Areas occupied by the ear conch are indicated by stippling. In part after Ridgway (1914); tracheae (dorsal aspect) were drawn to scale directly from preserved specimens, but scale is larger than that used for depicting heads and feet.

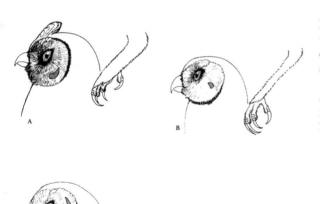

Figure 72. Head structures, foot structures, or both of (*A*) eastern screech-owl, (*B*) burrowing owl, (*C*) northern saw-whet owl, (*D*) northern pygmy-owl, (*E*) elf owl, (*F*) whiskered screech-owl, and (*G*) flammulated owl. Scale used is twice that of Figure 71. In part after Ridgway (1914).

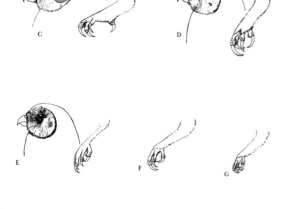

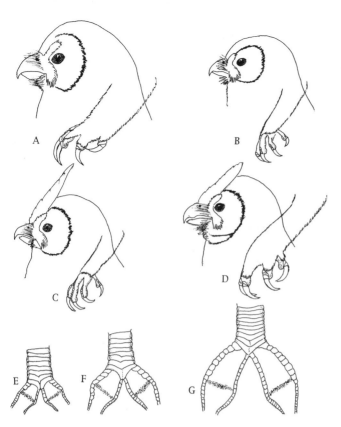

Figure 73. Head and foot structure of four Mexican owls, including (A) spectacled owl, (B) mottled owl, (C) crested owl and (D) striped owl. Scale approximately as in Figure 72. Partly after Ridgway (1914). Also shown are syringeal regions of female (E) and male (F) spotted owl, and (G) adult mottled owl, all drawn at 5× scale of heads and feet (after A. Miller, 1934, 1963).

Figure 74. Adult head patterns of indigenous Mexican owls: (A) spectacled owl, (B) black-and-white owl, (C), mottled owl (D) striped owl, (E) crested owl, (F) stygian owl, (G) Balsas screech-owl, (H) Central American (Guatemalan) screech-owl, (I) Central American (vermiculated) screech-owl, (J) Tamaulipas pygmy-owl, (K) Colima pygmy-owl, and (L) Central American pygmy-owl. Vocalizations of the pygmy-owls (J–L) are also shown (one-second intervals indicated, 2 kHz frequency range bracketed). Not drawn to uniform scale.

Appendix 2

Advertisement and Other Typical Calls of Non-Mexican North American Owls

Group 1.

Screeching, croaking, and other nonhooting or nonwhistling calls.
>A long, hoarse screech, *karr-r-r-r-r-ick* lasting about 2 seconds (Figure 75A), given at intervals of 1–20 seconds, and in series of up to 50 or more times while in flight (advertising song); also various snoring, croaking, and wheezing calls but never hoots. (Croaking, hissing, and screeching calls are uttered by many other owls, but not as primary advertising songs.) . . . . barn owl

Group 2.

Low-pitched hooting sounds, often in prolonged series of up to about 3 per second, but not rapidly pulsed or trilled, with variations in loudness and cadence but not pitch.
>A. A rather definite and consistent number of up to 9 notes that are distinctly accented or cadenced. Arranged below by increasing number of syllables in phrase.
>>1. Double-noted *coo-hoooo,* similar to a cuckoo clock, the second note much prolonged and sometimes rising slightly in pitch; the doublets often monotonously repeated for an hour or more (advertisement call); also a mellow, fluty 5-noted *whea-woo-who-woo-who* in courtship, with the last 4 notes slurred together. . . . . burrowing owl
>>2. Three or 4 low-pitched and cadenced notes, *who', who-whoo; whooo',* lasting nearly 2 seconds, the middle portion loudest and highest, the last prolonged and sometimes down slurred similar to barred owl (or "Who; who are youuuu?"); sometimes lacking the introductory note, and often with the last note distinctly emphasized (Figure 75G) (advertising call). Occasionally uttered as 2 long notes followed by 2 shorter ones; also diverse barking and sirenlike whistling noises. . . . . spotted owl
>>3. A variable series of 4–5 (rarely 3–9) low-pitched hooting notes, with no pitch variation but usually a distinct cadence, often *who'; hoo-hooo'; hoo-hooo',* sounding rather like "DON'T kill owls! Save owls!" to which a preliminary "Please!" and 1 or 2 additional "Save owls!" are often added (Figure 75E) (advertising call) . . . . great horned owl
>>4. A syncopated series of 2 short and closely spaced notes followed by 2–5 (usually 3) longer and equally spaced notes ("dot, dot, dash, dash, dash"), the series lasting about 1–1.5 seconds. Often repeated several times without pause, and ending with an extra long note (syncopated duet song). Also a series of about 6 notes, sometimes with a pause before the last, or the penultimate one emphasized (male song) . . . . whiskered screech-owl; see also 2AA6

249

5. Nine hooting notes in distinct 2-phrase cadence, the whole sequence lasting nearly 3 seconds and sounding like, "Who cooks for you; who cooks for you-all?" (Figure 75F) (advertising call); also diverse barking, chuckling and screaming notes .... barred owl

AA. A variable number of single or doublet hooting notes, not so distinctly accented (see also Group 3A). Arranged below by increasing pitch.

1. A series of up to 12 regularly spaced, very low-pitched (ca. 200 Hz) *boo* (sometimes double or triple) notes of equal duration (about 0.3 seconds) and uniform interval, the single units usually uttered at about 3 per 2 seconds (Figure 75H), but often becoming more rapid, lower, and softer toward the end of a calling sequence (advertising song). The female's notes are similar but harsher and are typically uttered in shorter series................................... great gray owl

2. Loud, hollow, and booming *hoot-hoo* notes, usually given in groups of 2 (range 1–6 or more), with 1–2 second intervals between the doublet calls (advertising song); frequency low-pitched but still unmeasured .... snowy owl

3. An indefinite series of prolonged, low (ca. 400 Hz), cooing *boo* sounds, each lasting about 0.5 seconds, the notes uttered at spaced intervals of about 2.5 seconds (range 2–8), the first usually lower in pitch and volume (Figure 76A). Sometimes uttered in flight (advertising song) .... long-eared owl

4. A single very low-pitched (to ca. 500 Hz) hoot, uttered monotonously and regularly 8–60 times (average about 25) per minute (Figure 75A), each hoot often preceded by 1–2 preliminary softer notes of even lower pitch (advertising song); also a similar but double *boo-boot'*, uttered about 40 per minute, with the emphasis on the second syllable (courtship song) .... flammulated owl

5. A low-pitched (ca. 500 Hz), indefinite series of spaced cooing or *boo* notes, each lasting about 0.1 seconds and recalling a distant steam engine, given at the rate of about 2–5 per second, with from 6 to 20 or more notes in each series (Figure 76B). The series may be repeated 5–6 times an hour; often uttered in flight (advertising song) .... short-eared owl

6. A series of about 6 (4–9, rarely to 16) rather evenly spaced *boo* notes (to ca. 800 Hz), the series usually lasting about 1 second or sometimes to 1.5 seconds (Figure 75D); often slowing toward the end, sometimes with the penultimate note emphasized (advertising song)........... .... whiskered screech-owl; see also 2A4

7. A variably long (6–30) series of short, mellow *took* notes, uttered at a uniform clocklike rate of about 5 every 2 seconds, each note lasting about 0.1 seconds, fairly high pitched (ca. 1000 Hz) (advertising song) .... ferruginous pygmy-owl; see also 3A3

Group 3.

Calls given as a series of variably rapid (to about 10 per second), generally higher pitched and mellow notes that sometimes approach pure whistles; or as nearly continuous trills, the sequence often markedly rising in pitch and/or volume.

A. A series of phrases of slower (up to about 5 per second), pulsed, single-noted (sometimes doublet) units separated at least in part by distinct intervals. (See Groups 3AA and 3AAA below for progressively faster note rates.) Pitch usually varied; arranged below by apparently increasing average pitch.

1. A series of 4–20 short (about 0.1 seconds) notes on same pitch (500–650 Hz) that begin slowly (to about 3.5 per second) but terminally accelerate (to about 11 per second) while declining in volume, recalling

a bouncing ball coming to a stop (Figure 75C, *left*) (advertising song) . . . . western screech-owl; see also 3AAA1

2. A series of mellow *too* notes uttered independently in a long series, at intervals of about 2 seconds (advertising song), or less often as a series of 5–8 notes that increase in speed and pitch (scale song). A low, rolling trill of numerous mellow and uninflected *to* notes, followed by a pause and then about 3 widely spaced *hoo* notes (these sounding something like, "Look, look, look!"). In southern Arizona (*gnoma*) the notes usually uttered as double *hoo-hoos* (Figure 76D), each doublet about a second apart, or in groups of 3 with interspersed single notes (advertising song) . . . . northern pygmy-owl

3. A long series of harsh, rapidly uttered and equally spaced "popping" or *poip* notes, each note with an upward inflection, uttered at the rate of about 2.5 notes per second and each lasting about 0.25 seconds (Figure 76E); sometimes interspersed with clear whistles . . . . ferruginous pygmy-owl; see also 2AA7

4. A series of 4–15 or more rapidly repeated (6–8 per second), excited, and high-pitched *chewk* notes that descend in pitch and have a cackling or yipping quality (Figure 76C); the series often uttered 3–4 times in succession. Also various other whining and barking sounds suggestive of small dogs or puppies . . . . elf owl

AA. A series of more rapid, usually monosyllabic toots, soft whistles, or metallic sounds uttered in extended phrases, sometimes in trilled or staccato fashion, at rates of about 1–8 notes per second. Arranged below by increasing rates of notes uttered per second.

1. An extended series of uniformly spaced and mellow *too* notes (about 1–2 per second), resembling dripping water (Figure 76H); the entire sequence lasting up to a minute or more, often becoming faster and ending quite rapidly (advertising song); also harsher *skreigh-aw* or *whurdle* notes, these often grouped in triplets, of varied pitch and cadence but recalling the filing of a saw . . . . northern saw-whet owl

2. A rapid series of whistled *hu* notes (about 5 per second), in long phrases lasting several seconds; 10–15 phrases per minute (advertising song) . . . . northern hawk-owl; variant of 3AAA3

3. A very rapid series (about 8 per second) of mellow and hollow *po* notes (range 11–23, average 16 in N. America), in rising and falling phrases about 1–3 seconds long (Figure 76G), resembling snipe winnowing. About 2–3 seconds between successive phrases, which may go on indefinitely (advertising song) . . . . boreal owl

AAA. A continuous or nearly continuous trill (at least 12 pulses per second) often lasting about 2 seconds or more and usually varying in pitch or loudness. Arranged below by increasing average phrase length.

1. A short burst of rapid notes (about 12 per second), lasting about 0.5 seconds, followed by a longer similar series, lasting about 1.0 seconds, forming a double trill (Figure 75C, *right*) (secondary and duetting song) . . . . western screech-owl; see also 3A1

2. A prolonged, continuous, descending or uniformly pitched "whinny" of quavering trilled quality, lasting nearly 2 seconds (advertising song) (Figure 75B, *left*). Also a trilled series of very short notes (about 14 per second) on same pitch that slowly get louder and then may fade (Figure 75B, *right*); lasting 2–4.5 seconds (secondary and duetting song) . . . . eastern screech-owl

3. A sonorous, trilling, vibrant, and rolling *hu-hu-hu-u-u-u* usually lasting 2–10 seconds (rarely to 14 seconds), with about 12 pulses per second (Figure 76F). Sometimes uttered as a bubbling, rising ripple of comparably pulsed notes; each phrase lasting 8–9 seconds, with a similar interval between phrases (advertising song) . . . . northern hawk-owl; variant of 3AA2

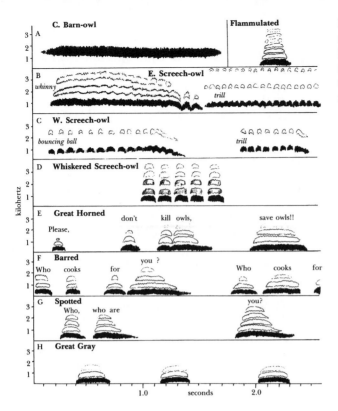

Figure 75. Diagrams of characteristic calls of nine North American owls, based on simplified sonographic representations of these calls. Major harmonics are indicated in lighter bands, minor ones are omitted.

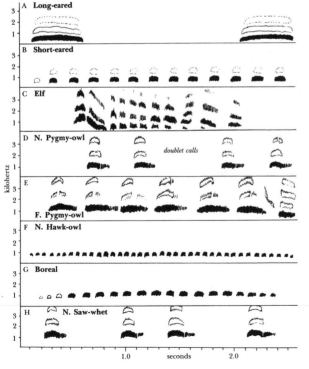

Figure 76. Diagrams of characteristic calls of eight North American owls, based on simplified sonographic representations of these calls. Harmonics shown as in Figure 75. From published sources and original sonograms.

Appendix 3

Origins of Scientific and Vernacular Names of North American Owls

This list includes all the extant genera and species, and nearly all currently recognized subspecies, of North American owls, including some names of fossil taxa and of taxa that are now considered synonyms. Self-explanatory vernacular names are excluded. The list is organized alphabetically by genus.

The English word "owl" derives from the old English *ule*, referring onomatopoeically to the bird's cries. (Compare "howl"; "howlets" are small owls.) Related words include the Latin *ulula*, "to cry out in pain"; the Greek *alala*, "an outcry"; the German *eule*, "an owl"; and even *uluka*, the Sanskrit word for an owl. The Greek *ololuzo*, "I call on the gods," has its English counterpart (via the Hebrew) in "hallelujah."

Aegolius: Probably from the Greek *aigolos*, "a nocturnal bird of prey." Perhaps originally from *aigos*, a goat, in reference to the reputed goat-sucking behavior of nocturnal birds.

    acadicus: A Latinism, "of Acadia" (a French colony of southeastern Canada, modern-day Nova Scotia).

      brooksi: After Major Allen Brooks (1869–1937), Canadian artist and ornithologist.

    funereus: Latin, "funereal," apparently in reference to a tolling, bell-like quality of the call or perhaps the association of its call with death. The North American name "boreal owl" reflects the species' generally northern distribution; the British name "Tengmalm's owl" refers to *Strix tengmalmi* Gmelin, a taxonomically invalid name for this species honoring P. G. Tengmalm, an early European naturalist.

      magnus: Latin, "large."

      richardsoni: Named for Sir John Richardson (1787–1865), Scottish naturalist on two of Sir John Franklin's (1786–1847) exploratory expeditions to arctic Canada.

Asio: Latin, "a horned owl." Pliny reportedly thus named it for its "ass-like" appearance, with its long "ears."

    flammeus: Latin, "flame-colored," in reference to the rusty plumage color.

    otus: From the Greek *otos*, "eared."

      wilsonianus: Named for Alexander Wilson (1766–1813), first notable American ornithologist.

      tufts: Named for R. W. Tufts, Canadian wildlife biologist.

    stygius: Latin, "of the River Styx," mythic abode of the dead. The English name has the same meaning.

Athene: Named after Athena, daughter of Zeus, the Greek goddess of wisdom, arts, and warfare, who traditionally was depicted with an owl on her shoulder and carrying an aegis (shield). The equivalent Roman goddess is Minerva, for whom a fossil Eocene owl has been named.

    cunicularia: From the Latin *cunicularius*, "a miner or burrower." This species is often placed in the genus *Speotyto,* coined from the Greek *speos,* "cave," and *tyto,* "owl."

    floridana: Latin, "of Florida."

    hypogaea: From the Greek *hypogeios,* "underground."

    rostrata: Latin, "beaked."

Bubo: Latin, "a horned or hooting owl," probably in reference to a low-pitched hooting call, especially one sounding like that of a bittern. Also perhaps related to the Greek *buzo,* "to hoot," and *buas,* "horned owl."

    virginianus: Latin, "from Virginia."

    algistus: From the Latin *algeo,* "to be cold."

    elachistus: From the Greek *elachistos,* "small."

    heterocnemis: From the Greek *heteros,* "different," and *cnem,* "legging or knee."

    lagophonus: From the Greek *lagos,* "hare," and *phonos,* "a killer."

    mayensis: Latin, "after the Mayan Indians."

    occidentalis: Latin, "western."

    pallescens: Latin, "becoming pale."

    saturates: Latin, "of full, rich color."

    scalariventris: From the Latin *scalaris,* "a ladder," and *ventris,* "belly."

    subarcticus: Latin, "of the low arctic."

    wapacuthu: Based on an aboriginal name used by the Eskimos of Hudson Bay.

Ciccaba: From the Greek *kikkaba,* a kind of owl mentioned by Aristophanes.

    nigrolineata: Latin, "black-lined."

    virgata: Latin, "twiggy" or "twiglike," probably referring to its striped plumage pattern.

    centralis: Latin, "centrally located."

    squamulata: Latin, "scaly" (referring to the scalloped feather pattern).

Glaucidium: Possibly from the Greek *glaukos* or *glaukidion,* meaning "gleaming" or "glaring," in reference to the eyes.

    brasilianum: A Latinism, "of Brazil" (the type locality). The vernacular name ferruginous refers to the rusty brown color of the plumage.

      cactorum: From the Greek *kaktos,* "a prickly plant."

      ridgwayi: Named for Robert Ridgway (1850–1929), premier American ornithologist and one-time president of the AOU (1899–1900).

    gnoma: From the Greek *gnomon* "to have knowledge, or be discerning and judicious," in reference to the bird's mythical role as an arbiter of destinies and its associated intelligence. In Latin *gnoma* refers to a subterranean spirit, and appears in English as "gnome."

      californicum: Latin, "of California."

      cobanense: After Coban, a famous Mayan ruin in Guatemala.

      grinnelli: Named for Joseph Grinnell (1877–1939), California ornithologist.

      hoskinsii: Named for Francis Hoskins, an assistant to the bird collector M. A. Frazer.

      swarthi: Named for H. S. Swarth, California ornithologist.

    griseiceps: Latin, "gray-headed."

    minutissimum: Latin, "very small."

    palmatum: Latin, "excellent."

    sanchezi: Named for Carlos Sanchez Mejorado Jr.

Lophostrix: Greek, "a crested owl."

    cristata: Latin, "crested."

Micrathene: From the Greek *mikros,* "small," and Athena, goddess of wisdom. The synonym *Micropallas* also refers to Pallas Athene, an alternate name for this goddess (and the source of such English words as palladium and athenaeum).

    whitneyi: Named for J. D. Whitney (1819–1896), American geologist and director of the geographical survey of California during which this species was discovered.

graysoni: Named for A. J. Grayson (1819–1869), ornithologist of California and Mexico.

idonea: Latin, "capable."

sanfordi: Named for Dr. L. C. Sanford, patron of ornithology and cosponsor of the Whitney-Sanford South Pacific Expedition of the American Museum of Natural History.

Minerva: In honor of Minerva, Roman goddess of wisdom.

antiqua: Latin, "ancient."

Nyctea: From the Greek *nycteus*, "nocturnal."

scandiaca: Latinism, "of Scandinavia."

Orgygoptynx: Greek, after Orgygos, mythic king of Thebes, and *ptynx*, an eagle-owl.

wetmorei: In honor of Alexander Wetmore (1896–1972), premier 20th-century American ornithologist and avian paleontologist.

Ornimegalonyx: Greek, from *ornith*, "bird," *megalo*, "largest," and *nyx*, "a puncture," perhaps referring to the bird's sharp talons.

Otus: Latin, "a horned or eared owl."

asio: Latin, "a horned owl" (see *Asio* above). The vernacular name "screech-owl" is not very appropriate; far better is the word "scops" for many forms in this genus, from the Greek *skopus*, "a watchman."

floridanus: Latin, "of Florida."

hasbroucki: Named for E. M. Hasbrouck, U.S. naturalist and screech-owl authority.

mccallii: Named for Colonel G. A. McCall, U.S. Army ornithologist.

maxwelliae: Named for Martha A. Maxwell, early Colorado taxidermist and naturalist.

naevius: Latin, "spotted."

swenki: Named for Myron H. Swenk, Nebraska naturalist.

atricapillus: Latin, "black-capped."

cassini: In honor of John Cassin (1813–1869), long-time curator of birds, Philadelphia Academy of Natural Sciences.

cooperi: In honor of Juan Cooper, a 19th-century collector of Neotropical birds.

lambi: In honor of Chester C. Lamb, California ornithologist and collector of Neotropical birds, who obtained the type.

flammeolus: Latin, "a small flame." The vernacular name flammulated is comparable in meaning.

borealis: Latin, "of the north."

frontalis: Latin, "with reference to the brow or forehead."

meridionalis: Latin, "southern."

guatemalae: Latin, "from Guatemala."

hastatus: Latin, "spear-shaped" (referring to its feather patterning).

kennicottii: Named for Robert Kennicott (1835–1866), first director of the Chicago Academy of Sciences.

aikeni: Named for Charles E. Aiken, early Colorado naturalist.

bendirei: Named for Captain Charles Bendire, U.S. Army naturalist.

cardonensis: Apparently referring to its nesting in the cardon cactus (*Pachycereus*).

gilmani: Named for Marshall F. Gilman (1871–?), California naturalist.

inyoensis: After Inyo, California.

macfarlanei: Named for Roderick R. MacFarlane (1833–1920), Canadian naturalist.

pacificus: Latin, "of the Pacific."

quercinus: Latin, "of oak leaves."

suttoni: Named for George M. Sutton (1898–1982), American bird artist and ornithologist.

vinaceous: Latin, "the color of red wine."

    xantusi: Named for L. John Xantus de Vesey, one-time U.S. consul to Mexico.

    yumanensis: After the Yuma, an Indian tribe on the lower Colorado River.

seductus: Latin, "distant."

trichopsis: From the Greek *trix,* "hair," and *opsis,* "having the appearance of," referring to the long rictal bristles. The vernacular name "whiskered" refers to the same trait.

    aspersus: Latin, "scattered."

    mesamericanus: Latin, "of Middle America."

vermiculatus: Latin, "vermiculated" (with small wormlike markings).

Pseudoscops: Greek, a false scops owl. Previously in *Rhinoptynx,* from the Greek *rhinos,* "nose," and *ptynx,* "eagle-owl," thus a large-beaked eagle-owl.

    clamator: Latin, "noisy" or "shrill."

        forbesi: Named for D. Forbes, of Veracruz, a collector of Mexican birds.

Pulsatrix: Latin, "a beater"; referring to its rhythmic calls. Its common name refers to the spectacle-like plumage pattern around the eyes.

    perspicillata: Latin, "spectacular."

    saturata: Latin, "saturated" (referring to the plumage color).

Sophiornis: Greek, from *sophia* "clever," and *ornis,* "bird."

Strix: Used in both Latin (*strix*) and Greek (*strigx*) to denote a kind of owl, especially a strident one. "Strigidae" has the same origin; *striges* is the plural of *strix.* (The term *striges* was applied in Rome to witches as well as to owls, as it was believed that witches could transform themselves into owls and suck the blood of sleeping children.)

    nebulosa: Latin, "clouded" (in reference to the plumage). This species has often been placed in the monotypic genus *Scotiaptex,* from the Greek *skotia,* "darkness or gloom," and perhaps the Greek *ptynx,* "an eagle-owl."

    occidentalis: Latin, "western."

        caurina: From the Latin *caurinus,* "northwestern."

        lucida: From the Latin *lucidus,* "clear."

    varia: Latin, "variegated" (in reference to the plumage).

        georgica: Latin, "of Georgia," where the type specimen was collected.

        helveola: From the Latin *helveolus,* "yellowish."

Surnia: Uncertain, but possibly based on the modern Greek *surnion,* a vernacular name for *Strix aluco.*

    ulula: Latin, "crying out as if in pain." Also related to the Greek *alala* and *ololuge,* "to howl, especially in pain." The American race *caparoch* is based on an aboriginal name used by the Inuit of the Hudson Bay area.

Tyto: From the Greek *tuto,* "a night owl." The New Latin form *tuton* is also suggestive of an owl's call.

    alba: Latin, "white."

        bondi: After James Bond, American ornithologist.

        furcata: Latin, "forked" (in reference to the tail).

        glaucous: From the Greek *glaukos,* "gleaming or silvery," and *opsis,* "appearance."

        guatemalae: Latin, "of Guatemala."

        lucayana: Latin, "of Lucaya, Bahama Islands."

        niveicauda: From the Latin *niveus,* "snowy," and *cauda,* "tail."

        pratincole: Latin, "inhabiting meadows."

Glossary

ABDOMEN. That section of the underparts between the breast and under tail coverts; sometimes called the belly.

ACCIDENTAL. An individual occurring well beyond its species' normal geographic range, sometimes called a vagrant.

ADAPTIVE RADIATION. The divergent patterns of evolution shown in a single phyletic line that result from varying speciation patterns and local evolutionary adaptations to differing environmental situations.

ADULT. A collective age category (composed of an indefinite number of age classes) of sexually mature individuals in their adult (definitive) plumage (q.v.).

ADULT (DEFINITIVE) PLUMAGE. Plumage attained and held by breeding adults. In owls there is apparently only a single annual molt and thus no distinction between breeding ("nuptial") and nonbreeding ("winter") plumages.

AERIAL. Capable of flight, as opposed to arboreal or terrestrial.

AGE CLASS (OR YEAR CLASS). A category of individuals including all those hatched or born during the same year (thus belonging to the same population cohort). See also COHORT.

AGONISTIC BEHAVIOR. Behavior associated with attack and escape, including intermediate stages of social dominance and submission.

ALLOPATRY (*adjectival form:* ALLOPATRIC). Occupation (by two or more populations) of completely separated geographic areas, at least during breeding. See also SYMPATRY.

ALLOSPECIES. Two or more populations (comprising a superspecies) that appear to have the necessary criteria to be considered separate species but are allopatric and thus cannot be tested for the presence of possible reproductive isolating mechanisms. Taxonomically, allospecies may be signified in trinomials by placing the name of the nominate form of the superspecies in parentheses between the names of the genus and the form(s) under consideration.

ALTRICIAL. Referring to species whose young are hatched blind, relatively helpless, and often naked.

ALULA. The group of miniature flight feathers (or bastard wing) associated with the wrist area (actually inserting on the first of the discernible digits, usually called the thumb).

ANISODACTYL. Referring to an avian foot arrangement in which one toe (the 1st) is oriented posteriorly, and the other three are normally directed anteriorly. See also ZYGODACTYL.

ANNUAL MORTALITY RATE. A statistic obtained by dividing a group's number of deaths during a year by the number of individuals in that group that had been alive at the start of the year. The group composition is often specified by age, sex, or both. See also MORTALITY RATE ("M").

ANNUAL SURVIVAL RATE. A statistic obtained by dividing the number of individuals alive at the end of the year by the number that had been alive at the beginning of the year. The group composition is often specified by age, sex, or both. See also SURVIVAL RATE ("S") .

ANTERIOR. Toward the front, as opposed to posterior.

AOU. Abbreviation for the American Ornithologists' Union.

ARBOREAL. Frequenting trees, as opposed to aerial or terrestrial.

ASYNCHRONOUS. Nonsimultaneous, such as the staggered hatching of eggs of a clutch over a period of several days.

ATTENUATED. Becoming slender toward the tip.

AUDITORY MEATUS. The external opening of the ear canal in birds, within which the tympanic membrane is located.

AXILLARS. Feathers of the "armpit" area between the underside of the wing and body.

BINOMIAL. A two-part name, such as a genus and species. See also TRINOMIAL.

BIOMASS. The total weight of organisms (of a specified species or collectively) in a given area of land.

BLUE LIST. A list compiled annually by the National Audubon Society of those species believed to be significantly declining over part or all of their ranges.

BRANCHING. A term applied to the behavior of unfledged owlets that leave the nest and begin clambering about on branches near the nest site.

BROODING. Parental sitting on or over the young, as opposed to incubation (sitting on eggs).

CERE. A skinlike or horny covering of the base of the upper mandible, typical of all owls, at the front of which the nostrils are located.

CLADE. A branching phyletic lineage; a cladistic classification is one based on shared derived traits.

CLINE. A graded geographic trend in one or more characters among members or demes (q.v.) of a species; such trends may be continuous (as in unbroken populations) or discontinuous (as in variously isolated populations).

CLUTCH. The number of eggs laid and simultaneously incubated by a female during nesting.

COEVOLVED. Traits that result from coevolution, those that are associated with common selective factors among different species.

COHORT. A component of a population consisting of individuals of the same age class.

COMMUNICATION. Behavior patterns of an individual that alter the probability of subsequent behavior by another individual in an adaptive manner.

COMMUNITY. The collective array of organisms occupying a particular habitat.

CONGENERIC. Members of the same genus.

CONSPECIFIC. Members of the same species.

CONVERGENT EVOLUTION. The evolution of structural or behavioral similarities in organisms that are not closely related, often because of analogous ecological adaptations. See also PARALLEL EVOLUTION.

COSMOPOLITAN. Referring to a taxon represented on all the major continental regions.

COURTSHIP. Communication between individuals of opposite sexes of a species that facilitates pair bonding, pair maintenance, or fertilization. See also DISPLAY.

COVERTS. Small feathers covering the wings (wing coverts) or tail (tail coverts); also sometimes used for feathers of other body areas, such as ear coverts (auriculars).

CREPUSCULAR. Active during the dawn and dusk hours, as opposed to strictly nocturnal (q.v.) or diurnal (q.v.).

CROWN. The top of the head.

CULMEN. The ridge of the upper mandible of the bill. The culmen length is normally measured as a straight line from the base of the bill (or, as sometimes defined, from the forehead feathering or the edge of the cere) to its tip, rather than as a curved line along the culmen's edge.

DEFINITIVE PLUMAGE. A species' final developmental plumage stage, after which no further significant changes occur.

DEME. A local population of a species that lacks sufficient unique attributes to merit formal taxonomic distinction as a subspecies.

DIMORPHISM (or DICHROMATISM). Occurring in two distinct and genetically determined forms or colors, either dependent upon sex (sexual dimorphism) or independent of it.

DISPERSAL. Movements (usually multidirectional and unpredictable) of organisms away from a point of origin or from centers of concentration; in predatory birds often occurring shortly after fledging (juvenile dispersal) or by adults after breeding (postbreeding dispersal).

DISPLACEMENT BEHAVIOR. Activities of an animal that appear to the human observer to be biologically inappropriate or irrelevant to the situation, but resembling activities appropriate to some other situation.

DISPLAY. Behavior patterns ("signals") that have been evolved ("ritualized") to provide communication functions for an organism, usually through stereotypical performance and exaggeration.

DISTAL. Toward the tip of the body or its appendages, as opposed to proximal.

DIURNAL. Active during the day, as opposed to nocturnal (q.v.) or crepuscular (q.v.).

DNA. Abbreviation for the genetic material (collective genome) of a species' chromosomes.

DNA HYBRIDIZATION. A biochemical technique for estimating phyletic and taxonomic relationships indicated by the chemical similarities of the genetic material (DNA) of two species. The technique determines the average temperature required to melt half of the hydrogen bonds formed in "hybrid DNA" that has been obtained by chemically combining single-stranded DNAs of the two component species; higher average melting temperatures indicate closer relationships between the species.

DORSAL. In the direction of or pertaining to the upperparts, as opposed to ventral.

EAR CONCH. The opening at either side of the feathered facial disk associated with hearing in owls.

EAR FLAPS. Skin-covered flaps located directly in front of the ear openings, behind them, or in both locations.

EAR TUFTS. Extended feathers on the heads of some owls that may resemble mammalian ear pinnae but are not related to hearing.

ECOLOGICAL (NICHE) SEGREGATION. The process by which competition between potentially competing individuals or populations is reduced by development of niche differences, such as behavioral or morphological differences associated with food getting. Ecological release is the opposite process, by which selection for such segregation is removed in areas where such competition does not occur, allowing for broader niche utilization than in areas of competitive interaction. See also HABITAT SEGREGATION.

EGG SUCCESS. An estimate of reproductive success based on the percentage of eggs in a breeding population that result in fledged young. See also FLEDGING SUCCESS, HATCHING SUCCESS, NESTING SUCCESS, and REPRODUCTIVE SUCCESS.

EMARGINATED. Either slightly notched or forked (when in reference to general tail shape), or abruptly narrowed or cut away toward the tip (when in reference to the web shape of individual remiges).

ENDANGERED. A conservation category, defined by the ICBP (q.v.) as including those taxa that are in danger of extinction and whose survival is unlikely if the factors causing their decline continue to operate. Legally defined by U.S. and Canadian federal environmental agencies (and by some states and provinces) in terms of total known surviving individuals of a taxon. See also RARE, THREATENED, and VULNERABLE.

ENDEMIC. Referring to a taxon that is native to and limited to a particular area.

EURYOECIOUS. Having a broad range of ecological tolerances, as opposed to stenoecious.

EXTINCT. Referring to a taxon that is no longer alive anywhere.

EXTIRPATED. Referring to a taxon that has been eliminated from part of a previously occupied range.

EXTRALIMITAL. Occurring beyond the stated limits of an area or range.

EYEBROW. Used in owl descriptions to refer to the distinctively shaped feathers at the top of the facial disk, above and between the eyes.

EYE RING. An area of bare, often colorful skin circumscribing the eye.

FAMILY. A taxonomic category representing a group of one or more related genera; consistently spelled with an "-idae" suffix, as in Strigidae.

FERRUGINOUS. Rusty brown.

FITNESS. The relative genetic contribution of an individual toward future generations of its species. Sometimes expanded to encompass inclusive fitness, the effect of the individual on the reproduction of all its genetic relatives.

FLEDGE. To attain the power of flight.

FLEDGING SUCCESS. An estimate of the percentage of hatched young that fledge successfully. See also EGG SUCCESS, HATCHING SUCCESS, NESTING SUCCESS, and REPRODUCTIVE SUCCESS.

FLEDGLING. A recently fledged bird.

FLIGHT FEATHERS. The collective primary and secondary feathers of the wing.

FOOD CHAIN. A sequence of energy transformations through successive "trophic levels" of a community by successive consumption of various of its members by one another. Food chains are actually only component parts of much more complex "food web" interactions, both of which end in "top-level" predators such as owls. Sometimes pesticides or other alien substances are biologically magnified during their passage through successive organisms comprising food chains or food webs, and thus accumulate in top-level predators.

FORM. A taxonomically neutral term for a species or some subdivision of a species; the term lacks nomenclatural significance and is usually used to avoid implying specific taxonomic meaning or rank when referring to a particular individual or population.

FRATRICIDE. The killing and consumption of younger or weaker nestmates by their siblings. Sometimes also called siblicide.

FUSCOUS. Brownish black.

GAPE. The lateral distance across the mouth at the base of the opened bill; also refers to the skin associated with the open mouth, as in gape color.

GENOTYPE. The genetic basis for an organism's trait or traits.

GENUS (*plural:* GENERA *adjectival form:* GENERIC). A taxonomic category representing one or more

species that are believed to be more closely related to one another than to any other species.
Consistently italicized and capitalized, as in
Strix, and comprising the first half of a binomial scientific name. See also SPECIES.

GRADUATED. Showing a progressive increase in
length, as in the feathers of a somewhat
pointed tail.

GULAR. Pertaining to the throat area.

GUTTATE. Having the shape of a tear drop.

HABITAT. The physical, chemical, and biotic characteristics of a specific environment.

HABITAT SEGREGATION. A physical subdivision of
available habitat resources by two or more of
the resource users (such as species, age classes,
or sexes); the resources may be subdivided by
space, time of usage, or both. See also ECOLOG-
ICAL (NICHE) SEGREGATION, of which habitat
segregation is one example.

HABITAT SELECTION. The ability of an organism to
assess environmental variations in such a way
as to be able to locate itself within desirable
habitats, whether by behavioral responses,
physiological tolerances, or both.

HATCHING SUCCESS. An estimate of the percent-
age of eggs in a sample that hatch successfully.
See also EGG SUCCESS, FLEDGING, SUCCESS,
NESTING SUCCESS, and REPRODUCTIVE SUCCESS.

HAWKING. Catching insects or other prey while in
full flight.

HERB. Any nonwoody plant, including broad-leaf
forbs.

HOLARCTIC. The land areas and islands of the
northern hemisphere including North America
north of Mexico's central highlands, Africa
north of the Sahara, and Asia north of the Hi-
malayas.

HOME RANGE. An area regularly used (but not
necessarily wholly or even partially defended)
by an individual or social group over some defined time period, such as a day (daily home
range) or a year (annual home range). The
home range may include a more restricted area
that is defended territorially against incursions
by other individuals or groups. See also TERRI-
TORY.

HYBRID. An individual produced by the crossing
of taxonomically different populations, usually
between subspecies (intraspecific), less often
between species (interspecific), and rarely between genera (intergeneric).

ICBP. International Council for Bird Protection.

IMMATURE. Referring to the period in a bird's life
from the time it fledges until it is sexually active. Visual distinction between juveniles (birds
still carrying their juvenal plumages) and older
but still sexually immature birds is often impossible for practical purposes.

INNATE. Inherited, such as instinctive patterns of
behavior.

INSTINCTIVE BEHAVIOR. Innate activities that typically are more complex than simple reflexes
and that are dependent upon specific external
stimuli ("releasers") as well as on variable and
specific internal states ("tendencies") for their
expression.

INTERGRADE. An individual or population having
transitional characteristics genetically and phe-
notypically linking different subspecies.

INTERSPECIFIC. Occurring between separate
species.

INTRASPECIFIC. Occurring within a single species.

IRRUPTIVE. Descriptive of nomadic or migratory
individuals or species the magnitudes of whose
movements differ markedly from time to time.

ISABELLINE. Brown, tinged with reddish yellow.

ISOLATING MECHANISM. An innate property of an
individual that prevents successful mating with
genetically unlike individuals. The term includes both premating and postmating mechanisms.

JUVENAL. The plumage stage typical of juveniles,
when the first generation of flight feathers
erupts and fledging occurs.

JUVENILE. A bird in its first (juvenal) plumage of
nondowny feathers.

K-SELECTED SPECIES. Those normally long-lived
and slowly reproducing species that have
evolved reproductive rates, mating systems,
and other adaptations that tend to keep their
populations fairly constant and near the carrying capacity ("K") of relatively stable environments. See also R-SELECTED SPECIES.

LATERAL. Toward the right or left side of the body,
as opposed to medial.

LEARNING. The adaptive modification of behavior
in an individual based on its past experience.

LONGEVITY. Records or estimates of lifespan. Includes maximum recorded longevity (in the
wild or captivity), mean or average longevity
(average life expectancy after hatching), and
mean afterlifetime (average expectancy of additional life for individuals that have attained
some stated age following hatching, such as
fledging, completion of the first year of life, or
some other appropriate starting point).

LORE (*adjectival form:* LORAL). The area between
the eye and the base of the upper mandible in
birds.

MANDIBLE. In plural, the upper and lower halves
of the bill; in singular, the lower half only (the
upper being the maxilla).

MATE SELECTION. The tendency for individuals of
a species selectively to "choose" appropriate
mates from the overall population in an adaptive manner, whether by innate tendencies or

learned clues as to individual differences in probable relative fitness among the available choices.

MATING SYSTEM. Patterns of mating within a population, including length and strength of a pair bond, the number of mates, degree of inbreeding, and the like.

MAXILLA. The upper half of the bill. See also MANDIBLE.

MEDIAL. Toward the midplane of the body, as opposed to lateral.

MIGRANT. Referring to an individual or population that regularly moves ("migrates") on a seasonal basis from one area to another and back, usually between breeding and nonbreeding areas.

MITOCHONDRIAL DNA. DNA that occurs only within the mitochondrial organelles of a cell and that passes from generation to generation without significant genetic changes, making it very useful for tracing phyletic lineages.

MOLT. The process by which feathers are individually or collectively lost and replaced, producing a change in plumage when done collectively. See also PLUMAGE.

MONOGAMY. A mating system involving the coordinated reproductive efforts of a single male and female through at least a single breeding cycle (single-brood and seasonal monogamy) and sometimes extended indefinitely (sustained or lifelong monogamy), even though the pairs may be separated during the nonbreeding season. See also POLYGAMY.

MONOTYPIC. Referring to a taxonomic category having only a single unit in the category immediately below it, such as a genus having only a single species, as opposed to polytypic.

MORPH. One of the recognized types of dimorphism or polymorphism characteristic of an organism (e.g., rufous morph, gray morph, etc.). As used here, essentially synonymous with "phase" (q.v.).

MORTALITY RATE ("M"). The estimated rate (usually given as a percentage or decimal fraction) of death among individuals of a population, usually calculated on an annual (12-month) basis and often defined by age or sex. See also SURVIVAL RATE ("S") and ANNUAL MORTALITY RATE.

NARES. The nasal cavities, opening via the nostrils.

NEARCTIC. The New World portion of the Holarctic (q.v.).

NEOSPECIES. Species having representatives that are still extant.

NEOTROPICAL. Referring to the land masses and associated islands of the New World south of the Nearctic, including South America and

Middle America north to Mexico's central highlands, plus the West Indies.

NESTING SUCCESS. An estimate of the proportion of initiated nests that successfully hatch one or more young (definition used here) or that produce fledged young (definition of Monroe, 2000). See also EGG SUCCESS, FLEDGING SUCCESS, HATCHING SUCCESS, and REPRODUCTIVE SUCCESS.

NESTLING. Descriptive of unfledged birds still in the nest ("nidicoles"). The nestling period is the length of time from hatching to fledging and is often shorter than the actual fledging period ("age at first flight") in owls, which may leave the nest while still unable to fly. See also BRANCHING, NIDICOLOUS, and NIDIFUGOUS.

NICHE. The sum of structural, physiological, and behavioral adaptations of a species to its environment.

NICHE SEGREGATION. See ECOLOGICAL (NICHE) SEGREGATION and HABITAT SEGREGATION.

NIDICOLOUS. Referring to species whose altricial young remain in the nest until attaining the ability to fly (fledging) or nearly so.

NIDIFUGOUS. Referring to species whose precocial young leave the nest very soon after hatching, usually within a day or so.

NOCTURNAL Active at night, as opposed to crepuscular (q.v.) or diurnal (q.v.).

NOMAD. A species or individual whose movements are relatively unpredictable as to their timing, direction, or duration.

NOMENCLATURE. The taxonomic procedure by which scientific names are applied to organisms. See also TAXONOMY.

NOMINATE. Referring to a taxon that is the nomenclatural basis for the name of the larger taxonomic group to which it belongs, for example, the genus *Strix* of the family Strigidae.

NUCHAL. Pertaining to the occipital (nape) area.

OBSOLETE. Nearly invisible or lacking, in reference to plumage pattern or structure.

OCCIPUT. The area of the head located at the rear of the skull (associated with the occipital bone).

OCHRACEOUS. The color of ochre, brownish yellow.

ORBITAL REGION. Used in owls to describe the area of the facial disk.

ORDER. A taxonomic category immediately subordinate to that of the class (or subclass), consisting of one or more related families and normally identified (at least in the taxonomic procedures advocated by A. Wetmore) by the suffix "-iformes," as in Strigiformes.

PAIR BOND. A prolonged and individualized social association between members of a mated pair in association with breeding. See also MONOGAMY.

PALATE. The upper roof of the mouth, including both the bony portion (bony palate) and its covering (soft or horny palate).

PALEARCTIC. The Old World component of the Holarctic (q.v.).

PARALLEL EVOLUTION. Traits or taxa that evolve in parallel fashion, showing neither convergent nor divergent trends. See also CONVERGENT EVOLUTION.

PHASE. A term traditionally used in morphological or taxonomic descriptions to designate nontransient plumage variations within a species, such as (usually) genetically determined pigmentary deviations from the norm of the species, that are typically independent of sex and age (unless sex-linked or age-related, respectively). See also MORPH, which is a more recently applied substitute term encompassing this phenomenon.

PHENOTYPE. The appearance of an organism, irrespective of the genetic basis of this appearance; the basis for phenetic analysis techniques of numerical taxonomy.

PHYLETIC. Referring to a pattern of evolutionary lineage, or phylogeny. A phylogeny is typically a hypothetical representation or diagram (phylogram) of evolutionary descent within a single phyletic line. It may be illustrated in the form of branching clades (cladograms), as diagrams showing degrees of phenotypic differences based on numerical analyses (phenograms), or in traditional treelike representations (dendrograms). See also CLADE and PHENOTYPE.

PLUMAGE. A collective generation of feathers produced by a molt (q.v.). See also JUVENAL, IMMATURE, SUBADULT, and ADULT (DEFINITIVE) PLUMAGE.

POLYGAMY. A mating system involving more than one individual of one sex mating with one individual of the other sex in conjunction with a single reproductive cycle; includes polygyny (the mating of a male with more than one female simultaneously or successively) and polyandry (the mating of a female with more than one male simultaneously or successively). Polygamy here includes bigamy; if no pair bonding occurs between members of copulating pairs polygamy may be defined as promiscuity. See also MONOGAMY.

POLYMORPHISM. The occurrence within one species of two or more forms of colors that are typically genetically controlled and often independent of sex (but are sometimes sex-limited).

POLYTYPIC. Referring to a taxon having more than one member in the category immediately subordinate to it, such as a genus with two or more species, as opposed to monotypic.

POSTOCULAR. Behind the eye. See also SUPERCILIARY.

PRECOCIAL. Referring to species whose young are able to feed and move about shortly after hatching.

PREDATION. The killing of one species by another for food. Predatory birds are often called birds of prey or raptors.

PRIMARIES. Those larger contour feathers attached to the bones of the hand and digits that, together with the secondaries, comprise the flight feathers (remiges).

PROXIMAL. Toward the main body axis, as opposed to distal.

RACE. An alternative name for the subspecies category.

RAPTOR. A predatory bird, typically one with sharp and strong talons and a pointed, curved bill. Also used in the adjectival form "raptorial" to describe these and associated predatory traits. See also PREDATION.

RARE. A conservation category defined by the ICBP (q.v.) as including those taxa having world populations that are small but not currently considered to be either endangered or vulnerable. See also ENDANGERED, THREATENED, and VULNERABLE.

RECOVERY. Defined by bird banders as the recapture of a bird by any means (trapping, shooting, etc.) at a point away from the original banding station. See also RETURN.

RECTRICES (*singular:* RECTRIX). The tail feathers.

REMIGES (*singular:* REMIX). The primary and secondary wing feathers, also called flight feathers.

REPRODUCTIVE ISOLATION. The result of a genetic barrier (anatomical, ecological, or behavioral) that helps to prevent matings between species (premating isolation) or, if such matings do occur, tends to prevent the hatching, survival, and reproduction of the resulting hybrid genotypes (postmating isolation). Such "intrinsic" (rather than extrinsic or environmental) barriers to gene exchange are called reproductive isolating mechanisms.

REPRODUCTIVE SUCCESS. As used here, an estimate of the percentage of eggs laid in a population that produces successfully fledged young. Termed "nest success" by Monroe (2000). Sometimes also defined as the average number of fledged young raised per reproductive pair, or as otherwise stipulated. See also EGG SUCCESS, FLEDGING SUCCESS, HATCHING SUCCESS, and NESTING SUCCESS.

RESIDENT (*adjectival form:* RESIDENTIAL). A sedentary (nonmigratory or nonnomadic) population.

RESOURCE. A particular feature of the environment, the control of which contributes to an organism's fitness (q.v.).

RETICULATE. Referring to a weblike or network-like pattern of rounded or hexagonal scales on the surface of the tarsus.

RETURN. Defined by bird banders as the recapture of a bird at a station at least 90 days after its previous capture or sighting there (called a repeat if recaptured in fewer than 90 days). See also RECOVERY.

REVERSED SEXUAL DIMORPHISM. The phenomenon, typical of raptorial birds, in which the female is the larger and more powerful sex, the reverse of the situation in most other birds.

RICTUS. The facial area at the base of the gape (q.v.), which in owls is typically very bristly ("rictal bristles").

RITUALIZATION. The evolutionary development of signaling ("display") behavior and associated signaling devices in an animal species, thereby providing an effective innate communication system.

R-SELECTED SPECIES. Those species that have evolved potentially high reproductive ("r") rates, flexible mating systems, and other adaptations such as efficient dispersal mechanisms that tend to allow for rapid changes of population sizes and locations, enabling them to exploit environments that are unpredictable as to their carrying capacity, duration of existence, and distribution. See also K-SELECTED SPECIES.

RUFESCENT. Tinged with rufous.

SCAPULARS. Those feathers located in the shoulder region (near the scapula bone), just medial to the upper wing coverts.

SCIENTIFIC NAME. The (usually binomial) combination of a general (generic) and a specific (species-level) name that collectively uniquely identify an organism, such as *Tyto alba*. To be complete, the name of the describer of the species and the year of its initial valid description should be added, such as *T. alba* (Scopoli) 1769—the parentheses around Scopoli's name indicating that he originally described it as a member of some genus other than the one in which it is currently being allocated. See also BINOMIAL and TRINOMIAL.

SCUTELLATE. Referring to a vertically aligned pattern of squarish scales on the surface (often only the front but sometimes front and rear edges) of the tarsus.

SECONDARIES. Those flight feathers attached to the forearm (ulna) of the wing.

SEDENTARY. Descriptive of nonmigratory and nonnomadic populations.

SEMISPECIES. A term sometimes used conveniently to designate allopatric populations that may be either subspecies or full species, there being no way of determining the exact level of speciation.

SENSU LATO. In the broad sense.

SEXUAL DIMORPHISM. The situation common in sexually reproducing animals for at least mature individuals of the sexes to differ in appearance ("dichromatism"), behavior ("diethism"), size, or all three. See also REVERSED SEXUAL DIMORPHISM.

SIBLICIDE. The killing and consumption of younger or weaker nestmates by their siblings. An alternate term for fratricide.

SPECIATION. The process of species proliferation through the gradual development of reproductive isolation between geographically separated populations.

SPECIES (*abbreviation:* SP.; *same spelling in plural but abbreviation:* SPP.; *adjectival form:* SPECIFIC). Taxonomically, the category below that of the genus and above that of the subspecies. It is written as the second and subsidiary component of a two-part (binomial) name, in italics but not capitalized, as in *Bubo virginianus*. Biologically, one or more populations of actually or potentially interbreeding organisms that are reproductively isolated from all other such populations. See also GENUS.

SPECIES GROUP. A group of two or more closely related species whose components usually have partially overlapping (sympatric) ranges. See also SUPERSPECIES.

STENOECIOUS. Having a narrow range of ecological tolerances, as opposed to euryoecious.

STRATEGY. The evolved niche adaptations of a population that are associated with its adaptation to a particular environment. Includes mating strategies, foraging strategies, etc.

STRIGES. Taxonomic term used in some classifications to represent the order Strigiformes.

STRIGINE. An adjective referring to members of the Striginae, or to birds having strigidlike characteristics.

SUBADULT. Referring to a late immature developmental stage of those species that require several years to attain sexual maturity, sometimes marked by a distinctive plumage or soft-part traits different from both younger immature stages and adult, or definitive, plumages.

SUBFAMILY. A taxonomic category representing an initial subdivision of a family (above that of an infrafamily or tribe); identified by the suffix "-inae," as in Striginae.

SUBGENUS. A taxonomic category of convenience below that of the genus, used to associate the more closely related members of some more inclusive genus.

SUBORDER. A taxonomic category occasionally interpolated between the order and family (or superfamily) levels of classification, having no generally agreed upon suffix ending other than those of normal Latinized plurals.

SUBSPECIES (*abbreviation:* SSP. *in singular;* SSPP. *in plural*). A taxonomic category that is defined as a recognizable geographic subdivision of a species; also called a race. Taxonomically identified as the final part of a three-part (trinomial) scientific name, for example, *Bubo virginianus virginianus.*

SUPERCILIARY. Above the eye, such as a superciliary stripe. See also POSTOCULAR.

SUPERFAMILY. A taxonomic category immediately higher than that of the family but below that of order or suborder; identified by its characteristic "-oidea" suffix.

SUPERSPECIES. Two or more species, with largely or entirely nonoverlapping ranges ("allospecies"), that are clearly derived from a common ancestor but are too distinct to be considered a single species. If significant sympatry is present among the included species they are usually instead called a species group. See also ALLOSPECIES.

SURVIVAL RATE ("S"). The probability (given as a percentage or decimal fraction, the latter equal to 1- M) of an individual surviving for a given period; usually defined as a 12-month period (annual survival) and often differentiated as to sex or age. See also MORTALITY RATE ("M") and ANNUAL SURVIVAL RATE.

SYMPATRY (*adjectival form:* SYMPATRIC). Coexistence by two or more populations in the same area, especially during the breeding season. See also ALLOPATRY.

SYRINX (*plural:* SYRINGES). The sound-producing organ of birds, located in the junction of the trachea and bronchi.

SYSTEMATICS. The practices of taxonomy concerned with the erection of taxonomic classifications and phylogenies. See also TAXONOMY and NOMENCLATURE.

TAIL. As used ornithologically, the collective rectrices of a bird. Measured (unless otherwise indicated) from the point of insertion of the central rectrices to their tips.

TALONS. The sharply pointed and curved claws of a raptorial bird.

TARSUS. A term collectively applied to the tarsometatarsus of birds; sometimes also called the leg or foot, but actually consisting of the fused ankle and foot bones.

TAXON (*plural:* TAXA). As used here, any taxonomic unit (category) or a particular example of that category.

TAXONOMY. The science of biological classification, which is the basis for providing appropriate biological names (nomenclature) and the establishment of systematic hierarchies (systematics) believed to best reflect evolutionary relationships (lines of phyletic descent). Various contemporary taxonomic techniques include cladistics (the study and analysis of definable monophyletic units, or clades) and numerical taxonomies (the use of "operational taxonomic units" for estimating degrees of phenotypic differences in related groups).

TERRESTRIAL. Frequenting or associated with the ground, as opposed to aerial or arboreal.

TERRITORIALITY. The advertisement and agonistic behavior associated with territorial establishment and defense.

TERRITORY. A definable area having resources that are consistently controlled or defended by an animal against others of its species (intraspecific territories) or, less often, against individuals of other species (interspecific territories), at least for some part of the year. Among owls, often only the area immediately around the nest is strongly defended. Such territories comprise part of more inclusive home ranges. See also HOME RANGE.

THREATENED. A legal category of the U.S. and Canadian wildlife agencies (and of some individual states and provinces) for designating those taxa that are not yet believed to be endangered but whose known numbers place them at risk of falling into that category. Similar to the "vulnerable" category of the ICBP (q.v.). See also ENDANGERED, RARE, and VULNERABLE.

TRIBE. A taxonomic category representing a suprageneric subdivision of a subfamily and containing one or more genera; identified by the suffix "-ini," as in Strigini.

TRINOMIAL. A three-part name, typically consisting of the names of a genus, species, and subspecies. See also BINOMIAL.

TYMPANIFORM (or TYMPANIC) MEMBRANES. Paired membranes of the syrinx that, when set into motion by the passage of air, are the basis for bird vocalizations.

VAGRANT. An individual occurring well outside its population's normal migratory or nomadic limits. Also called an accidental.

VENTRAL. In the direction of or pertaining to the underside, as opposed to dorsal.

VERNACULAR NAME. The "common" name of a taxon (usually a species) or morph in some particular language.

VULNERABLE. A conservation category defined by the ICBP (q.v.) as including those taxa believed likely to move into the endangered category in the near future if the causal factors responsible for their declines continue to operate. See also ENDANGERED, RARE, and THREATENED.

WING. The arm and associated feathers of birds, measured from the bend (wrist) of the folded wing to the tip of the longest primary, usually done with the feathers flattened, unless stated as being the chord (unflattened) distance. The less commonly used wingspan distance is measured from tip to tip of the extended wings.

ZYGODACTYL. Referring to an avian foot arrangement in which two toes (1st and 4th) are facultatively (as in owls) or permanently oriented posteriorly and the other two (2nd and 3rd) anteriorly. See also ANISODACTYL.

References

NOTE: The following references include many post-1988 citations that are not specifically cited in the text but that seem important enough to be included, since no recent published bibliography of the North American owls exists. The Raptor Information System Web site (http://ris.wr.usgs.gov/) provides a searchable online database of about 29,000 references.

[ABA.] 1999. 1998–1999 ABA Check-list report. *Birding* 31 (6): 518–524. (Stygian owl in Texas.)

Abbruzzese, C. M., and G. Ritchison. 1997. The hunting behavior of eastern screech-owls. *In* Duncan, Johnson, and Nicholls, 1997, pp. 21–32.

Adam, C. I. G. 1987. Status of the eastern screech owl in Saskatchewan with reference to adjacent areas. *In* Nero et al., 1987, pp. 268–276.

Adamcik, R. S., A. W. Todd, and L. B. Keith. 1978. Demographic and dietary responses of great horned owls during a snowshoe hare cycle. *Canadian Field-Nat.* 92:156–166.

Allaire, P. N., and D. F. Landrum. 1975. Summer census of screech-owls in Brethitt County. *Kentucky Warbler* 51:23–29.

Allen, A. A. 1924. A contribution to the life history and economic status of the screech owl (*Otus asio*). *Auk* 41:1–16.

Allen, H. L., and L. W. Brewer. 1985. A review of current northern spotted owl (*Strix occidentalis caurina*) research in Washington State. *In* Gutiérrez and Carey, 1985, pp. 55–57.

Allen, H. L., T. Hamer, and L. W. Brewer. 1985. Range overlap of the spotted owl (*Strix occidentalis caurina*) and the barred owl (*Strix varia*) in Washington and implications for the future. Proc. Raptor Research Found. Symp., 9–10 Nov., Sacramento (abstract).

Allen, J. A. 1893. Hasbrouck on "Evolution and dichromatism in the genus *Megascops.*" *Auk* 10:347–353.

American Ornithologists' Union (AOU). 1998. *Check-list of North American birds.* 7th. ed. American Ornithologists' Union, Washington, D.C. (6th ed., 1983; 5th ed., 1957).

Andrle, R. F., and J. R. Carroll, eds. 1988. *The atlas of the breeding birds in New York State.* Ithaca: Cornell Univ. Press.

Andrusiak, D. P., and K. M. Cheng. 1997. Breeding biology of the barn owl (*Tyto alba*) in the lower mainland of British Columbia. *In* Duncan, Johnson, and Nicholls, 1997, pp. 38–46.

Anweiler, G. 1960. The boreal owl influx. *Blue Jay* 18:61–63.

Apfelbaum, S. I., and P. Seelbach. 1983. Nest tree, habitat selection, and productivity of seven North American raptor species based on the Cornell Univ. nest record program. *Raptor Research* 17:97–113.

Armstrong, E. A. 1970. *The folklore of birds.* London: Collins.

Armstrong, W. H. 1958. Nesting and food habits of the long-eared owl in Michigan. *Pub. Mich. State Univ. Biol. Series* 1:63–96.

Arrendondo, O. 1976. The great predatory birds of the Pleistocene of Cuba. *Smithsonian Contr. Paleobiol.* 27:169–187.

———. 1982. Los Strigiformes fossiles dei Pleistoceneo cubano. *Boletin Soc. Venez. Cienc. Nat.* 37:33–35.

Arsenault, D. P. 1999. The ecology of flammulated owls: Nest-site preferences, spatial structure, and mating system. Master's thesis, Univ. Nevada, Reno.

Arsenault, D. P., A. Hodgson, and P. C. Stacey. 1997. Dispersal movements of juvenile Mexican spotted owls (*Strix occidentalis lucida*) in New Mexico. *In* Duncan, Johnson, and Nicholls, 1997, pp. 47–57.

Atkinson, E. 1989. Great gray owl (*Strix nebulosa*) surveys on the Payette National Forest. Idaho Department of Fish and Game report, Pocatello.

Atkinson, E., and M. L. Atkinson. 1990. Distribution and status of flammulated owls (*Otus flammeolus*) in the Salmon National Forest. Idaho Department of Fish and Game, Boise.

Austing, G. R., and J. B. Holt Jr. 1966. *The world of the great horned owl.* Philadelphia: Lippincott.

Axelrod, M. 1980. Diet of a Minnesota hawk owl. *Loon* 52:117–118.

Baekken, B. T., J. O. Nybo, and G. A. Sonerud. 1987. Home-range size of hawk owls: Dependence on calculation method, number of tracking days, and number of plotted perchings. *In* Nero et al., 1987, pp. 145–148.

Bailey, A. M., and R. J. Niedrach. 1965. *Birds of Colorado.* 2 vols. Denver: Denver Museum of Nat. Hist.

Balda, R. P., B. C. McKnight, and C. D. Johnson. 1975. Flammulated owl migration in the southwestern United States. *Wilson Bull.* 87:520–533.

Balgooyen, T. G. 1969. Pygmy owl attacks California quail. *Auk* 86:358.

Banks, R. C. 1964. An experiment on a flammulated owl. *Condor* 66:79.

Barlow, J. C., and R. Johnson. 1967. Current status of the elf owl in the southwestern United States. *Southwest Nat.* 12:331–332.

Barrowclough, G. F., and S. L. Coats. 1985. The demography and population genetics of owls, with special reference to the conservation of the spotted owl (*Strix occidentalis*). *In* Gutiérrez and Carey, 1985, pp. 74–85.

Barrowclough, G. F., and R. J. Gutiérrez. 1990. Genetic variation and differentiation in the spotted owl (*Strix occidentalis*). *Auk* 107:737–744.

Barrows, C. 1981. Roost selection by spotted owls. An adaptation to heat stress. *Condor* 83:302–309.

———. 1985. Breeding success relative to fluctuations in diet for spotted owls in California. *In* Gutiérrez and Carey, 1985, pp. 50–54.

———. 1989. Diets of five species of desert owls. *West. Birds* 20:1–10.

Bart, J., and E. D. Forsman. 1992. Dependence of northern spotted owls (*Strix occidentalis*) on old-growth forests in the western USA. *Conserv. Biol.* 62:95–100.

Batey, K. M., H. H. Batey, and I. O. Buss. 1980. First boreal owl fledglings for Washington State. *Murrelet* 61:80.

Baumgartner, F. M. 1938. Courtship and nesting of the great horned owl. *Wilson Bull.* 50:274–285.

———. 1939. Territory and population in the great horned owls. *Auk* 56:274–282.

Beddard, F. E. 1888. On the classification of Striges. *Ibis,* ser. 6, 5:335–344.

Belcher, C., and G. D. Smooker. 1936. Birds of the colony of Trinidad and Tobago. *Ibis* 13th ser. 6:1–35.

Belthoff, J. R. 1986. Roost site selection by juvenile eastern screech-owls during the postfledging period. Proc. ann. meeting, Raptor Research Found., 20–23 Nov., Gainesville, Fla. (abstract).

Belthoff, J. R., and A. M. Duffy Jr. 1995. Locomotor activity levels and the dispersal of western screech-owls, *Otus kennicottii. Anim. Behav.* 50:558–561.

Belthoff, J. R., and G. Ritchison. 1986. Natal dispersal and mortality of juvenile eastern screech-owls in central Kentucky. Proc. ann. meeting, Raptor Research Found., 20–23 Nov., Gainesville, Fla. (abstract).

———. 1989. Natal dispersal of eastern screech-owls. *Condor* 91:254–265.

———. 1990a. Nest-site selection by eastern screech-owls in central Kentucky. *Condor* 92:982–990.

———. 1990b. Roosting behavior of postfledging eastern screech-owls. *Auk* 107:567–579.

Belthoff, J. R., E. J. Sparks, and G. Ritchison. 1993. Home ranges of adult and juvenile eastern screech-owls: Size, seasonal variation, and extent of overlap. *J. Raptor Res.* 27:8–15.

Bendire, C. E. 1892. Life histories of North American birds. *U.S. Natl. Mus. Spec. Bull.* 1:1–425.

Bent, A. C. 1938. Life histories of North American birds of prey. Pt. 2. *U.S. Natl. Mus. Bull.* 170:1–482.

Bergeron, D., C. Jones, D. L. Genter, and D. Sullivan. 1992. P. D. Skaar's Montana bird distribution. 6th ed. Spec. Pub. no. 2, Montana Nat. Heritage Program, Helena.

Berggren, V., and J. Wahlstedt. 1977. (The sound repertoire of the great gray owl.) *Vår Fågelvärld* 36:243–249. (In Swedish, English summary.)

Bergman, C. A. 1983. Flaming owl of ponderosa. *Audubon* 85 (6): 66–71.

Bergmann, H.-H., and M. Ganso. 1965. (On the biology of the pygmy owl.) *J. Ornith.* 106:255–84. (In German, English summary.)

Bevis, D. F., K. E. Duffy, D. M. Whalen, B. D. Watts, and K. M. Dodge. 1997. Autumn migration of northern saw-whet owls (*Aegolius acadicus*) in the Middle Atlantic and Northeastern United States: What observations from 1995 suggest. *In* Duncan, Johnson, and Nicholls, 1997, pp. 74–89.

Bibles, B. D. 1992. Is there competition between exotic and native cavity-nesting birds in the Sonoran Desert: An experiment. Master's thesis, Univ. of Arizona, Tucson.

Binford, L. C. 1989. *A distributional survey of the birds of the Mexican State of Oaxaca*. Ornithological Monographs 43, American Ornithologists' Union, Washington, D.C.

Bird, D. M., D. E. Varland, and J. Negro, eds. 1995. *Raptors in human landscapes*. New York: Academic Press.

Blakesley, J. A., A. B. Franklin, and R. J. Gutiérrez. 1990. Sexual dimorphism in northern spotted owls from northwestern California. *J. Field Ornith.* 61:320–327.

———. 1992. Spotted owl roost and nest site selection in northwestern California. *J. Wildl. Manage.* 56:388–392.

Bloom, P. H. 1979. Ecological studies of the barn owl in California. *In* Schaeffer and Ehlers, 1979, pp. 36–39.

———. 1983. Notes on the distribution and biology of the flammulated owl in California. *West. Birds* 14:49–52.

Blus, L. J. 1996. Effects of pesticides on owls in North America. *J. Raptor Res.* 30:198–206.

Boal, C. W., B. D. Bibles, and R. W. Mannan. 1997. Nest defense and mobbing behavior of elf owls. *J. Raptor Res.* 31:286–287.

Bohmke, B. W., and M. Macek. 1994. Breeding the spectacled owl. *AFA Watchbird* 21 (5): 4–7.

Bolen, E. C. 1978. Long-distance displacement of two southern barn owls. *Bird-Banding* 49:78–79.

Bondrup-Nielsen, S. 1976. First boreal owl nest for Ontario with notes on development of the young. *Canadian Field-Nat.* 90:477–479.

———. 1977. Thawing of frozen prey by boreal and saw-whet owls. *Canadian J. Zool.* 55:595–601.

———. 1978. Vocalizations, nesting, and habitat preferences of the boreal owl (*Aegolius funereus*) in North America. Master's thesis, Univ. of Toronto.

———. 1983. Ambivalence of the concealing pose of owls. *Canadian Field-Nat.* 97:329–330.

———. 1984. Vocalizations of the boreal owl, *Aegolius funereus richardsoni*, in North America. *Canadian Field-Nat.* 98:191–197.

Bonnot, P. 1922. Notes on the voice of the California screech owl. *Condor* 24:30–31.

Bosakowski, T. 1984. Roost selection and behavior of the long-eared owl (*Asio otus*) wintering in New Jersey. *Raptor Research* 18:137–142.

———. 1986. Short-eared owl winter roosting strategies. *Amer. Birds* 40:237–240.

Bosakowski, T., R. Speiser, and J. Benzinger. 1987. Distribution, density, and habitat relationships of the barred owl in northern New Jersey. *In* Nero et al., 1987, pp. 135–143.

Bosakowski, T., R. Speiser, and D. G. Smith. 1989. Nesting ecology of forest-dwelling great horned owls in the Eastern Deciduous Forest Biome. *Canadian Field-Nat.* 103:65–69.

Bouchart, M. L. 1991. Great gray owl habitat use in southeastern Manitoba and the effects of forest resource management. Natural Resources Inst. Practicum, Univ. of Manitoba, Winnipeg.

Boula, K. 1982. Food habits and roost sites of northern saw-whet owls in northeastern Oregon. *Murrelet* 63:92–93.

Bowen, P. J. 2000. Demographic, distribution, and metapopulation analyses of the burrowing owl (*Athene cunicularia*) in Florida. Master's thesis, Univ. Central Florida, Orlando.

Bowers, R. K. 1998. Western screech-owl. *In* Glinski, 1998a, pp. 145–148.

Boxall, P. C., and M. R. Lein. 1982a. Feeding ecology of snowy owls (*Nyctea scandiaca*) wintering in S. Alberta. *Arctic* 35:282–290.

———. 1982b. Possible courtship behavior by snowy owls in winter. *Wilson Bull.* 94:79–81.

———. 1982c. Territoriality and habitat selection of female snowy owls (*Nyctea scandiaca*) in winter. *Canadian J. Zool.* 50:2344–2350.

Brauning, D. W., ed. 1992. *The atlas of the breeding birds in Pennsylvania*. Pittsburgh: Univ. of Pittsburgh Press.

Brewer, L. W., and H. L. Allen. 1985. Home range size and habitat use of northern spotted owls (*Strix occidentalis caurina*) in Washington. Proc. Raptor Research Found. Symp., 9–10 Nov., Sacramento (abstract).

Brewer, R., G. A. McPeek, and R. J. Adams Jr., eds. 1991. *The atlas of breeding birds of Michigan*. East Lansing: Michigan State Univ. Press.

Bridges, D. 1992. Northern saw-whet owls vs. boreal owls above 10,000 feet in the Wet, Sangre de Cristo, and Culebra mountains of south-central Colorado: A preliminary report. *Colo. Field Ornith. J.* 26:29–31.

Briggs, M. A. 1954. Apparent neoteny in the saw-whet owls of Mexico and Central America. *Proc. Biol. Soc. Wash.* 67:79–182.

Brinker, D. F., K. E. Duffy, D. M. Wahlen, B. D. Watts, and K. M. Dodge. 1997. Autumn migration of northern saw-whet owls (*Aegolius acadicus*) in the Middle Atlantic and Northeastern United States: What observations from

1995 might suggest. *In* Duncan, Johnson, and Nicholls, 1997, pp. 74–89.

Brodkorb, P. 1971. Catalogue of fossil birds. Pt. 4. *Bull. Florida State Mus.* 15:163–266.

Browning. N. R. 1990. Erroneous emendations to names proposed by Hekstra (Strigidae: *Otus*). *Proc. Biol. Soc. Wash.* 103:452.

Browning, N. R., and R. C. Banks. 1990. The identity of Pennant's "Wapacuthu owl" and the subspecific name of the population of *Bubo virginianus* from west of Hudson Bay. *J. Raptor Res.* 24:80–83.

Brunton, D. F., and R. Pittaway Jr. 1971. Observations of the great gray owl in its winter range. *Canadian Field-Nat.* 85:315–322.

Buchanan, J. H., L. L. Irwin, and E. L. McCutcheon. 1994. Characteristics of spotted owl nest trees in the Wenatchee National Forest. *J. Raptor Res.* 27:1–7.

Bühler, P., and W. Epple. 1980. (The vocalizations of the barn-owl). *J. Ornith.* 121:36–71. (In German, English summary.)

Bull, E. L., and R. G. Anderson. 1978. Notes on flammulated owls in northeastern Oregon. *Murrelet* 59:26–28.

Bull, E. L., and J. R. Duncan. 1993. Great gray owl. No. 41 in *The birds of North America,* ed. A. Poole, P. Stettenheim, and F. Gill. Philadelphia: Academy of Natural Sciences; Washington, D.C.: American Ornithologists' Union.

Bull, E. L., and M. G. Henjum. 1985. Ecology of great gray owls in northeastern Oregon. Proc. Raptor Research Found. Symp., 9–10 Nov., Sacramento (abstract).

———. 1987. Ecology of great gray owls in northeastern Oregon. Paper presented at Symposium on Biology and Conservation of Northern Forest Owls, 3–7 Feb., Winnipeg, Manitoba.

———. 1990. *Ecology of the great gray owl.* USDA, For. Serv., Gen. Tech. Rep., PNW-GTR-265, Portland, Oreg.

Bull, E. L., M. G. Henjum, and R. S. Rohweder. 1988a. Home range and dispersal of great gray owls in northeastern Oregon. *J. Raptor Res.* 22:101–106.

———. 1988b. Nesting and foraging habitat of great gray owls in northeastern Oregon. *J. Raptor Res.* 22:107–115.

———. 1989a. Diet and optimal foraging of great gray owls. *J. Wildl. Manage.* 53:47–50.

———. 1989b. Reproduction and mortality of great gray owls in Oregon. *Northwest Sci.* 63:38–43.

Bull, E. L., A. L. Wright, and M. G. Henjum. 1989. Nesting and diet of long-eared owls in conifer forests, Oregon. *Condor* 91:908–912.

———. 1990. Nesting habitat of flammulated owls in Oregon. *J. Raptor Res.* 24:52–55.

Bunn, D. S., A. B. Warburton, and R. D. S. Wilson. 1982. *The barn owl.* Berkhamsted, Eng.: T. and A. D. Poyser.

Burnham, K. P., D. R. Anderson, and G. C. White. 1996. Meta-analysis of vital rates of the northern spotted owl. *Studies in Avian Biol.* 17:92–101.

Burton, J. A., ed. 1973. *Owls of the world.* New York: E. P. Dutton.

Butts, K. O. 1973. Life history and habitat requirements of burrowing owls in western Oklahoma. Master's thesis, Oklahoma State Univ., Stillwater.

Byrkjedal, I., and G. Langhelle. 1986. Sex and age biased mobility in hawk owls *Surnia ulula*. *Ornis Scand.* 17:306–308.

Cahn, A., and J. T. Kemp. 1930. On the food of certain owls in east-central Illinois. *Auk* 47:323–328.

Campbell, R. W., N. K. Dawe, I. McT. Cowan, J. M. Cooper, G. W. Keiser, and M. C. E. McNall. 1990. *The birds of British Columbia*. Vol. 2. Victoria, B.C.: Royal Brit. Col. Mus.

Campbell, R. W., E. D. Forsman, and B. M. van der Ray. 1984. An annotated bibliography of literature on the spotted owl. Land Management Report no. 24, Ministry of Forests, British Columbia, Victoria.

Campbell, R. W., and M. D. MacColl. 1978. Winter foods of snowy owls in southwestern British Columbia. *J. Wildl. Manage.* 42:190–192.

Campbell, R. W., D. A. Manuwal, and A. S. Harestad. 1987. Food habits of the common barn owl in British Columbia. *Canadian J. Zool.* 65:578–586.

Cannell, P. F. 1985. The syrinx, systematics, and biogeography of wood owls. Paper presented at 103rd meeting, American Ornithologists' Union, 7–10 Oct., Tempe, Arizona (abstract).

Cannings, R. J. 1987. The breeding biology of northern saw-whet owls in southern British Columbia. *In* Nero et al., 1987, pp. 193–198.

———. 1993. Northern saw-whet owl. No. 42 in *The birds of North America,* ed. A. Poole, P. Stettenheim, and F. Gill. Philadelphia: Academy of Natural Sciences; Washington, D.C.: American Ornithologists' Union.

Cannings, R. J., and T. Angell. 2001. Western screech-owl. No. 597 in *The birds of North America,* ed. A. Poole, P. Stettenheim, and F. Gill. Philadelphia: Academy of Natural Sciences; Washington, D.C.: American Ornithologists' Union.

Cannings, R. J., and S. R. Cannings. 1982. A flammulated owl nests in a nest box. *Murrelet* 63:66–68.

Cannings, R. J., S. R. Cannings, J. M. Cannings, and G. P. Sirk. 1978. Successful breeding of the flammulated owl in British Columbia. *Murrelet* 59:74–75.

Canova, L. 1989. Influence of snow cover on prey selection by long-eared owls *Asio otus. Ethol. Ecol. and Evol.* 1:367–372.

Carey, A. B. 1985. A summary of the scientific basis for spotted owl management. *In* Gutiérrez and Carey, 1985, pp. 100–114.

Carey, A. B., and K. C. Peeler. 1995. Spotted owls: Resource and space use in mosaic landscapes. *J. Raptor Res.* 29:223–239.

Carey, A. B., J. A. Reed, and S. P. Horton. 1990. Spotted owl home range and habitat use in southern Oregon coast ranges. *J. Wildl. Manage.* 54:11–17.

Carleson, R. D., and W. I. Haight. 1985. Status of spotted owl management in Oregon as perceived by Oregon Department of Fish and Wildlife. *In* Gutiérrez and Carey, 1985, pp. 27–30.

Carlsson, B-G. 1991. Recruitment of mates and deceptive behavior by male Tengmalm's owls. *Behav. Ecol. Sociobiol.* 28:321–328.

Carlsson, B-G., B. Hornfeldt, and O. Lofgren. 1987. Bigyny in Tengmalm's owl *Aegolius funereus. Ornis Scand.* 18:237–243.

Carpenter, T. W. 1987. Effects of environmental variables on responses of eastern screech owls to playback. *In* Nero et al., 1987, pp. 277–280.

———. 1992. Utility of wing length, tail length, and tail barring in determining the sex of barred owls collected in Michigan and Minnesota. *Condor* 94:794–795.

Cartron, J-L. E., W. S. Richardson, and G. A. Proudfoot. 1999. The cactus ferruginous pygmy-owl: Taxonomy, distribution, and natural history. In *Ecology and conservation of the cactus ferruginous pygmy-owl in Arizona,* ed. J-L. E. Cartron and D. M. Finch, pp. 5–16. Gen. Tech. Rep. RMRS-GTR-43, Ogden, Utah.

Catling, P. M. 1971. Spring migration of saw-whet owls at Toronto, Ontario. *Bird-Banding* 42:110–114.

———. 1972a. A behavioral attitude of saw-whet and boreal owls. *Auk* 89:194–196.

———. 1972b. Food and pellet analysis of the saw-whet owl. *Ont. Field Biol.* 26:72–85.

———. 1972c. A study of the boreal owl in southern Ontario with particular reference to the irruption of 1968/69. *Canadian Field-Nat.* 86:223–232.

Cavanagh, P. M., and G. Ritchison. 1987. Variation in the bounce and whinny songs of the eastern screech-owl. *Wilson Bull.* 99:620–627.

Chamberlin, M. I. 1980. Winter hunting behavior of a snowy owl in Michigan. *Wilson Bull.* 92:116–120.

Chandler, R. M. 1982. A reevaluation of the Pliocene owl *Lechusia stirtoni* Miller. *Auk* 99:580–581.

Christie, D. A., and A. M. van Woudenberg. 1997. Modeling critical habitat for flammulated owls (*Otus flammeolus*). *In* Duncan, Johnson, and Nicholls, 1997, pp. 97–106.

Cink, C. L. 1975. Population densities of screech owls in northeastern Kansas. *Kansas Ornith. Soc. Bull.* 26:13–16.

Clark, R. J. 1975. A field study of the short-eared owl *Otus flammeus* (Pontoppidan) in North America. *Wildl. Monogr.* 47:1–67.

———. 1987. Distributional status and literature of northern forest owls. *In* Nero et al., 1987, pp. 47–55.

Clark, R. J., D. G. Smith, and L. H. Kelso. 1978. *Working bibliography of owls of the world.* Washington, D.C.: National Wildlife Federation.

Clayton, K. M., and J. K. Schmutz. 1997. Burrowing owl (*Speotyto cunicularia*) survival in prairie Canada. *In* Duncan, Johnson, and Nicholls, 1997, pp. 107–110.

Coats, S. 1979. Species status and phylogenetic relationships of the Andean pygmy-owl, *Glaucidium jardinii. Amer. Zool.* 19:892.

Coats, S., and P. F. Cannell. 1985. Systematics of the Strigidae. Proc. Raptor Research Found. Symp., 9–10 Nov., Sacramento (abstract).

Collins, C. T. 1963. Notes of the feeding behavior, metabolism, and weight of the saw-whet owl. *Condor* 65:528–530.

———. 1979. The ecology and conservation of burrowing owls. *In* Schaeffer and Ehlers, 1979, pp. 6–17.

———. 1993. A threat display of the northern saw-whet owl (*Aegolius acadicus*). *J. Raptor Res.* 27:113–115.

Collins, C. T., and R. E. Landry. 1977. Artificial nest burrows for burrowing owls. *N. Amer. Bird-Bander* 2:151–154.

Collins, K. M. 1980. Aspects of the biology of the great gray owl. Master's thesis, Univ. of Manitoba, Winnipeg.

Collister, D. M. 1995. Prey caching by non-breeding northern hawk owls in Alberta. *Blue Jay* 53:203–204.

———. 1997. Seasonal distribution of the great gray owl (*Strix nebulosa*) in southwestern Alberta. *In* Duncan, Johnson, and Nicholls, 1997, pp. 119–122.

Colvin, B. A. 1985. Common barn-owl population decline in Ohio and the relationship to agricultural trends. *J. Field Ornith.* 56:224–235.

Colvin, B. A., and P. L. Hegdal. 1985. A comprehensive research effort on the common barn owl (*Tyto alba*). Proc. Raptor Research Found. Symp., 9–10 Nov., Sacramento (abstract).

Colvin, B., A., and S. R. Spaulding. 1983. Winter

foraging behavior of short-eared owls (*Asio flammeus*) in Ohio. *Amer. Midl. Nat.* 110:124–128.

Comeau, N. A. 1923. *From the life and sport of the North Shore.* Quebec: Telegraphic Printing Co.

Conners, V. A. 1982. Differential roost site selection in screech owls (*Otus asio*). Master's thesis, Southern Connecticut State College, New Haven.

Cook, W. E. 1997. *Avian desert predators.* Berlin: Springer-Verlag.

Cooksey, M. 1998. A pre-1996 North American record of Stygian owl. *Field Notes* 52:265–266.

Coulombe, H. N. 1971. Behavior and population ecology of the burrowing owl (*Speotyto cunicularia*) in the Imperial Valley of California. *Condor* 73:162–176.

Courser, W. D. 1972. Variability of tail molt in the burrowing owl. *Wilson Bull.* 84:93–95.

Craig, E. H., T. H. Craig, and L. R. Powers. 1988. Activity patterns and home-range use of nesting long-eared owls. *Wilson Bull.* 100:204–213.

Craighead, F. C., Jr., and D. P. Mindell. 1981. Nesting raptors in western Wyoming, 1947 and 1975. *J. Wildl. Manage.* 45:865–872.

Craighead, J. J., and F. C. Craighead Jr. 1956. *Hawks, owls, and wildlife.* Harrisburg, Pa.: Stackpole; Washington, D.C.: Wildl. Manage. Inst. (Reprinted 1969 by Dover Pub., New York.)

Cramp, S., ed. 1985. *Handbook of the birds of Europe, the Middle East and North Africa.* Vol. 4. Oxford: Oxford Univ. Press.

Curtis, W. 1952. Quantitative studies of echolocation in bats (*Myotis l. lucifugus*), studies of vision of bats (*Myotis l. lucifugus* and *Eptesicus f. fuscus*), and quantitative studies of vision of owls (*Tyto alba pratincola*). Ph.D. diss., Cornell Univ., Ithaca.

Dark, S. J., R. J. Gutiérrez, and G. I. Gould Jr. 1998. The barred owl (*Strix varia*) invasion in California. *Auk* 115:50–56.

Dawson. J. W. 1998. Great horned owl. *In* Glinski, 1998a, pp. 152–155.

Dechant, D. J., M. L. Sondreal, D. H. Johnson, L. D. Igl, C. M. Goldade, M. P. Nenneman, and B. R. Euliss. 1999a. Effects of management practices on grassland birds: Burrowing owl. Northern Prairie Wildlife Research Center, Jamestown, N.Dak.

———. 1999b. Effects of management practices on grassland birds: Short-eared owl. Northern Prairie Wildlife Research Center, Jamestown, N.Dak.

de Gubernatis, A. 1872. *Zoological mythology.* London: Trubner. (Reprinted 1968 by Singing Tree Press, Detroit.)

DeSimone, P., M. Root, and D. Roddy. 1985. Barred owl (*Strix varia*) nesting and behavior in northwestern Connecticut. Proc. Raptor Research Found. Symp., 9–10 Nov., Sacramento (abstract).

De Smet, K. D. 1997. Burrowing owl (*Speotyto cunicularia*) monitoring and management activities in Manitoba, 1987–1996. *In* Duncan, Johnson, and Nicholls, 1997, pp. 123–130.

Desmond, M. J. 1997. Evolutionary history of the genus Speotyto: A genetic and morphological perspective. Ph.D. diss., Univ. of Nebraska–Lincoln.

Desmond, M. J, and J. A. Savidge. 1995. Spatial patterns of burrowing owl (*Speotyto cunicularia*) nests within black-tailed prairie dog (*Cynomys ludovicianus*) towns. *Canadian J. Zool.* 73:1375–1379.

———. 1996. Factors influencing burrowing owl (*Speotyto cunicularia*) nest densities and numbers in western Nebraska. *Amer. Midl. Nat.* 136:143–148.

Desmond, M. J., J. A. Savidge, and K. M. Eskridge. 2000. Correlations between burrowing owl and black-tailed prairie dog declines: A seven-year analysis. *J. Wildl. Manage.* 64:1067–1075.

Devereux, J. C., and J. A. Moser. 1984. Breeding ecology of barred owls in the central Appalachians. *Raptor Research* 18:49–58.

deVos, J. C., Jr. 1998a. Burrowing owl. *In* Glinski, 1998a, pp. 166–169.

———. 1998b. Northern saw-whet owl. *In* Glinski, 1998a, pp. 182–184.

Dice, L. R. 1945. Minimum intensities of illumination under which owls can find dead prey by sight. *Amer. Nat.* 79:385–416.

Dickerman, R. W. 1993. The subspecies of the great horned owls of the central Great Plains, with notes on adjacent areas. *Kansas Ornith. Soc. Bull.* 44:17–21.

Diller, L. V., and D. M. Thome. 1999. Population density of northern spotted owls in managed young-growth forests in coastal northern California. *J. Raptor Res.* 33:275–286.

Doak, D. 1989. Spotted owls and old growth logging in the Pacific Northwest. *Conserv. Biol.* 3:389–396.

Dorn, J. L., and R. D. Dorn. 1994. Further data on screech-owl distribution and habitat use in Wyoming. *West. Birds* 25:35–42.

Duffy, K. E., and P. E. Matheny. 1997. Northern saw-whet owls (*Aegolius acadicus*) captured at Cape May Point, N.J., 1980–1994: Comparison of two capture techniques. *In* Duncan, Johnson, and Nicholls, 1997, pp. 131–137.

Dunbar, D. L., B. P. Booth, E. D. Forsman, A. E. Hetherton, and D. Wilson. 1991. Status of the spotted owl (*Strix occidentalis*) and barred owl (*Strix varia*) in southwestern British Columbia. *Canadian Field-Nat.* 105:464–468.

Dunbar, D. L., and D. J. Wilson. 1987. The status of spotted owls in southwestern British Columbia. Paper presented at Symposium on Biology and Conservation of Northern Forest Owls, 3–7 Feb., Winnipeg, Manitoba.

Duncan, J. R. 1987. Movement strategies, mortality, and behavior of radio-marked great gray owls in southeastern Manitoba and northern Minnesota. *In* Nero et al., 1987, pp. 101–107.

———. 1992. Influence of prey abundance and snow cover on great gray owl breeding dispersal. Ph.D. diss., Univ. of Manitoba, Winnipeg.

———. 1997. Great gray owls (*Strix nebulosa*) and forest management in North America: A review. *J. Raptor Res.* 31:160–166.

Duncan, J. R., and P. A. Duncan. 1997. Increase in distribution records of owl species in Manitoba based on a volunteer nocturnal survey using boreal owl (*Aegolius funereus*) and great gray owl (*Strix nebulosa*) playback. *In* Duncan, Johnson, and Nicholls, 1997, pp. 519–524.

———. 1998. Northern hawk owl. No. 356 in *The birds of North America,* ed. A. Poole, P. Stettenheim, and F. Gill. Philadelphia: Academy of Natural Sciences; Washington, D.C.: American Ornithologists' Union.

Duncan, P. A., and W. C. Harris. 1997. Northern hawk owls (*Surnia ulula caparoch*) and forest management in North America: A review. *J. Raptor Res.* 31:187–190.

Duncan, P. A., D. H. Johnson, and T. H. Nicholls, eds. 1997. *Biology and conservation of owls of the northern hemisphere: Symposium proceedings,* 5–9 Feb., Winnipeg, Manitoba. Tech. Rep. NC-190. USDA Forest Service, North Central Forest Experiment Station. St. Paul, Minn.

Duncan, P. A., and A. E. Kearns. 1997. Habitat associated with barred owl (*Strix varia*) locations in southeastern Manitoba: A review of a habitat model. *In* Duncan, Johnson, and Nicholls, 1997, pp. 138–147.

Dunham, S., L. Butcher, D. A. Charlet, and J. M. Reed. 1996. Breeding range and conservation of flammulated owls (*Otus flammeolus*) in Nevada. *J. Raptor Res.* 30:189–193.

Dunning, J. B., Jr. 1985. Owl weights in the literature: A review. *Raptor Research* 19:113–121.

———, ed. 1993. *CRC handbook of avian body masses.* Baton Rouge: CRC Press.

Dunstan, T. C., and S. D. Sample. 1972. Biology of barred owls in Minnesota. *Loon* 44:111–115.

Dunstan, T. C., and T. E. M. Varchmin. 1985. Behavioral and physical development of barred owls, *Strix varia.* Proc. Raptor Research Found. Symp., 9–10 Nov., Sacramento (abstract).

Earhart, C. M., and N. K. Johnson. 1970. Sex dimorphism and food habits of North American owls. *Condor* 72:251–264.

Eckert, K. 1979. First boreal owl nesting record south of Canada: A diary. *Loon* 51:20–27.

———. 1984. A record invasion of great gray owls. *Loon* 56:143–147.

Edberg, R. 1955. (The irruption of hawk-owls [*Surnia ulula*] in northwestern Europe, 1950–1951.) *Vår Fågelvärld* 14:10–21. (In Swedish, English summary.)

Ekstein, R. 1999. Local and landscape factors affecting burrowing owl nest site selection and nest success in western Nebraska. Master's thesis, Univ. of Nebraska–Lincoln.

Elderkin, M. F. 1987. The breeding and feeding ecology of a barred owl *Strix varia* population in Kings County, Nova Scotia. Master's thesis, Acadia Univ., Wolfsville, Nova Scotia.

Ellison, P. T. 1980. Habitat use by resident screech owls (*Otus asio*). Master's thesis, Univ. of Massachusetts, Amherst.

Elody, B. I., and N. F. Sloan. 1985. Movements and habitat use of barred owls in the Huron Mountains of Marquette County, Michigan, as determined by radiotelemetry. *Jack-Pine Warbler* 63:3–8.

Emlen, J. T., Jr. 1973. Vocal stimulation in the great horned owl. *Condor* 75:126–127.

Enriquez-Rocha, P. 1995. Abundancia relativa, uso de habitat y conocimiento popular de los Strigiformes en un bosque humedo tropical en Costa Rica. Master's thesis, Universidad Nacional Costa Rica, Heredia.

———. 1997. Seasonal records of the burrowing owl in Mexico. *J. Raptor Res. Report* 9:49–51.

Enriquez-Rocha, P., and J. L. Rangel-Salazar. 1997. Intra- and interspecific calling in a tropical owl community. *In* Duncan, Johnson, and Nicholls, 1997, pp. 525–532.

Enriquez-Rocha, P., J. L. Rangel-Salazar, and D. Holt. 1993. Presence and distribution of Mexican owls: A review. *J. Raptor Res.* 27:154–160.

———. 1994. The distribution of Mexican owls. *In* Meyburg and Chancellor, 1994, pp. 567–574.

Epple, W. 1985. (Ethological adaptations in the reproductive system of the barn owl *Tyto alba* Scop., 1769.) *Ökol. Vögel* 7:1–95. (In German, English summary.)

Erdman, T. C., T. O. Meyer, J. H. Smith, and D. M. Erdman. 1997. Autumn populations and movements of migrant northern saw-whet owls (*Aegolius acadicus*) at Little Suamico, Wisconsin. In Duncan, Johnson, and Nicholls, 1997, pp. 167–174.

Erdoes, R., and A. Ortiz. 1984. *American Indian myths and legends.* New York: Pantheon Books.

Errington, P. L. 1932a. Food habits of southern Wisconsin raptors. Pt. 1. Owls. *Condor* 34:176–186.

──────. 1932b. Studies on the behavior of the great horned owl. *Wilson Bull.* 44:212–220.

Errington, P. L., and L. J. Bennett. 1935. Food habits of burrowing owls in northwestern Iowa. *Wilson Bull.* 47:125–128.

Errington, P. L., F. Hamerstrom, and F. N. Hamerstrom Jr. 1940. The great horned owl and its prey in north-central United States. *Iowa Agric. Exp. Sta. Res. Bull.* 277:758–850.

Evans, D. L. 1980. Vocalizations and territorial behavior of wintering snowy owls. *Amer. Birds* 34:748–751.

──────. 1982. Status reports on twelve raptors. *U.S. Fish and Wildl. Serv. Spec. Sci. Rep., Wildl.* no. 238:1–68.

Evans, D. L., and R. N. Rosenfield. 1987. Remigial molt in fall migrant long-eared and northern saw-whet owls. *In* Nero et al., 1987, pp. 209–214.

Everett, M. 1977. *A natural history of owls.* London: Hamlyn.

Fast, S. J., and H. W. Ambrose III. 1976. Prey preference and hunting habitat selection in the barn owl. *Amer. Midl. Nat.* 96:503–507.

Feduccia, A. 1996. *The origin and evolution of birds.* New Haven: Yale Univ. Press.

Feduccia, A., and C. E. Ferree. 1978. Morphology of the bony stapes (columella) in owls: Evolutionary implications. *Proc. Biol. Soc. Wash.* 91:431–438.

Feusier, S. 1989. Distribution and behavior of western screech-owls (*Otus kennicottii*) of the Starr Ranch Audubon Sanctuary, Orange County, California. Master's thesis, Humboldt State Univ., Arcata, Calif.

Ffrench, R. 1973. *A guide to the birds of Trinidad and Tobago.* Wynnewood, Pa.: Livingston Pub.

Fisher, A. K. 1893. The hawks and owls of the United States in their relation to agriculture. *U.S. Dept. Agr. Div. Ornith. and Mammal. Bull.* 3:1–210.

Fitch, H. S. 1947. Predation by owls in the Sierran foothills of California. *Condor* 49:137–151.

──────. 1958. Home ranges, territories, and seasonal movements of vertebrates of the natural history reservation. *Univ. Kansas Pub. Mus. Nat. Hist.* 11:63–226.

Fite, K. V. 1973. Anatomical and behavioral correlates of visual acuity in the great horned owl. *Vision Res.* 13:219–30.

Fitton, S. 1993. Screech-owl distribution in Wyoming. *West. Birds* 24:182–188.

Fleig, M. 1971. Tytonidae × Strigidae cross produces fertile eggs. *Auk* 88:178.

Fletcher, A. C. 1900–1901. The Hako, a Pawnee ceremony. *Ann. Rep., U.S. Bur. Amer. Ethnol.* 22:1–372.

Forbes, J. E., and D. W. Warner. 1974. Behavior of a radio-tagged saw-whet owl. *Auk* 91:783–795.

Ford, N. L. 1967. A systematic study of the owls based on comparative osteology. Ph.D. diss., Univ. of Michigan, Ann Arbor.

Forsman, E. D. 1980. Ageing and moult in western Palaearctic hawk owls *Surnia u. ulula* L. *Ornis Fenn.* 57:173–175.

Forsman, E. D., S. DeStefano, M. G. Rapheal, and R. J. Gutiérrez, eds. 1996. Demography of the spotted owl. *Studies in Avian Biol.* 17:1–122.

Forsman, E. D., and E. C. Meslow. 1985. Old-growth forest retention for spotted owls: How much do they need? *In* Gutiérrez and Carey, 1985, pp. 58–59.

Forsman, E. D., E. C. Meslow, and M. J. Strub. 1977. Spotted owl abundance in young versus old-growth forests, Oregon. *Wildl. Soc. Bull.* 5:43–47.

Forsman, E. D., E. C. Meslow, and H. M. Wight. 1984. Distribution and biology of the spotted owl in Oregon. *Wildl. Monogr.* no. 87:1–64.

Forsman, E. D., and H. M. Wight. 1979. Allopreening in owls: What are its functions? *Auk* 96:525–531.

Frank, R. 1997. Population ecology of great horned owls at the Rocky Mountain Arsenal National Wildlife Refuge. Master's thesis, Univ. of Wisconsin, Madison.

Frank, R., and R. S. Lutz. 1997. Great horned owl (*Bubo virginianus*) productivity and home range characteristics. *In* Duncan, Johnson, and Nicholls, 1997, pp. 185–189.

Franklin, A. B. 1985. Breeding ecology of the great gray owl (*Strix nebulosa*) in the Grand Teton region of Idaho and Wyoming. Proc. Raptor Research Found. Symp., 9–10 Nov., Sacramento (abstract).

──────. 1987. Breeding biology of the great gray owl in southeastern Idaho and northwestern Wyoming. Master's thesis, Humboldt State Univ., Arcata, Calif.

──────.1988. Breeding biology of the great gray owl in southeastern Idaho and northwestern Wyoming. *Condor* 90:689–696.

──────. 1997. Factors affecting spatial variation in northern spotted owl populations in northwest California. Ph.D. diss., Colo. State Univ., Fort Collins.

Franklin, A. B., S. R. Anderson, E. D. Forsman, K. P. Burnham, and F. W. Wagner. 1996. Methods for collecting and analyzing demographic data on the northern spotted owl. *Studies in Avian Biol.* 17:12–20.

Franklin, A. B., R. I. Gutiérrez, B. R. Noon, and J. P. Ward. 1996. Demographic characteristics and trends of northern spotted owl populations in northwestern California. *Studies in Avian Biol.* 17:83–91.

Freeman, P. 2000. Identification of individual barred owls using spectrographic analysis and auditory cues. *J. Raptor Res.* 34:85–92.

Freethy, R. 1992. *Owls: A guide for ornithologists.* Hildenborough, Kent, Eng.: Bishopgate Press.

Frost, B. J., P. J. Baldwin, and M. Csizy. 1989. Auditory localization in the northern saw-whet owl, *Aegolius acadicus. Canadian J. Zool.* 67:1955–1959.

Frylestam, B. 1972. (Movements and mortality rates of Scandinavia-ringed barn-owls *Tyto alba.*) *Ornis Scand.* 3:45–54. (In German, English summary.)

Fuller, M. E. 1979. Spatiotemporal ecology of four sympatric raptors. Ph.D. diss., Univ. of Minnesota, Minneapolis.

Gamel, C. M. 1997. Habitat selection, population density, and home range of the elf owl, *Micrathene whitneyi,* at Santa Ana National Wildlife Refuge, Texas. Master's thesis, Univ. of Texas–Pan American, Edinburg, Tex.

Ganey, J. L. 1988. Distribution and habitat ecology of Mexican spotted owls in Arizona. Master's thesis, Northern Arizona Univ., Flagstaff.

———. 1990. Calling behavior of spotted owls in northern Arizona. *Condor* 92:485–490.

———. 1992. Food habits of Mexican spotted owls in Arizona. *Wilson Bull.* 104:521–426.

———. 1998. Spotted owl. *In* Glinski, 1998a, pp. 170–174.

Ganey, J. L., and R. P. Balda. 1985. Distribution of Mexican spotted owls in Arizona. Proc. Raptor Research Found. Symp., 9–10 Nov., Sacramento (abstract).

Ganey, J. L., and W. M. Block. 2000. Roost sites of radio-marked Mexican spotted owls in Arizona and New Mexico: Sources of variability and descriptive characteristics. *J. Raptor Res.* 34:270–228.

Garcia, E. R. 1979. A survey of the spotted owl in Washington. *In* Schaeffer and Ehlers, 1979, pp. 18–28.

Gates, J. M. 1972. Red-tailed hawk populations and ecology in east-central Wisconsin. *Wilson Bull.* 84:421–433.

Gaunt, A. S., and S. L. L. Gaunt. 1985. Syringeal structure and avian phonation. *Current Ornith.* 2:213–245.

Gaunt, A. S., S. L. L. Gaunt, and R. M. Casey. 1982. Syringeal mechanics reassessed: Evidence from *Streptopelia. Auk* 99:474–494.

Gehlbach, F. R. 1986. Odd couple of suburbia. *Natural History* 95 (6): 56–66.

———. 1989. Screech owl. In *Lifetime reproduction in birds,* ed. I. Newton, pp. 315–326. London: Academic Press.

———. 1994a. *The eastern screech-owl: Life history, ecology, and behavior in the suburbs and the countryside.* College Station: Texas A&M Univ. Press.

———. 1994b. Nest-box and natural-cavity nests of the eastern screech-owl: An exploratory study. *J. Raptor Res.* 28:154–157.

———. 1994c. Recruitment in an eastern screech-owl, *Otus asio,* population: On components of fitness and inheritance. *In* Meyburg and Chancellor, 1994, pp. 507–509.

———. 1995a. Biogeographic controls of avifaunal richness in isolated forests of the U.S.-Mexican borderlands. In *Storm Over a Mountain Island,* ed. C. A. Istock and R. S. Hoffman, pp. 135–150. Tucson: Univ. of Ariz. Press.

———. 1995b. Eastern screech-owl. No. 165 in *The birds of North America,* ed. A. Poole, P. Stettenheim, and F. Gill. Philadelphia: Academy of Natural Sciences; Washington, D.C.: American Ornithologists' Union.

———. 1995c. Eastern screech-owls in suburbia: A model of raptor urbanization. In Bird, Varland, and Negro, 1995, pp. 69–74.

———. In press. Body size variation and evolutionary ecology of eastern and western screech-owls. *Southwestern Naturalist.*

Gehlbach, F. R., and N. Y. Gehlbach. 2000. Whiskered screech-owl. No. 507 in *The birds of North America,* ed. A. Poole, P. Stettenheim, and F. Gill. Philadelphia: Academy of Natural Sciences; Washington, D.C.: American Ornithologists' Union.

Gehlbach, F. R., and J. S. Leverett. 1995. Avian mobbing of eastern screech-owls: Predatory cues, risk to mobbers, and degree of threat. *Condor* 97:831–834.

Gerber, R. 1960. *Die Sumpforeule* Asio flammeus. Neue Brehm-Bücherei 259. Wittenberg Lutherstadt: A. Ziemsen Verlag.

Gerhardt, R. P. 1991. Response of the mottled owl (*Ciccaba virgata*) to broadcast conspecific signals. *J. Field Ornith.* 62:239–244.

Gerhardt, R. P., and D. M. Gerhardt. 1997. Size, dimorphism, and related characteristics of *Ciccaba* owls from Guatemala. *In* Duncan, Johnson, and Nicholls, 1997, pp. 190–196.

Gerhardt, R. P., D. M. Gerhardt, C. J. Flatten, and G. N. Bonilla. 1994a. Breeding biology and home ranges of two *Ciccaba* owls. *Wilson Bull.* 106:629–639.

———. 1994b. The food habits of sympatric *Ciccaba* owls in northern Guatemala. *J. Field Ornith.* 65:258–264.

Gervais, J. A., and D. K. Rosenberg. 1999. Western burrowing owls in California produce second broods of chicks. *Wilson Bull.* 111:569–571.

Gessaman, J. A. 1972. Bioenergetics of the snowy

owl (*Nyctea scandiaca*). *Arctic and Alpine Res.* 4:223–238.

Getz, L. L. 1961. Hunting areas of the long-eared owl. *Wilson Bull.* 73:79–82.

Giese, A. R. 1999. Habitat selection by northern pygmy-owls on the Olympic Peninsula. Master's thesis, Oregon State Univ., Corvallis.

Gilbert, R. 1981. Radiotelemetry study of home range, habitat use, and roost site selection of the eastern screech owl (*Otus asio*). Master's thesis, Southern Connecticut State College, New Haven.

Gilkey, A. K., W. D. Loomis, B. M. Breckenridge, and C. H. Richardson. 1943. The incubation period of the great horned owl. *Auk* 60:272–273.

Gill, M., and R. J. Cannings. 1997. Habitat selection of northern saw-whet owl (*Aegolius acadicus brooksi*) on the Queen Charlotte Islands, British Columbia. *In* Duncan, Johnson, and Nicholls, 1997, pp. 197–204.

Gilligan, J., M. Smith, D. Rogers, and A. Contreras, eds. 1994. *Birds of Oregon: Status and distribution.* McMinnville, Oreg.: Cinclus Pub.

Gleason, R., and T. H. Craig. 1979. Food habits of burrowing owls in southeastern Idaho. *Great Basin Nat.* 39:274–276.

Gleason, R., and D. R. Johnson. 1985. Factors influencing nesting success of burrowing owls in southeastern Idaho. *Great Basin Nat.* 45:81–84.

Glinski, R. L., ed. 1998a. *The raptors of Arizona.* Tucson: Univ. of Arizona Press; Phoenix: Arizona Game and Fish Dept.

———. 1998b. Short-eared owl. *In* Glinski, 1998a, pp. 195–77.

Glover, F. A. 1953. Summer foods of the burrowing owl. *Condor* 55:275.

Glue, D. E. 1977a. Breeding biology of long-eared owls. *Brit. Birds* 70:318–331.

———. 1977b. Feeding ecology of the short-eared owl in Britain and Ireland. *Bird Study* 24:70–78.

Glutz von Blotzheim, U. N., and K. M. Bauer, eds. 1980. *Handbuch der Vögel Mitteleuropas.* Band 9. Wiesbaden: Akademische Verlagsgesellschaft.

Goad, M. S. 1985. Summer habitat and nest site selection of elf owls at Saguaro National Monument, Arizona. Master's thesis, Univ. of Arizona, Tucson.

Goad, M. S., and R. W. Mannan. 1987. Nest site selection by elf owls in Saguaro National Monument, Arizona. *Condor* 89:659–662.

Goggans, R. 1985. Flammulated owl habitat use in northeast Oregon. Master's thesis, Oregon State Univ., Corvallis.

Gomez de Silva, M. Perez-Villafana, and J. A. Santos-Moreno. 1997. Diet of spectacled owl. (*Pulsatrix perspicillata*) during the rainy season in northern Oaxaca, Mexico. *J. Raptor Res.* 31:385–387.

Goodman, A. E., and E. J. Fisk. 1973. Breeding behaviour of captive striped owls (*Rhinoptynx clamator*). *Avic. Mag.* 79:158–162.

Gottfred, J., and A. Gottfred. 1996. Copulatory behavior in the great horned owl. *Blue Jay* 54:180–184.

Gould, G. I., Jr. 1974. The status of the spotted owl in California. Calif. Dept. of Fish and Game, Sacramento.

———. 1977. Distribution of the spotted owl in California. *West. Birds* 8:131–146.

———. 1979. Status and management of elf and spotted owls in California. *In* Schaeffer and Ehlers, 1979, pp. 86–97.

———. 1985. Current and future distribution and abundance of spotted owls in California. Proc. Raptor Research Found. Symp., 9–10 Nov., Sacramento (abstract).

Graber, R. R. 1962. Food and oxygen consumption in three species of owls (Strigidae). *Condor* 64:473–487.

Grant, J. 1966. The barred owl in British Columbia. *Murrelet* 47:39–45.

Grant, R. A. 1965. The burrowing owl in Minnesota. *Loon* 37:2–17.

Green, G. A., and R .G. Anthony. 1989. Nesting success and habitat relationships of burrowing owls in the Columbia River basin, Oregon. *Condor* 91:347–354.

Gross, A. O. 1944. Food of the snowy owl. *Auk* 61:1–18.

———. 1948. Cyclic invasions of the snowy owl and the migrations of 1945–1946. *Auk* 64:584–601.

Grossman, M. L., and J. Hamlet. 1964. *Birds of prey of the world.* New York: Bonanza Books.

Gutiérrez, R. J. 1985. An overview of recent research on the spotted owl. *In* Gutiérrez and Carey, 1985, pp. 39–49.

———. 1994. Changes in the distribution of spotted owls during the past century. *Studies in Avian Biol.* 15:293–300.

———. 1996. Biology and distribution of the northern spotted owl. *Studies in Avian Biol.* 17:2–5.

Gutiérrez, R. J,. and A. B. Carey, eds. 1985. *Ecology and management of the spotted owl in the Pacific Northwest.* USDA, Gen. Tech. Rep. PNW-185, Portland, Oreg.

Gutiérrez, R. J., E. D. Forsman, A. B. Franklin, and E. C. Nieslow. 1996. History of demographic studies in the management of the northern spotted owl. *Studies in Avian Biol.* 17:6–11.

Gutiérrez, R. J., A. B. Franklin, and W. S. LaHaye. 1995. Spotted owl. No. 179 in *The birds of North America,* ed. A. Poole, P. Stettenheim, and F. Gill. Philadelphia: Academy of Natural Sciences; Washington, D.C.: American Ornithologists' Union.

Gutiérrez, R. J., A. B. Franklin, W. S. LaHaye, V. J. Meretsky, and J. P. Ward. 1985. Juvenile spotted owl dispersal in northwestern California: Preliminary analysis. *In* Gutiérrez and Carey, 1985, pp. 60–65.

Gutiérrez, R. J., J. E. Hunter, G. Chavez-Leon, and J. Price. 1998. Characteristics of spotted owl habitat in landscapes disturbed by timber harvest in northwestern California. *J. Raptor Res.* 32:104–110.

Gutiérrez, R. J., and J. Pritchard. 1990. Distribution, density, and age structure of spotted owls on two southern California habitat islands. *Condor* 92:491–495.

Hagar, D. C., Jr. 1957. Nesting populations of red-tailed hawks and horned owls in central New York State. *Wilson Bull.* 69:263–272.

Hagen, Y. 1956. The irruption of hawk owls (*Surnia ulula* L.) in Fennoscandia, 1950–51. *Sterna* 24:1–22.

———. 1960. (Snowy owl studies on Hardangervidda, summer 1959.) *Papers Norwegian Game Research* 2 (7): 1–25. (In Norwegian.)

Halterman, M. D., S. A. Laymon, and M. J. Whitfield. 1989. Status and distribution of the elf owl in California. *West. Birds* 20:71–80.

Hamer, T. E. 1988. Home range size of the northern barred and northern spotted owl in northwestern Washington. Master's thesis, Western Wash. Univ., Bellingham.

Hamer, T. E., and H. L. Allen. 1985. Continued range expansion of the barred owl (*Strix varia*) in western North America. Proc. Raptor Research Found. Symp., 9–10 Nov., Sacramento (abstract).

Hamer, T. E., E. D. Forsman, A. D. Fuchs, and M. L. Walters. 1994. Hybridization between barred and spotted owls. *Auk* 111:487–492.

Hamer, T. E., D. L. Hays, C. M. Senger, and E. D. Forsman. 2001. Diets of northern barred owls and northern spotted owls in an area of sympatry. *J. Raptor Res.* 35:221–227.

Hamer, T. E., F. B. Samson, K. A. O'Halloran, and L. W. Brewer. 1987. Activity patterns and habitat use of barred and spotted owls in northwestern Washington. Paper presented at Symposium on Biology and Conservation of Northern Forest Owls, 3–7 Feb., Winnipeg, Manitoba.

Handy, E. L. 1918. Zuni tales. *J. American Folk-lore* 31:451–471.

Hanson, W. C. 1971. Snowy owl incursions in south-eastern Washington and the Pacific Northwest, 1966–67. *Condor* 73:114–116.

Hardin, K. I., and D. E. Evans. 1977. *Cavity-nesting bird habitat in oak hickory forests: A review.* USDA, Forest Service, Gen. Tech. Rep. NC-30.

Hardy, J. W., B. B. Coffey, and G. B. Reynard, compilers. 1999. *Voices of the New World owls.* (Revised version.) Gainesville, Fla.: ARA Records.

Hardy, P. C. 1997. Habitat selection by elf owls and western screech-owls in the Sonoran Desert. Master's thesis, Univ. of Ariz., Tucson.

Hardy, P. C., and M. L. Morrison. 2001. Nest site selection by elf owls in the Sonoran desert. *Wilson Bull.* 113:25–32.

Hardy, P. C., M. L. Morrison, and R. X. Barry. 1999. Abundance and habitat associations of elf owls and western screech-owls in the Sonoran Desert. *Southwest Nat.* 44:311–323.

Harris, W. C. 1984. Great gray owls in Saskatchewan, 1974–1983. *Blue Jay* 42:152–160.

———. 1987. Habitat use by northern forest owls in central Saskatchewan. Paper presented at Symposium on Biology and Conservation of Northern Forest Owls, 3–7 Feb., Winnipeg, Manitoba.

Harrison, C. J. O., and C. A. Walker. 1975. The Bradycnemidae, a new family of owls from the Upper Cretaceous of Romania. *Paleontology* 18:563–570.

Hasbrouck, E. M. 1893a. Evolution and dichromatism in the genus *Megascops*. *Amer. Nat.* 27:521–533.

———. 1893b. The geographical distribution of the genus *Megascops* in North America. *Auk* 10:250–264.

Haug, E. A. 1985. The breeding ecology of burrowing owls in Saskatchewan. Proc. Raptor Research Found. Symp., 9–10 Nov., Sacramento (abstract).

Haug, E. A., B. A. Millsap, and M. S. Martell. 1993. Burrowing owl (*Athene cunicularia*). No. 61 in *The birds of North America,* ed. A. Poole, P. Stettenheim, and F. Gill. Philadelphia: Academy of Natural Sciences; Washington, D.C.: American Ornithologists' Union.

Haug, E. A., and L. W. Oliphant. 1990. Movements, activity patterns, and habitat use of burrowing owls in Saskatchewan. *J. Wildl. Manage.* 54:27–35.

Haverschmidt, F. 1968. *Birds of Surinam.* Edinburgh: Oliver and Boyd.

Hayward, G. D. 1983. Resource partitioning among six forest owls in the River of No Return Wilderness, Idaho. Master's thesis, Univ. of Idaho, Moscow.

———. 1984. Roost habitat selection by three small forest owls. *Wilson Bull.* 96:690–692.

———. 1989. Habitat use and population biology of boreal owls in the northern Rocky Mountains. Ph.D. diss., Univ. Idaho, Moscow.

———. 1997. Forest management and conservation of boreal owls in North America. *J. Raptor Res.* 31:114–124.

Hayward, G. D., and E. O. Garton. 1983. First nesting record for boreal owl in Idaho. *Condor* 85:501.

———. 1988. Resource partitioning among forest owls in the River of No Return Wilderness, Idaho. *Oecologia* 75:253–265.

Hayward, G. D., and P. H. Hayward 1991. Body measurements of boreal owls in Idaho and a discriminant model to determine sex of live specimens. *Wilson Bull.* 103:497–500.

———. 1993. Boreal owl. No. 63 in *The birds of North America,* ed. A. Poole, P. Stettenheim, and F. Gill. Philadelphia: Academy of Natural Sciences; Washington, D.C.: American Ornithologists' Union.

Hayward, G. D., P. H. Hayward, and E. O. Garton. 1987. Movements and home range use by boreal owls in central Idaho. *In* Nero et al., 1987, pp. 175–184.

———. 1993. Ecology of boreal owls in the northern Rocky Mountains, USA. *Wildl. Monogr.* no. 124:1–59. (Rev. RRJ 29:148)

Hayward, G. D., P. H. Hayward, E. O. Garton, and R. Escano. 1987. Revised breeding distribution of the boreal owl in the northern Rocky Mountains. *Condor* 89:431–432.

Hayward, G. D., and J. Verner, eds. 1994. *Flammulated, boreal, and great gray owls in the United States: A technical conservation assessment.* U.S. Forest Service, Gen. Tech. Rep. RM 253. Rocky Mtn. Forest and Range Exp. Station, Fort Collins, Colo.

Heinrich, B. 1987. *One man's owl.* Princeton: Princeton Univ. Press.

Hekstra, C. P. 1982. Description of twenty-four new subspecies of American *Otus. Bull. Zool. Mus. Univ. Amsterdam* 9:49–63.

Henny, C. J. 1969. Geographic variation in mortality rates and production requirements of the barn owl (*Tyto alba* sspp.). *Bird-Banding* 40:277–290.

———. 1972. An analysis of the population dynamics of selected avian species. *U.S. Fish and Wildl. Serv., Wildl. Res. Rept.* 1.

Henny, C. J., and L. J. Blus. 1981. Artificial burrows provide new insight into burrowing owl nesting biology. *Raptor Research* 15:82–85.

Henny, C. J., and L. F. VanCamp. 1979. Annual weight cycle in wild screech owls. *Auk* 96:795–796.

Henrioux, F. 2000. Home range and habitat use by the long-eared owl in northwestern Switzerland. *J. Raptor Res.* 34:93–101.

Henry, S. G. 1998. Elf owl. *In* Glinski, 1998a, pp. 162–165.

Henry, S. G., and F. R. Gehlbach. 1999. Elf owl. No. 413 in *The birds of North America,* ed. A. Poole, P. Stettenheim, and F. Gill. Philadel-phia: Academy of Natural Sciences; Washington, D.C.: American Ornithologists' Union.

Herren, V., S. H. Anderson, and L. F. Ruggiero. 1996. Boreal owl mating habitat in the northwestern United States. *J. Raptor Res.* 30:123–129.

Herrera, C. M., and F. Hiraldo. 1976. Food-niche and trophic relationships among European owls. *Ornis Scand.* 7:29–41.

Herter, D. R., and L. L. Hicks. 2000. Barred owl and spotted owl populations and habitat in the central Cascade range of Washington. *J. Raptor Res.* 34:279–286.

Herting, B. L., and J. R. Belthoff. 1997. Testosterone, aggression, and territoriality in male western screech-owls (*Otus kennicottii*): Results from preliminary experiments. *In* Duncan, Johnson, and Nicholls, 1997, pp. 213–217.

Hilty, S., and W. L. Brown. 1986. *A guide to the birds of Colombia.* Princeton: Princeton Univ. Press.

Hocking, B., and B. L. Mitchell. 1961. Owl vision. *Ibis* 103a:284–288.

Hoffman, W. J. 1892–1893. The Menomini Indians. *Ann. Rep., U.S. Bur. Amer. Ethnol.* 14:1–328.

Hoffmeister, D. F., and H. W. Setzer. 1947. The postnatal development of two broods of great horned owls (*Bubo virginianus*). *Univ. Kans. Pub. Mus. Nat. Hist.* 1:157–173.

Holmberg, T. 1982. (Breeding density and site tenacity of Tengmalm's owl, *Aegolius funereus.*) *Vår Fågelvärld* 41:265–267. (In Swedish, English summary.)

Holmgren, V. C. 1988. *Owls in folklore and natural history.* Santa Barbara: Capra Press.

Holroyd, G. L., and T. I. Wellicome. 1997. Report on the western burrowing owl (*Speotyto cunicularia*) conservation workshop. *In* Duncan, Johnson, and Nicholls, 1997, pp. 612–615.

Holroyd, G. L., and J. G. Woods. 1975. Migration of the saw-whet owl in eastern North America. *Bird-Banding* 46:101–105.

Holt, D. W. 1992. Notes on short-eared owl (*Asio flammeus*) nest sites, reproduction, and territory size in coastal Massachusetts. *Can. Field-Nat.* 106:352–356.

———. 1993. Trophic niche of Nearctic short-eared owls. *Wilson Bull.* 105:497–503.

———. 1996. A banding study of Cincinnati area great horned owls. *J. Raptor Res.* 30:194–197.

———. 1997. The long-eared owl (*Asio otus*) and forest management: A review of the literature. *J. Raptor Res.* 32:175–186.

Holt, D. W., E. Andrews, and N. Claflin. 1991. Nonbreeding season diet of northern saw-whet owls, *Aegolius acadicus,* on Nantucket Island, Massachusetts. *Canadian Field-Nat.* 105:382–385.

Holt, D. W., and J. M. Hillis. 1987. Current status and habitat associations of forest owls in western Montana. *In* Nero et al., 1987, pp. 281–288.

Holt, D. W., and S. M. Leasure. 1993. Short-eared owl (*Asio flammeus*). No. 62 in *The birds of North America,* ed. A. Poole and F. Gill. Philadelphia: Academy of Natural Sciences; Washington, D.C.: American Ornithologists' Union.

Holt, D. W., and L. A. Leroux. 1996. Diets of northern pygmy-owls and northern saw-whet owls in west-central Montana. *Wilson Bull.* 108:123–128.

Holt, D. W., and S. M. Melvin. 1986. Population dynamics, habitat use, and management needs of the short-eared owl in Massachusetts: Summary of 1985 research. Mass. Div. Fish Wildl. Nat. Her. Prog., Boston.

Holt, D. W., and J. L. Petersen. 2000. Northern pygmy-owl (*Glaucidium gnoma*). No. 494 in *The birds of North America,* ed. A. Poole and F. Gill. Philadelphia: Academy of Natural Sciences; Washington, D.C.: American Ornithologists' Union.

Holzinger, J., M. Mickley, and K. Schilhansl. 1973. (Studies on the breeding and foraging biology of the short-eared owl [*Otus flammeus*] in a south-German breeding area with observations on the movements of the species in central Europe.) *Anz. Orn. Ges. Bayern* 12:176–197. (In German, English summary.)

Hough, F. 1960. Two significant calling periods of the screech owl. *Auk* 77:227–228.

Houston, C. S. 1971. Brood size of the great horned owl in Saskatchewan. *Bird-Banding* 42:103–105.

————. 1975. Reproductive performance of great horned owls in Saskatchewan. *Bird-Banding* 46:302–304.

————. 1978. Recoveries of Saskatchewan-banded great horned owls. *Canadian Field-Nat.* 92:61–66.

————. 1987. Nearly synchronous cycles of the great horned owl and snowshoe hare in Saskatchewan. *In* Nero et al., 1987, pp. 56–58.

————. 1997. Banding of *Asio* owls in south-central Saskatchewan. *In* Duncan, Johnson, and Nicholls, 1997, pp. 237–242.

Houston, C. S., and C. M. Francis. 1995. Survival of great horned owls in relation to the snowshoe hare cycle. *Auk* 112:44–59.

Houston, C. S., D. G. Smith, and C. Rohner. 1998. Great horned owl. No. 372 in *The birds of North America,* ed. A. Poole, P. Stettenheim, and F. Gill. Philadelphia: Academy of Natural Sciences; Washington, D.C.: American Ornithologists' Union.

Howell, S. N. G., and M. B. Robbins. 1995. Species limits in the least pygmy-owl (*Glaucidium minutissimum*) complex. *Wilson Bull.* 107:7–25.

Howell, S. N. G., and S. Webb. 1995. *A guide to the birds of Mexico and northern Central America.* Oxford: Oxford Univ. Press.

Howie, R. R. 1980a. The burrowing owl in British Columbia. In *Threatened and endangered species and habitats in British Columbia and the Yukon,* ed. R. Stace-Smith, L. Johns, and P. Joslin, pp. 88–95. Victoria, B.C.: Ministry of Environment.

————. 1980b. The spotted owl in British Columbia. In *Threatened and endangered species and habitats in British Columbia and the Yukon,* ed. R. Stace-Smith, L. Johns, and P. Joslin, pp. 96–105. Victoria, B.C.: Ministry of Environment.

Howie, R. R., and R. Ritcey. 1987. Distribution, habitat selection, and densities of flammulated owls in British Columbia. *In* Nero et al., 1987, pp. 249–254.

Hoyo, J. del, A. Elliott, and J. Sargatal. 1999. *Handbook of the birds of the world.* Vol. 5. *Barn-owls to hummingbirds.* Barcelona: Lynx Edicions. (Owls, pp. 34–242.)

Hudson, G. E. 1937. Studies on the muscles of the pelvic appendage of birds. *Amer. Midl. Nat.* 1:1–108.

Hughes, A. J. 1993. Breeding density and habitat preference of the burrowing owl in northeastern Colorado. Master's thesis, Colo. State Univ., Fort Collins.

Huhtala, K., E. Korpimäki, and E. Pullianien. 1987. Foraging activity and growth of nestlings in the hawk owl: Adaptive strategies under northern conditions. *In* Nero et al., 1987, pp. 152–156.

Hume, R. 1991. *Owls of the world.* Limpsfield, U.K.: Dragon's World Ltd.

Ibanez, C., C. Ramo, and B. Busto. 1992. Notes on food habits of the black-and-white owl. *Condor* 94:529–531.

Isaach, J. P., M. S. Bo, and M. M. Martinez. 2000. Food habits of the striped owl (*Asio clamator*) in Buenos Aires Province, Argentina. *J. Raptor Res.* 34:235–237.

Jacot, E. C. 1931. Notes on the spotted and flammulated screech owls in Arizona. *Condor* 33:8–11.

Jaksic, F. M., and J. H. Carothers. 1985. Ecological, morphological, and bioenergetic correlates of hunting mode in hawks and owls. *Ornis Scand.* 16:165–172.

James, P. C., and T. J. Ethier. 1988. Trends in the winter distribution and abundance of burrowing owls. *Amer. Birds* 43:1224–1225.

James, R. D., and S. V. Nash. 1983. An upright posture of young northern hawk-owls. *Ont. Field Biol.* 37:90–93.

James, T. R., and R. W. Seabloom. 1968. Notes on the burrow ecology and food habits of the burrowing owl in southwestern North Dakota. *Blue Jay* 26:83–84.

Jannson, F. 1964. (Notes on a breeding pair of pygmy owls [*Glaucidium passerinum*] in central Sweden.) *Vår Fågelvärld* 23:209–222. (In Swedish, English summary.)

Janzen, D. H., and C. M. Pond. 1974. Food and feeding behavior of a captive Costa Rican least pygmy-owl (*Glaucidium minutissimum rarum*). *Brenesia* 9:71–78.

Jensen, W. F., W. L. Robinson, and N. L. Heitman. 1982. Breeding of the great gray owl on Neebish Island, Michigan. *Jack-Pine Warbler* 60:27–28.

Johns, S., G. R. A. Ebel, and A. Johns. 1978. Observations on the nesting behaviour of the saw-whet owl in Alberta. *Blue Jay* 36:36–38.

Johnsgard, P. A. 2001. *Prairie birds: Fragile splendor in the Great Plains.* Lawrence: Univ. Press of Kansas.

Johnson, D. H. 1987. Barred owls and nest boxes: Results of a five-year study in Minnesota. *In* Nero et al., 1987, pp. 129–134.

———. 1997. Wing loading in 15 species of North American owls. *In* Duncan, Johnson, and Nicholls, 1997, pp. 553–561.

Johnson, D. R. 1978. *The study of raptor populations.* Moscow: Univ. Press of Idaho.

Johnson, E. D., and P. J. Zwank. 1990. Flammulated owl biology on the Sacramento Unit of the Lincoln National Forest. MS, U.S. Forest Serv., Lincoln National Forest, Alamogordo, N.Mex.

Johnson, N. K. 1963. The supposed migratory status of the flammulated owl. *Wilson Bull.* 75:174–178.

Johnson, N. K., and W. C. Russell. 1962. Distribution data on certain owls in the western Great Basin. *Condor* 64:513–515.

Johnson, R. R., B. T. Brown, L. T. Haight, and J. M. Simpson. 1981. Playback recordings as a special avian censusing technique. *Studies in Avian Biol.* 6:68–71.

Johnson, R. R., and L. T. Haight. 1985. Status of the ferruginous pygmy-owl in the southwestern United States. Paper presented at 103rd meeting, American Ornithologists' Union, 7–10 Oct., Tempe, Arizona (abstract).

Johnson, R. R., L. T. Haight, and J. M. Simpson. 1979. Owl populations and species status in the southwestern states. *In* Schaeffer and Ehlers, 1979, pp. 40–59.

Jones, S. R. 1991. Distribution of small forest owls in Boulder County, Colorado. *J. Colo. Field Ornith.* 25:55–70.

Karalus, K. E., and A. W. Eckert. 1974. *The owls of North America.* New York: Doubleday.

Karr, J. R. 1978. Weights of some Central American birds. *Brenesia* 14–15:249–257.

Keith, L. B. 1963. *Wildlife's ten-year cycle.* Madison: Univ. of Wisconsin Press.

———. 1964. Territoriality among wintering snowy owls. *Canadian Field-Nat.* 78:17–24.

Kellomaki, E., E. Heinonen, and H. Tiainen. 1977. (Two successful nestings of Tengmalm's owl in one summer.) *Ornis Fenn.* 54:134–135. (In Finnish, English summary.)

Kelso, L. H. 1940. Variation of the external ear-opening in the Strigidae. *Wilson Bull.* 52:24–29.

Kerlinger, P., and M. R. Lein. 1986. Differences in winter range among age-sex classes of snowy owls (*Nyctea scandiaca*) in North America. *Ornis Scand.* 17:1–7.

———. 1988a. Causes of mortality, fat condition, and weights of wintering snowy owls. *J. Field Ornith.* 59:7–12.

———. 1988b. Population ecology of snowy owls during winter on the Great Plains of North America. *Condor* 90:866–874.

Kerlinger, P., M. R. Lein, and B. J. Sevick. 1985. Distribution and population fluctuations of wintering snowy owls (*Nyctea scandiaca*) in North America. *Canadian J. Zool.* 63:1829–1834.

Kertell, K. 1977. The spotted owl at Zion National Park, Utah. *West. Birds* 8:147–150.

———. 1986. Reproductive biology of northern hawk-owls in Denali National Park. *J. Raptor Res.* 20:91–101.

Kimball, H. H. 1925. Pygmy owl killing a quail. *Condor* 27:209–210.

Kingery, H., ed. 1998. *Colorado breeding bird atlas.* Denver: Colorado Bird Atlas Partnership and Colorado Div. of Wildlife.

Kirk, D. A., and C. Hyslop. 1998. Population status and recent trends in Canadian raptors: A review. *Biol. Conserv.* 83:92–118.

Klimkiewicz, M. K., and A. G. Futcher. 1989. Longevity records of North American birds. Suppl. 1. *J. Field Ornith.* 60:469–494.

Knight, R. L., and A. W. Erickson. 1977. Ecological notes on long-eared and great horned owls along the Columbia River. *Murrelet* 58:2–6.

Knight, R. L., and R. E. Jackman. 1984. Food-niche relationships between great horned owls and common barn-owls in eastern Washington. *Auk* 101:175–179.

Knudsen, E. I. 1981. The hearing of the barn owl. *Sci. Amer.* 245:112–125.

Koivula, N., E. Korpimäki, and J. Viijala. 1997. Do Tengmalm's owls see vole scent marks visible in ultraviolet light? *Anim. Behav.* 54:873–877.

König, C. 1968. (Vocalizations of Tengmalm's owl

[*Aegolius funereus*] and the pygmy owl [*Glaucidium passerinum*].) *Vogelvelt* suppl. 1:115–138. (In German.)

———. 1994. Biological patterns in owl taxonomy, with emphasis on bioacoustical studies on Neotropical pygmy (*Glaucidium*) and screech owls (*Otus*). *In* Meyburg and Chancellor, 1994, pp. 1–19.

König, C., F. Weick, and J. H. Becking. 1999. *Owls: A guide to the owls of the world.* New Haven: Yale Univ. Press.

Konishi, M. 1973. How the owl tracks its prey. *Amer. Scientist* 61:414–424.

———. 1983. Night owls are good listeners. *Natural History* 92 (9): 56–59.

———. 1993. Listening with two ears. *Sci. Amer.* 268:66–73.

Konrad, P. M., and D. S. Gilmer. 1984. Observations on the nesting ecology of burrowing owls in central North Dakota. *Prairie Nat.* 16:129–130.

Korpimäki, E. 1981. On the ecology and biology of Tengmalm's owl (*Aegolius funereus*) in southern Ostrobothnia and Suonmenseklä, western Finland. *Acta Univ. Ouluensis* (A). 118:1–84.

———. 1986. Reversed size dimorphism in birds of prey, especially in Tengmalm's owl *Aegolius funereus*: A test of the "starvation hypothesis." *Ornis Scand.* 17:326–332.

———. 1987a. Clutch size, breeding success, and brood size experiments in Tengmalm's owl *Aegolius funereus*: A test of the "starvation hypothesis." *Ornis Scand.* 18:277–284.

———. 1987b. Sexual size dimorphism and life-history traits of Tengmalm's owl: A review. *In* Nero et al., 1987, pp. 157–161.

———. 1988a. Effects of age on breeding performance of Tengmalm's owl *Aegolius funereus* in western Finland. *Ornis Scand.* 19:21–26.

———. 1988b. Effects of territory quality on occupancy, breeding performance, and breeding dispersal in Tengmalm's owl. *J. Anim. Ecol.* 57:97–108.

———. 1991. Poor reproductive success of polygynously mated female Tengmalm's owls: Are better options available? *Anim. Behav.* 41:37–47.

———. 1992a. Diet composition, prey choice, and breeding success of long-eared owls: Effects of multiannual fluctuations in food abundance. *Canadian J. Zool.* 70:2373–2381.

———. 1992b. Fluctuating food abundance determines the lifetime reproductive success of male Tengmalm's owls. *J. Anim. Ecol.* 61:103–111.

Korpimäki, E., and H. Hakkarainen. 1991. Fluctuating food supply affects the clutch size of Tengmalm's owl independent of laying date. *Oecologia* 85:543–552.

Korpimäki, E., and K. Norrdahl. 1991. Numerical and functional responses of kestrels, short-eared owls, and long-eared owls to vole densities. *Ecology* 72:814–826.

Krahe, R. G. 1981. Breeding the striped owl *Rhinoptynx clamator*. *Avic. Mag.* 87:242–248.

Kumar, T. S. 1984. Man-owl: A superstitious concept. Proc. Raptor Research Found. Symp., 9–10 Nov., Sacramento (abstract).

Laidig, K. J., and D. S. Dobkin. 1995. Spatial overlap and habitat associations of barred owls and great horned owls in southern New Jersey. *J. Raptor Res.* 29:151–157.

Land, H. C. 1970. *Birds of Guatemala.* Wynnewood, Pa.: Livingston Press.

Lande, R. 1988. Demographic models of the northern spotted owl (*Strix occidentalis caurina*). *Oecologia* 75:601–607.

———. 1991. Population dynamics and extinction in heterogeneous environments: The northern spotted owl. *In Bird population studies: Relevance to conservation and management,* ed. C. M. Perrins, J.-D. Lebreton, and G. J. M. Hirons, pp. 566–580. Oxford: Oxford Univ. Press.

Landry, R. E. 1979. Growth and development of the burrowing owl, *Athene cunicularia*. Master's thesis, Calif. State Univ., Long Beach.

Lane, W. H., D. E. Andersen, and T. H. Nicholls. 1997a. Distribution, abundance, and habitat use of territorial male boreal owls (*Aegolius funereus*) in northeast Minnesota. *In* Duncan, Johnson, and Nicholls, 1997, pp. 246–247.

———. 1997b. Habitat use and movements of breeding male boreal owls (*Aegolius funereus*) in northeast Minnesota as determined by radio telemetry. *In* Duncan, Johnson, and Nicholls, 1997, pp. 248–249.

Lasley, G. H., C. Sexton, and D. Hillsman. 1988. First record of the mottled owl in the United States. *Amer. Birds* 42:23–24.

Lauff, R. F. 1997. Range expansion of northern hawk owls (*Surnia ulula* [L.]) and boreal owls (*Aegolius funereus* [L.]) in Nova Scotia. *In* Duncan, Johnson, and Nicholls, 1997, pp. 569–571.

Laughlin, S. H., and D. P. Kibbe, eds. 1985. *The atlas of the breeding birds of Vermont.* Hanover, Vt.: Univ. Press of New England.

Lawless, S. G., G. Ritchison, P. H. Klatt, and D. Westneal. 1997. The mating strategies of eastern screech-owls: A genetic analysis. *Condor* 99:213–217.

Laymon, S. A. 1985. General habitats and movements of spotted owls in the Sierra Nevada. *In* Gutiérrez and Carey, 1985, pp. 66–68.

———. 1988. Ecology of the spotted owl in the central Sierra Nevada, California. Ph.D. diss., Univ. of Calif., Berkeley.

Leder, J. E., and M. L. Walters. 1980. Nesting observations of the barred owl in western Washington. *Murrelet* 61:111–112.

Lemkuhl, J. F., and M. G. Raphael. 1993. Habitat pattern around northern spotted owl locations on the Olympic Peninsula. *J. Wildl. Manage.* 57:302–315.

Levad, R. 1989. Western screech-owls in the Rio Grande Valley. *Colo. Field Ornith. J.* 23:107–109.

Ligon, J. D. 1963. Breeding range expansion of the burrowing owl in Florida. *Auk* 80:367–368.

————. 1968. The biology of the elf owl, *Micrathene whitneyi. Misc. Pub. Mus. Zool., Univ. Mich.* no. 136.

Lindblad, J. 1967. *I ugglemarker.* Stockholm: Bonniers.

Lindhe, U. 1966. (An investigation into the prey selection of the Tengmalm's owl in southwestern Lapland.) *Vår Fågelvärld* 25:40–48. (In Swedish, English summary.)

Linkhart, B. D. 1984. Range, activity, and habitat use by nesting flammulated owls in a Colorado ponderosa pine forest. Master's thesis, Colo. State Univ., Fort Collins.

————. 1987. Brood division and postnesting behavior of flammulated owls. *Wilson Bull.* 99:240–243.

Linkhart, B. D., and R. T. Reynolds. 1985. Breeding biology of nesting flammulated owls (*Otus flammeolus*). Proc. Raptor Research Found. Symp., 9–10 Nov., Sacramento (abstract).

————. 1997. Territories of flammulated owls (*Otus flammeolus*): Is occupancy a measure of habitat quality? *In* Duncan, Johnson, and Nicholls, 1997, pp. 250–254.

Linkhart, B. D., R. T. Reynolds, and R. A. Ryder. 1999. Home range and breeding habitat of flammulated owls in Colorado. *Wilson Bull.* 11:342–351.

Lockie, J. D. 1955. The breeding habits and foods of short-eared owls after a vole plague. *Bird Study* 2:53–69.

Longland, W. S., and M. V. Price. 1991. Direct observations of owls and heteromyid rodents: Can predation risk explain microhabitat use? *Ecology* 72:2261–2273.

Lowery, G. H., and W. W. Dahlquest. 1951. Birds from the State of Veracruz, Mexico. *Univ. of Kansas Pub., Museum of Nat. Hist.* 3 (4): 1–119.

Luce, R., R. Oakleaf, A. Cerovski, L. Hunter, and J. Priday. 1997. *Atlas of birds, mammals, reptiles, and amphibians in Wyoming.* Lander: Wyoming Game and Fish Dept.

Lundberg, A. 1979. Residency, migration, and a compromise: Adaptations to nest-site scarcity and food specialization in three Fennoscandian owl species. *Oecologia* 41:273–281.

Lundsten, J. A . 1993. Survey of the northern pygmy-owl in the Oregon Coast Range. *Oreg. Birds* 19:75–76.

Lutz, R. S., and D. L. Plumpton. 1997. Metapopulation dynamics of a burrowing owl population in Colorado. *In* Duncan, Johnson, and Nicholls, 1997, pp. 255–259.

————. 1999. Philopatry and nest site reuse by burrowing owls: Implications for productivity. *J. Raptor Res.* 33:149–153.

Lynch, P. J., and D. G. Smith. 1984. Census of eastern screech-owls (*Otus asio*) in urban open-space areas using tape-recorded song. *Amer. Birds* 38:388–391.

MacCracken, J. G., D. W. Uresk, and R. M. Hansen. 1985. Vegetation and soils of burrowing owl nest sites in Conata basin, South Dakota. *Condor* 87:152–154.

Manning, T. H., E. O. Höhn, and A. H. Macpherson. 1956. The birds of Banks Island. *Bull. Nat. Mus. Canada* 143:1–144.

Marcot, B. G. 1995. *Owls of old forests of the world.* Gen. Tech. Rep. RNW-343. Portland, Oreg.: USDA, Forest Service, Pacific Northwest Research Station.

Marcot, B. G., and R. Hill. 1980. Flammulated owls in northwestern California. *West. Birds* 11:141–149.

Marks, J. S. 1984. Feeding ecology of breeding long-eared owls in southwestern Idaho. *Canadian J. Zool.* 62:1528–1533.

————. 1986. Nest site characteristics and reproductive success of long-eared owls (*Asio otus*) in southwestern Idaho. *Wilson Bull.* 98:547–560.

Marks, J. S., J. L. Dickinson, and J. Haydock. 1999. Genetic monogamy in long-eared owls. *Condor* 101:844–859.

Marks, J. S., and J. H. Doremus. 1988. Breeding-season diet of northern saw-whet owls in southwestern Idaho. *Wilson Bull.* 100:690–694.

————. 2000. Are northern saw-whet owls nomadic? *J. Raptor Res.* 34:299–304.

Marks, J. S., J. H. Doremus, and R. J. Cannings. 1989. Polygyny in the northern saw-whet owl. *Auk* 106:732–734.

Marks, J. S., D. L. Evans, and D. W. Holt. 1994. Long-eared owl (*Asio otus*). No. 133 in *The birds of North America*, ed. A. Poole, P. Stettenheim, and F. Gill. Philadelphia: Academy of Natural Sciences; Washington, D.C.: American Ornithologists' Union.

Marks, J. S., and C. D. Marti. 1984. Feeding ecology of sympatric barn owls and long-eared owls in Idaho. *Ornis. Scand.* 15:135–143.

Marks, J. S., and A. E. Perkins. 1999. Double brooding in the long-eared owl. *Wilson Bull.* 111:273–276.

Marr, T. G., and D. W. McWhirter. 1982. Differential hunting success in a group of short-eared owls. *Wilson Bull.* 94:82–83.

Marshall, J. R., Jr. 1939. Territorial behavior of the flammulated screech owl. *Condor* 41:71–78.

———. 1957. Birds of pine-oak woodland in southern Arizona and adjacent Mexico. *Pacific Coast Avifauna* 31:1–125.

———. 1967. Parallel variation in North and Middle American screech owls. *Monogr. West. Found. Vert. Zool.* 1:1–72.

———. 1968. Systematics of smaller Asian night birds based on voice. *Ornith. Monogr.* 25:1–58.

Marshall, J. R., Jr., R. Behrstock, and C. König. 1991. Review of *Voices of the New World nightjars and their allies* (Caprimulgiformes) and *Voices of the New World owls* (Strigiformes). *Wilson Bull.* 103:311–314.

Martell, M. S. 1985. The current status of the burrowing owl in Minnesota. Proc. Raptor Research Found. Symp., 9–10 Nov., Sacramento (abstract).

Marti, C. D. 1969. Some comparisons of the feeding ecology of four owls in north-central Colorado. *Southwest. Nat.* 14:163–170.

———. 1974. Feeding ecology of four sympatric owls. *Condor* 76:45–61.

———. 1976. A review of prey selection by the long-eared owl. *Condor* 78:331–336.

———. 1979. Status of owls in Utah. *In* Schaeffer and Ehlers, 1979, pp. 29–35.

———. 1988. A long-term study of food-niche dynamics in the common barn owl: Comparisons within and between populations. *Canadian J. Zool.* 66:1803–1812.

———. 1992. Barn owl. No. 1 in *The birds of North America,* ed. A. Poole, P. Stettenheim, and F. Gill. Philadelphia: Academy of Natural Sciences; Washington, D.C.: American Ornithologists' Union.

———. 1997a. Flammulated owls (*Otus flammeolus*) breeding in deciduous forests. *In* Duncan, Johnson, and Nicholls, 1997, pp. 262–266.

———. 1997b. A twenty-year study of barn owl (*Tyto alba*) reproduction in northern Utah. *In* Duncan, Johnson, and Nicholls, 1997, p. 261.

———. 1999. Natal and breeding dispersal in barn owls. *J. Raptor Res.* 33:181–189.

Marti, C. D., and M. N. Kochert. 1995. Are red-tailed hawks and great horned owls diurnal-nocturnal dietary counterparts? *Wilson Bull.* 107:615–628.

———. 1996. Diet and trophic characteristics of great horned owls in southwestern Idaho. *J. Field Ornith.* 67:499–506.

Marti, C. D., E. Korpimäki, and F. M. Jaksic. 1993. Trophic structure of raptor communities: A continent comparison and synthesis. In *Current Ornith.,* vol. 10, ed. D. M. Power. New York: Plenum Press.

Marti, C. D., and J. S. Marks. 1989. Medium-sized owls. In *Proceedings of the Western Raptor Management Symposium and Workshop,* ed. B. G. Pendleton, pp. 124–133. Natl. Wildl. Fed., Sci. Tech. Ser. no. 12, Washington, D.C.

Marti, C. D., K. Steenhof, M. N. Kochert, and J. S. Marks. 1993. Community trophic structure: The roles of diet, body size, and activity time in vertebrate predators. *Oikos* 67:6–18.

Martin, D. J. 1973a. Selected aspects of burrowing owl ecology and behavior. *Condor* 75:446–456.

———. 1973b. A spectrographic analysis of burrowing owl vocalizations. *Auk* 90:564–578.

———. 1974. Copulatory and vocal behavior of a pair of whiskered owls. *Auk* 91:619–624.

———. 1986. Sensory capacities and the nocturnal habit of owls (Strigiformes). *Ibis* 128:266–277.

Martin, G. R. 1982. An owl's eyes: Schematic optics and visual performance in *Strix aluco. J. Comp. Physiol.* 145:341–349.

Martinez, M. M., J. P. Isaach, and F. Donatti. 1996. (Distribution and reproductive biology of the striped owl in Buenos Aires Province, Argentina.) *Ornitologia Neotropical* 7:157–161. (In Spanish.)

Maser, C., and E. D. Brodie Jr. 1966. A study of owl pellets from Linn, Benton, and Polk counties, Oregon. *Murrelet* 47:9–14.

Matthiae, T. M. 1981. A nesting boreal owl from Minnesota. *Loon* 54:212–214.

Mayr, E., and M. Mayr. 1954. The tail molt of small owls. *Auk* 71:172–178.

Mays, J. L. 1996. Population size and distribution of cactus ferruginous pygmy-owls in Brooks and Kenedy counties, Texas. Master's thesis, Texas A&M Univ., Kingsville.

Mazur, K. N., S. D. Frith, and P. C. James. 1998. Barred owl home range and habitat selection in the boreal forest of central Saskatchewan. *Auk* 115:746–754.

Mazur, K. N., and P. C. James. 2000. Barred owl. No. 508 in *The birds of North America,* ed. A. Poole, P. Stettenheim, and F. Gill. Philadelphia: Academy of Natural Sciences; Washington, D.C.: American Ornithologists' Union.

Mazur, K. N., P. C. James, M. J. Fitzsimmons, C. Langen, and R. H. M. Espie. 1997. Habitat associations of the barred owl in the boreal forest of Saskatchewan, Canada. *J. Raptor Res.* 31:253–259.

Mazur, K. N., P. C. James, and S. D. Frith. 1997. Barred owl (*Strix varia*) nest characteristics in the boreal forest of Saskatchewan, Canada. *In*

Duncan, Johnson, and Nicholls, 1997, pp. 267–271.

McCallum, D. A. 1994a. Conservation status of the flammulated owl in the United States. *In* Hayward and Verner, 1994, pp. 74–79.

———. 1994b. Flammulated owl. No. 93 in *The birds of North America,* ed. A. Poole, P. Stettenheim, and F. Gill. Philadelphia: Academy of Natural Sciences; Washington, D.C.: American Ornithologists' Union.

———. 1994c. Review of technical knowledge: Flammulated owls. *In* Hayward and Verner, 1994, pp. 14–46.

McCallum, D. A., and F. R. Gehlbach. 1988. Nest site preferences of flammulated owls in western New Mexico. *Condor* 90:653–666.

McCallum, D. A., F. R. Gehlbach, and S. W. Webb. 1995. Life history and ecology of flammulated owls in a marginal New Mexico population. *Wilson Bull.* 107:530–537.

McGarigal, K., and J. D. Fraser. 1985. Barred owl responses to recorded vocalizations. *Condor* 87:552–553.

McGillivray, W. B. 1989. Geographic variation in size and reverse size dimorphism of the great horned owl in North America. *Condor* 91:777–786.

McInvaille, W. B., Jr., and L. B. Keith. 1974. Predator-prey relations and breeding biology of the great horned owl and red-tailed hawk in central Alberta. *Canadian Field-Nat.* 88:1–20.

McKeever, K. 1997. Remaining choices. *In* Duncan, Johnson, and Nicholls, 1997, pp. 6–10.

McNair, D. B. 1994. Caching by an irruptive hawk-owl. *Blue Jay* 52:216–217.

McQueen, L. B. 1972. Observations on copulatory behavior of a pair of screech owls (*Otus asio*). *Condor* 74:101.

Meehan, R. H. 1980. Behavioral significance of boreal owl vocalizations during the breeding season. Master's thesis, Univ. of Alaska, Fairbanks.

Meehan, R. H., and R. J. Ritchie. 1982. Habitat requirements of boreal and hawk owls in interior Alaska. In *Raptor management and biology in Alaska and western Canada,* ed. W. N. Ladd and P. F. Schempgf, pp. 188–196. Anchorage: U.S. Fish and Wildl. Serv.

Mendall, H. L. 1944. Food of hawks and owls in Maine. *J. Wildl. Manage.* 8:198–208.

Meyburg, B.-U., and R. D. Chancellor, eds. *Raptor conservation today.* London: World Working Group on Birds of Prey and Owls.

Meyer, J. S., L. L. Irwin, and M. S. Boyce. 1998. Influence of habitat abundance and fragmentation on northern spotted owls in western Oregon. *Wildl. Monogr.* 139.:1–51.

Mikkola, H. 1972. Hawk owls and their prey in northern Europe. *Brit. Birds* 65:453–460.

———. 1981. Der Barthaus *Strix nebulosa.* Neue Brehm-Bücherei 538. Wittenberg: A. Ziemsen Verlag.

———. 1983. *Owls of Europe.* Vermillion, S. Dak: Buteo Books.

Miller, A. H. 1934. The vocal apparatus of some North American owls. *Condor* 36:204–213.

———. 1935. The vocal apparatus of the elf owl and spotted screech owl. *Condor* 37:288.

———. 1947. The structural basis of the voice of the flammulated owl. *Auk* 64:133–135.

———. 1952. Supplementary data on the tropical avifauna of the Upper Magdalena Valley of Colombia. *Auk* 69:450–457.

———. 1963. The vocal apparatus of two South American owls. *Condor* 65:440–441.

Miller, A. H., and L. Miller. 1951. Geographic variation of the screech owls of the deserts of western North America. *Condor* 53:171–177.

Miller, G. S., and E. C. Meslow. 1985. Dispersal data for juvenile spotted owls: The problem of small sample size. *In* Gutiérrez and Carey, 1985, pp. 69–73.

Miller, G. S., S. K. Nelson, and W. C. Wright. 1985. Two-year-old female spotted owl breeds successfully. *West. Birds* 16:93–94.

Miller, L. 1930. The territorial concept in the horned owl. *Condor* 32:290–291.

Mills, L. S., R. J. Fredrickson, and B. B. Moorhead. 1993. Characteristics of old-growth forests associated with spotted owls in northwest California. *J. Wildl. Manage.* 57:315–321.

Millsap, B. A. 1998a. Barn owl. *In* Glinski, 1998a, pp. 136–139.

———. 1998b. Long-eared owl. *In* Glinski, 1998a, pp. 175–177.

Millsap, B. A., and C. Bear. 1992. Mate and territory fidelity and natal dispersal in an urban population of Florida burrowing owls (*Athene cunicularia*). Burrowing Owl Symposium, Raptor Research Found. ann. meeting, Seattle, Wash. (abstract).

Millsap, B. A., and R. R. Johnson. 1988. Ferruginous pygmy-owl. Natl. Wildl Fed. Sci. Tech. Ser. no. 11:137–139.

Mindell, D. P. 1983. *Nesting raptors in southwestern Alaska: Status, distribution, and aspects of biology.* BLM-Alaska Tech. Rep. 8., U.S. Dept. of Interior, Bur. Land Manage., Alaska State Office, Anchorage.

Mindell, D. P., M. D. Sorenson, C. J. Huddleston, H. C. Miranda Jr., A. Knight, S. J. Savvchuk, and T. Yuri. 1997. Phylogenetic relationships among and within select avian orders based on mitochondrial DNA. In *Avian molecular evolution and systematics,* ed. D. P. Mindell, pp. 213–247. San Diego and London: Academic Press.

Mineau, P., et al. 1999. Poisoning of raptors with organophosphorus and carbamate pesticides, with emphasis on Canada, U.S., and U.K. *J. Raptor Res.* 33:1–37.

Moen, C. A., A. B. Franklin, and R. J. Gutiérrez. 1991. Age determination of subadult northern spotted owls in northwest California. *Wildl. Soc. Bull.* 19:489–493.

Monroe, B. L., Jr. 1968. *A distributional survey of the birds of Honduras.* Ornith. Monogr. no. 7. American Ornithologists' Union.

———. 2000. Measuring annual reproductive success. *Condor* 102:470–473.

Monson, G. 1998a. Ferruginous pygmy owl. *In* Glinski, 1998a, pp. 159–161.

———. 1998b. Whiskered screech-owl. *In* Glinski, 1998a, pp. 149–151.

Mooney, J. 1995. *Myths of the Cherokee.* New York: Dover Publications.

Moore, R. T., and J. T. Marshall. 1959. A new race of screech owl from Mexico: *Otus asio lambi.* *Condor* 61:224–225.

Morrell, T. E., and R. H. Yahner. 1994. Habitat characteristics of great horned owls in south-central Pennsylvania. *J. Raptor Res.* 28:164–170.

Moser, J. A., and C. J. Henry. 1976. Thermal adaptiveness of plumage color in screech owls. *Auk* 93:614–619.

Motta-Junior, J. C., and S. A. Talamoni. 1994. Observations on feeding ecology of striped owls in southeastern Brazil. *J. Raptor Res.* 28:62 (abstract).

Mourer-Chauvire, C. 1987. *Minerva antiqua* (Aves, Strigiformes), an owl mistaken for an edentate mammal. *Amer. Mus. Novitates* 2773:1–11.

Mueller, H. C. 1986. The evolution of reversed sexual dimorphism in owls: An empirical analysis of possible selective factors. *Wilson Bull.* 98:387–406.

Mueller, H. C., and D. D. Berger. 1967. Observations on migrating saw-whet owls. *Bird-Banding* 38:120–125.

Mueller, H. C., and K. Meyer. 1985. The evolution of reversed sexual dimorphism in size: A comparative analysis of the Falconiformes of the western Palearctic. In *Current ornithology,* vol. 2, ed. R. F. Johnston, pp. 65–101. New York: Plenum Press.

Mullay, G. A. 1976. Geographic variation in the clutch sizes of seven owl species. *Auk* 93:602–613.

Muller, K. A. 1970. Exhibiting and breeding elf owls *Micrathene whitneyi* at Washington Zoo. *Inter. Zoo Yearbook* 10:33–36.

Mumford, R. E., and R. L. Zusi. 1958. Notes on movements, territory, and habitat of wintering saw-whet owls. *Wilson Bull.* 70:188–191.

Munro, J. A. 1925. Notes on the economic relations of Kennicott's screech owl (*Otus asio kennicotti*) in the Victoria region. *Canadian Field-Nat.* 39:166–167.

Murphy, C. J., and H. C. Howland. 1983. Owl eyes: Accommodation, corneal curvature, and refractive state. *J. Comp. Physiol.* 151:277–284.

Murray, G. A. 1976. Geographic variation in the clutch size of seven owl species. *Auk* 93:602–613.

Navarro, S. A. G. 1992. Altitudinal distribution of birds in the Sierra Madre del Sur, Guerrero, Mexico. *Condor* 94:29–39.

Nero, R. W. 1969. The status of the great gray owl in Manitoba, with special reference to the 1968–69 influx. *Blue Jay* 27:191–209.

———. 1980. *The great gray owl: Phantom of the northern forest.* Washington, D.C.: Smithsonian Inst. Press.

Nero, R. W., R. J. Clark, R. J. Knapton, and R. H. Hamre. 1987. *Biology and conservation of northern forest owls: Symposium proceedings.* USDA, Forest Service, Gen. Tech. Rep. RM-142.

Nero, R. W., and H. W. R. Copland. 1981. High mortality of great gray owls in Manitoba, winter 1980–81. *Blue Jay* 39:158–165.

Nero, R. W., H. W. R. Copland, and J. Mezibroski. 1984. The great gray owl in Manitoba, 1968–83. *Blue Jay* 43:130–151.

Newton, I., I. Wylie, and L. Dale. 1997. Mortality causes in British barn owls (*Tyto alba*), based on 1,101 carcasses examined during 1963–1996. *In* Duncan, Johnson, and Nicholls, 1997, pp. 299–307.

Nicholls, T. H. 1970. Ecology of barred owls as determined by an automatic radio-tracking system. Ph.D. diss., Univ. of Minnesota, Minneapolis.

Nicholls, T. H., and M. R. Fuller. 1987. Territorial aspects of barred owl home range and behavior in Minnesota. *In* Nero et al., 1987, pp. 121–128.

Nicholls, T. H., and D. W. Warner. 1972. Barred owl habitat use as determined by radiotelemetry. *J. Wildl. Manage.* 36:213–224.

Nicholson, C. P., ed. 1997. *Atlas of the breeding birds of Tennessee.* Knoxville: Univ. of Tennessee Press.

Nilsson, I. N., and T. von Schantz. 1982. The reversed size dimorphism in birds of prey: A reply. *Oikos* 38:388.

Norberg, A. 1964. (Studies on the ecology and ethology of Tengmalm's owl *Aegolius funereus.*) *Vår Fågelvärld* 23:228–244. (In Swedish, English summary.)

———. 1968. Physical factors in directional hearing in *Aegolius funereus* with special reference to

the significance of the asymmetry of the external ears. *Ark. Zool.* 20:181–204.

———. 1970. Hunting technique of Tengmalm's
owl *Aegolius funereus* (L.). *Ornis Scand.* 1:49–64.

———. 1978. Skull asymmetry, ear structure and
function, and auditory localization in Tengmalm's owl *Aegolius funereus* (Linne): *Philos.
Trans. R. Soc. Lond., Biol. Sci.* 282 (991): 325–
410.

———. 1987. Evolution, structure, and ecology of
northern forest owls. *In* Nero et al., 1987,
pp. 9–43.

Nowicki, T. 1974. A census of screech owls (*Otus
asio*) using tape-recorded calls. *Jack-Pine Warbler*
52:98–101.

Oberholser, H. C. 1974. *The bird life of Texas.* 2 vols.
Austin: Univ. of Texas Press.

O'Connell, M. W. 1987. Occurrence of the boreal
owl in northeastern Washington. *In* Nero et al.,
1987, pp. 185–188.

Oeming, A. F. 1955. A preliminary study of the
great gray owl (*Scotiaptex nebulosa nebulosa*
Forster) in Alberta, with observations on some
other species of owls. Master's thesis, Univ. of
Alberta, Edmonton.

Olenick, B. E. 1990. Breeding biology of burrowing owls using artificial nest burrows in southeastern Idaho. Master's thesis, Idaho State
Univ., Pocatello.

Oliphant, L. W., M. R. Robinson, C. Murphy, and
H. Howland. 1983. The musculature and pupil-
lary response of the great horned owl iris. *Exp.
Eye Res.* 37:583–595.

Olson, B. T. 1999. Breeding habitat ecology of the
barred owl (*Strix varia*) at three spacial scales in
the boreal mixedwood forest of north-central
Alberta. Master's thesis, Univ. of Alberta, Edmonton.

Olson, S. L. 1995. The genera of owls in the
Asioninae. *Bull. Brit Ornith. Club* 115:35–39.

Olson, S. L., and W. B. Hilgartner. 1982. Fossil
and subfossil birds from the Bahamas. *Smithsonian Contr. Paleobiol.* 48:22–60.

Olson, S. L., and H. F. James. 1991. *Descriptions of
thirty-two new species of birds from the Hawaiian Islands.* Part 1. *Non-passerines.* Ornithological
Monographs 45, American Ornithologists'
Union, Washington, D.C.

Orians, G., and F. Kuhlman. 1956. Red-tailed
hawk and horned owl populations in Wisconsin.
Condor 58:371–385.

Osborne, T. O. 1987. Biology of the great gray owl
in interior Alaska. *In* Nero et al., 1987, pp. 91–
95.

Osterlof, S. 1969. Report for 1962 of the Bird-
ringing Office, Swedish Museum of Natural
History. *Vår Fågelvärld* suppl. 5:1–159.

Otteni, L. C., E. G. Bolen, and C. Cottam. 1972.

Predator-prey relationships and reproduction
of the barn owl in southern Texas. *Wilson Bull.*
84:434–448.

Owen, D. F. 1963a. Polymorphism in the screech
owl in eastern North America. *Wilson Bull.*
75:183–190.

———. 1963b. Variation in North American
screech owls and the subspecies concept. *Syst.
Zool.* 12:8–14.

Palmer, D. A. 1986. Habitat selection, movements, and activity of boreal and saw-whet
owls. Master's thesis, Colo. State Univ., Fort
Collins.

———. 1987. Annual, seasonal, and nightly variation in calling activity of boreal and northern
saw-whet owls. *In* Nero et al., 1987, pp. 162–
168.

Palmer, D. A., and R. A. Ryder. 1984. The first
documented breeding of the boreal owl in Colorado. *Condor* 86:215–217.

Parker, G. R. 1974. A population peak and crash
of lemmings and snowy owls on Southampton
Island, Northwest Territories. *Canadian Field-
Nat.* 88:151–156.

Parkes, K. C., and A. R. Phillips. 1978. Two new
Caribbean subspecies of barn owl (*Tyto alba*),
with remarks on variation in other populations.
Ann. Carn. Mus. 47:479–492.

Parmelee, D. F. 1992. Snowy owl. No. 10 in *The
birds of North America,* ed. A. Poole, P. Stettenheim, and F. Gill. Philadelphia: Academy of
Natural Sciences; Washington, D.C.: American
Ornithologists' Union.

Patuxent Wildlife Research Center. [Data on
longevity.] http://www.pwrc.usgs.gov/bbl/
homepage/longvrec.htm.

Payne, R. 1962. How the barn owl locates its prey
by hearing. *Living Bird* 1:151–159.

Peck, G. K., and R. D. James. 1983. *Breeding birds of
Ontario: Nidiology and distribution.* Vol. 1 *Non-
passerines.* Toronto: Royal Ontario Museum.

Perkins, J. P., J. A. Thrailkill, W. J. Ripple, and
K. T. Hershey. 1997. Landscape patterns
around northern spotted owl (*Strix occidentalis
caurina*) nest sites in Oregon's central coast
ranges. *In* Duncan, Johnson, and Nicholls, 1997,
p. 314.

Perrone, M. 1981. Adaptive significance of ear
tufts in owls. *Condor* 83:383–384.

Peters, D. C. 1992. Zoogeographic relationships of
the Eocene from Messel (Germany). In *Acta
20th Congressus Internationalis Ornithologici,* ed.
B. D. Bell et al., 1:572–577. Wellington, New
Zealand.

Peters, J. L. 1940. *Check-list of birds of the world.* Vol.
4. (Cuculiformes, Strigiformes, Apodiformes.)
Cambridge: Harvard Univ. Press.

Petersen, L. 1979. Ecology of great horned owls

and red-tailed hawks in southeastern Wisconsin. *Wisc. Dept. Nat. Resour. Tech. Bull.* no. 111:1–63.

———. 1991. Mixed woodland owls. In *Proceedings of the Midwest Raptor Management Symposium and Workshop,* ed. B. G. Pendleton and D. L. Krahe, 85–95. Natl. Wildl. Fed. Sci. Tech. Ser. no. 15, Washington, D.C.

Peterson, R. T. 1963. *The birds.* New York: Time, Inc.

Pezzolesi, L. S. 1994. The western burrowing owl: Increasing prairie dog abundance, foraging theory, and nest site fidelity. Master's thesis, Texas Tech. Univ., Lubbock.

Phillips, A. R. 1942. Notes on the migrations of the elf and flammulated screech owls. *Wilson Bull.* 54:132–137.

Phillips, A. R., J. Marshall Jr., and G. Monson. 1964. *The birds of Arizona.* Tucson: Univ. of Arizona Press.

Pitelka, F. A., P. Q. Tomich, and G. W. Trichel. 1955a. Breeding behavior of jaegers and owls near Barrow, Alaska. *Condor* 57:3–18.

———. 1955b. Ecological relations of jaegers and owls as lemming predators near Barrow, Alaska. *Ecol. Monogr.* 25:85–117.

Plumpton, D. L., and R. S. Lutz 1993. Nesting habitat use by burrowing owls in Colorado. *J. Raptor Res.* 27:175–179.

———. 1994. Sexual size dimorphism, mate choice, and productivity of burrowing owls. *Auk* 111:724–727.

Poole, E. L. 1938. Weights and wing areas in North American birds. *Auk* 55:511–517.

Portenko, L. A. 1972. Die Schnee-eule *Nyctea scandiaca.* Neue Brehm-Bücherei 454. Wittenberg: A. Ziemsen Verlag.

Postupalsky, S., J. M. Papp, and L. Scheller. 1997. Nest sites and reproductive success of barred owls (*Strix varia*) in Michigan. *In* Duncan, Johnson, and Nicholls, 1997, pp. 325–337.

Powers, L. R., A. Dale, P. A. Gaede, C. Rhodes, L. Nelson, J. J. Dean, and J. D. May. 1996. Nesting and food habits of flammulated owls (*Otus flammeolus*) in southcentral Idaho. *J. Raptor Res.* 30:15–20.

Price, J., S. Droege, and A. Price. 1995. *The summer atlas of North American birds.* San Diego: Academic Press.

Proudfoot, G. A. 1996. Natural history of the cactus ferruginous pygmy-owl. Master's thesis, Texas A&M Univ., Kingsville.

Proudfoot, G. A., and S. L. Beasom. 1996. Responsiveness of cactus ferruginous pygmy-owls to broadcasted conspecific calls. *Wildl. Soc. Bull.* 24:294–297.

———. 1997. Food habits of nesting ferruginous pygmy-owls in southern Texas. *Wilson Bull.* 109:741–748.

Proudfoot, G. A., and R. R. Johnson. 2000. Ferruginous pygmy-owl (*Glaucidium brasilianum*). No. 496 in *The birds of North America,* ed. A. Poole, P. Stettenheim, and F. Gill. Philadelphia: Academy of Natural Sciences; Washington, D.C.: American Ornithologists' Union.

Proudfoot, G. A., J. L. Mays, S. L. Beasom, and R. Bingham. 1997. Effectiveness of broadcast surveys in determining habitat use of ferruginous pygmy-owls (*Glaucidium brasilianum*) in southern Texas. *In* Duncan, Johnson, and Nicholls, 1997, p. 338.

Pyle, P. 1997a. Flight-feather molt patterns and age in North American owls. *Monogr. Field Ornith.* 2:1–32.

———. 1997b. *Identification guide to North American birds.* Pt. 1. Bolinas, Calif.: Slate Creek Press.

Rains, C. 1997. Comparison of food habits of the northern saw-whet owl (*Aegolius*) and the western screech-owl (*Otus kennicottii*) in southwestern Idaho. *In* Duncan, Johnson, and Nicholls, 1997, pp. 339–346.

Randi, E., G. Fusco, R. Lorenzini, and F. Spina. 1991. Allozyme divergence and phylogenetic relationships within the Strigiformes. *Condor* 93:295–301.

Randle, W., and R. Austin. 1952. Ecological notes on long-eared and saw-whet owls in southwestern Ohio. *Ecology* 33:422–426.

Raptor Research Foundation, and Univ. of California Davis Raptor Center. 1985. Abstracts of the symposium on the biology, status, and management of owls. Part of International Symposium on the Management of Birds of Prey, 1–10 Nov., Sacramento.

Rashid, S. 1999. Northern pygmy-owl (*Glaucidium gnoma*) in Rocky Mountain National Park. *J. Colo. Field Ornith.* 32:94–101.

Ratcliffe, B. D. 1986. The Manitoba burrowing owl survey, 1982–1984. *Blue Jay* 44:31–37.

Read, M., and J. Allsop. 1994. *The barn owl.* London: Blandford.

Reese, J. G. 1972. A Chesapeake barn owl population. *Auk* 89:106–114.

Reynolds, R. T., and B. D. Linkhart. 1984. Methods and materials for capturing and monitoring flammulated owls. *Great Basin Nat.* 44:49–51.

———. 1985. Pair bonding and site tenacity in flammulated owls. Proc. Raptor Research Found. Symp., 9–10 Nov., Sacramento (abstract).

———. 1987a. Fidelity to territory and mate in flammulated owls. *In* Nero et al., 1987, pp. 234–238.

———. 1987b. The nesting biology of flammulated owls in Colorado. *In* Nero et al., 1987, pp. 239–248.

———. 1990a. Extra-pair copulation and extra-range movements in flammulated owls. *Ornis Scand.* 21:71–77.

———. 1990b. Longevity records for male and female flammulated owls. *J. Field Ornith.* 61:243–244.

———. 1998. Flammulated screech-owl. *In* Glinski, 1998a, pp. 140–144.

Reynolds, R. T., R. S. Ryder, and B. D. Linkhart. 1988. Small forest owls. In *Proceedings of the Western Raptor Management Symposium and Workshop,* ed. B. G. Pendleton, 134–145. Natl. Wildl. Fed. Sci. Tech. Ser. no. 12, Washington, D.C.

Rice, W. R. 1982. Acoustical location of prey by the marsh hawk: Adaptation to concealed prey. *Auk* 99:403–413.

Rich, P. V. 1982. Tarsometatarsus of *Protostrix* from the mid-Eocene of Wyoming. *Auk* 99:576–579.

Rich, P. V., and D. J. Bohaska. 1976. The world's oldest owl: A new strigiform from the Paleocene of southwestern Colorado. *Smithsonian Contr. Paleobiol.* 27:87–93.

———. 1981. The Ogygoptyngidae, a new family of owls from the Paleocene. *Alcheringa* 5:95–102.

Rich, R. 1986. Habitat and nest-site selection by burrowing owls in the sagebrush steppe of Idaho. *J. Wildl. Manage.* 50:548–555.

Richmond, M. L., L. R. Deweese, and R. E. Pillmore. 1980. Brief observations on the breeding biology of the flammulated owl in Colorado. *West. Birds* 11:35–46.

Ricklefs, R. E. 1983. Comparative avian demography. *Current Ornith.* 1:1–32.

Ridgely, R. S., and J. A. Gwynne Jr. 1989. *A guide to the birds of Panama.* 2nd ed. Princeton: Princeton Univ. Press.

Ridgway, R. 1914. Birds of North and Middle America. *U.S. Natl. Mus. Bull.* 50, pt. 6:1–882.

Righter, R. 1995. Description of a northern pygmy-owl vocalization from the southern Rocky Mountains. *Colo. Field Ornithol J.* 29:21–23.

Rinkevich, S. E., and R. J. Gutiérrez. 1996. Mexican spotted owl habitat characteristics in Zion National Park. *J. Raptor Res.* 30:74–78.

Ritchison, G., J. R. Belthoff, and E. I. Sparks. 1992. Dispersal restlessness: Evidence for innate dispersal by juvenile eastern screech-owls. *Anim. Behav.* 43:57–65.

Ritchison, G., and P. M. Cavanagh. 1986. Response of eastern screech-owls to playback of the bounce songs of neighboring and non-neighboring individuals. Proc. ann. meeting, Raptor Research Found., 20–23 Nov., Gainesville, Fla. (abstract).

———. 1992. Prey use by eastern screech-owls: Seasonal variation in eastern Kentucky and a review of previous studies. *J. Raptor Res.* 26:66–73.

Ritchison, G., P. M. Cavanagh, J. R. Belthoff, and E. J. Sparks. 1988. The singing behavior of eastern screech-owls: Seasonal timing and response to playback of conspecific song. *Condor* 90:648–652.

Robbins, C. S., and E. A. T. Blom, eds. 1996. *Atlas of the breeding birds of Maryland and the District of Columbia.* Pittsburgh: Univ. of Pittsburgh Press.

Robiller, F. 1982. (Behavior of a breeding pair of northern hawk owls *Surnia ulula* in Torne Lapmark.) *Beitr. Vogelkd.* 28:366–368. (In German.)

Robinson, M., and C. D. Becker. 1986. Snowy owls in Fetlar. *Brit. Birds* 78:228–242.

Robinson, W. D., J. D. Brown, and S. K. Robinson. 2000. Forest bird community structure in central Panama: Influence of spacial scale and biogeography. *Ecol. Monogr.* 70:209–235.

Rohner, C. 1995. Great horned owls and snowshoe hares: What causes the time lag in the numerical response of predators to cyclic prey? *Oikos* 74:61–68.

———. 1996. The numerical response of great horned owls to the snowshoe hare cycle: Consequences of non-territorial "floaters" on demography. *J. Anim. Ecol.* 65:359–370.

———. 1997. Non-territorial "floaters" in great horned owls: Space use during a cyclic peak of snowshoe hares. *Anim. Behav.* 53:901–912.

Rohner, C., and D. B. Hunter. 1996. First-year survival of great horned owls during a peak and decline of the snowshoe hare cycle. *Canadian J. Zool.* 74:1092–1097.

Rohner, C., and C. J. Krebs. 1996. Owl predation on snowshoe hares: Consequences of antipredator behaviour. *Oecologia* 198:303–310.

Rohner, C., and J. N. M. Smith. 1996. Brood size manipulations in great horned owls *Bubo virginianus:* Are predators food limited at the peak of prey cycles? *Ibis* 138:236–242.

Rohner, C., J. N. M. Smith, J. Stroman, and M. Joyce. 1995. Northern hawk-owls in the Narcotic boreal forest: Prey selection and population consequences of multiple prey cycles. *Condor* 97:208–220.

Rohweder, R. 1978. Barred owl expanding into northeastern Oregon. *Oreg. Birds* 4:41–42.

Root, T. 1988. *Atlas of wintering North American birds.* Chicago: Univ. of Chicago Press.

Ross, A. 1969. Ecological aspects of the food habits of insectivorous screech owls. *Proc. West. Found. Vert. Zool.* 1:301–344.

Roulin, A., C. Riols, C. Dijkstra, and A. Ducrest. 2001. Female plumage spottiness signals parasite resistance in the barn owl (*Tyto alba*). *Behav. Ecol.* 12:1–3–10.

Rowland, B. 1978. *Birds with human souls: A guide to bird symbolism.* Knoxville: Univ. of Tennessee Press.

Rowley, J. S. 1984. Breeding records of land birds from Oaxaca, Mexico. *Proc. Found. Vert. Zool.* 2:73–224.

Rudolph, S. G. 1978. Predation ecology of coexisting great horned and barn owls. *Wilson Bull.* 90:134–137.

Russell, F. 1904–1905. The Pima Indians. *Ann. Rep., U.S. Bur Amer. Ethnol.* 26:1–389.

Russell, S. M. 1964. *A distributional survey of the birds of British Honduras.* Ornith. Monogr. no. 1. American Ornithologists' Union.

Russell, S. M., and G. Monson. 1998. *The birds of Sonora.* Tucson: Univ. of Arizona Press.

Ryder, R. A., D. A. Palmer, and J. J. Rawinski. 1985. Status of the boreal owl (*Aegolius funereus*) in Colorado. Proc. Raptor Research Found. Symp., 9–10 Nov., Sacramento (abstract).

———. 1987. Distribution and status of the boreal owl in Colorado. Paper presented at Symposium on Biology and Conservation of Northern Forest Owls, 3–7 Feb., Winnipeg, Manitoba.

Schaeffer, P., and S. Ehlers, eds. 1979. *Owls of the West: Their ecology and conservation.* Tiburon, Calif: Western Education Center, National Audubon Soc.

Schaldach, W. J., Jr. 1963. The avifauna of Colima and adjacent Jalisco, Mexico. *Proc. West. Found. Vert. Zool.* 1:1–100.

Scherzinger, W. 1970. Zum Aktionssystem des Sperlingskauzes (*Glaucidium passerinum* L.). *Zoologica* (Stuttgart) 118:1–120.

———. 1971a. (Observations on the nestling development of some owls [Strigidae].) *Z. Tierpsychol.* 28:494–504. (In German, English summary.)

———. 1971b. (Predator reactions of some owls [Strigidae].) *Z. Tierpsychol.* 29:165–174. (In German, English summary.)

———. 1974. (On the ecology of the Eurasian pygmy-owl in Bayerischer Wald Nationalpark.) *Anz. Orn. Ges. Bayern* 13:121–156. (In German, English summary.)

———. 1977. Small owls in aviaries. *Avic. Mag.* 83:18–21.

Schifferli, A. 1957. Alter und Sterblichkeit bei Waldkauz (*Strix aluco*) und Schleiereule (*Tyto alba*) in der Schweiz. *Orn. Beob.* 54:50–56.

Schlatter, R. P., J. Yanez, H. Nunez, and F. M. Jaksic. 1980. The diet of the burrowing owl in central Chile and its relation to prey size. *Auk* 97:616–619.

Schmutz, J. K., G. Wood, and D. Wood. 1991. Spring and summer prey of burrowing owls in Alberta. *Blue Jay* 49:93–97.

Schmutz, S. M., and J. S. Moker. 1991. A cytogenetic comparison of some North American owl species. *Genome* 34:714–717.

Schönn, S. 1978. Der Sperlingkauz *Glaucidium passerinum passerinum.* Neue Brehm-Bücherei 513. Wittenberg: A. Ziemsen Verlag.

Schönwetter, M. 1964. *Handbuch der Oologie.* Acad.-Verlag Berlin. Lfg. 10:577–640.

Schulz, T. A., and D. Yasuda. 1985. Ecology and management of the common barn owl (*Tyto alba*) in the California Central Valley. Proc. Raptor Research Found. Symp., 9–10 Nov., Sacramento (abstract).

Schwartzkopff, J. 1955. On the hearing of birds. *Auk* 72:340–347.

———. 1963. Morphological and physiological properties of the auditory system in birds. *Proc. 13th Internat. Ornith. Congress* 2:1059–1068. Ithaca: American Ornithologists' Union.

Scott, D. 1997. *The long-eared owl.* London: Hawk and Owl Trust.

Sealy, S. G., and K. A. Hobson. 1987. On the "concealing pose" of the northern saw-whet owl. *Blue Jay* 45:33–37.

Seaman, D. E. 1997. Abundance and population characteristics of northern spotted owls (*Strix occidentalis caurina*) in Olympic National Park, Washington. *In* Duncan, Johnson, and Nicholls, 1997, p. 381.

Seamans, M. E., and R. J. Gutiérrez. 1999. Diet composition and reproductive success of Mexican spotted owls. *J. Raptor Res.* 33:143–148.

Servos, M. C. 1987. Summer habitat use by great gray owls in southeastern Manitoba. *In* Nero et al., 1987, pp. 108–114.

Sharrock, J. T. R. 1976. *The atlas of breeding birds in Britain and Ireland.* Berkhamsted, Eng.: T. and A. D. Poyser.

Shawyer, C. R. 1998. *The barn owl.* Chelmsford, Eng.: Alrequin Press.

Sheffield, S. R. 1997. Current status, distribution, and conservation of the burrowing owl (*Speotyto cunicularia*) in midwestern and western North America. *In* Duncan, Johnson, and Nicholls, 1997, pp. 399–408.

Sherman, A. R. 1911. Nest life of the screech owl. *Auk* 28:155–168.

Shuffeldt, R. W. 1900. Professor Collett on the morphology of the cranium and the auricular openings in the north-European species of the family Strigidae. *J. Morph.* 17:119–176.

Sibley, C. G., and J. Ahlquist. 1985. The relationships of some groups of African birds, based on comparisons of the genetic material, DNA. In *Proceedings Int. Symp. on African Vertebrates,* ed. K.-L. Schuchmann, pp. 115–161. Bonn: Zool. Forschunginstitut and Mus. A. Koenig.

Sibley, C. G., and B. L. Monroe. 1990. *Distribution and taxonomy of birds of the world.* New Haven: Yale Univ. Press.

Simpson, M. B., Jr. 1972. The saw-whet owl population of North Carolina's southern Great Balsam Mountains. *Chat* 36:39–47.

Sisco, C., and D. Sharp. 1986. The occurrence of spotted and barred owls in Olympic National Park, Washington. Proc. Raptor Research Found. Symp., 9–10 Nov., Sacramento (abstract).

Slud, P. 1980. The birds of Hacienda Paloverde, Guanacaste, Costa Rica. *Smithsonian Contr. Zool.* 292:1–92.

Smith, D. G. 1969. Nesting ecology of the great horned owl *Bubo virginiana. Brigham Young Univ. Sci. Bull. Biol. Ser.* 10 (4): 16–25.

———. 1971. Population dynamics, habitat selection, and partitioning of breeding raptors in the eastern Great Basin of Utah. Ph.D. diss., Brigham Young Univ., Provo.

———. 1981. Winter roost site fidelity by long-eared owls in central Pennsylvania. *Amer. Birds* 35:339.

Smith, D. G., A. Devine, and D. Gendron. 1982. An observation of copulation and allopreening of a pair of whiskered owls. *J. Field Ornith.* 53:51–52.

Smith, D. G., and H. H. Frost. 1974. History and ecology of a colony of barn owls in Utah. *Condor* 76:131–136.

Smith, D. G., and R. Gilbert. 1984. Eastern screech-owl home range and use of suburban habitats in southern Connecticut. *J. Field Ornith.* 55:322–329.

Smith, D. G., and E. Hiestand.1990. Alloparenting at an eastern screech-owl nest. *Condor* 92:246–247.

Smith, D. G., and J. R. Murphy. 1973. Breeding ecology of raptors in the eastern Great Basin of Utah. *Brigham Young Univ., Biol Ser.* 18 (3): 1–76.

Smith, D. G., D. Walsh, and A. Devine. 1987. Censusing eastern screech-owls in southern Connecticut. *In* Nero et al., 1987, pp. 255–267.

Smith, D. G., and S. M. Wiemeyer. 1992. Determining sex of eastern screech-owls using discriminant function analysis. *J. Raptor Res.* 26:24–26.

Smith, D. G., and C. R. Wilson. 1971. Notes on the winter food of screech-owls. *Great Basin Nat.* 31:83–84.

Smith, N. 1997. Observations of wintering snowy owl (*Nyctea scandiaca*) at Logan Airport, East Boston, Massachusetts, from 1981–1997. *In* Duncan, Johnson, and Nicholls, 1997, pp. 591–596.

Smith, S. M. 1982. Raptor "reverse" dimorphism revisited: A new hypothesis. *Oikos* 29:118–122.

Smithe, F. 1966. *The birds of Tikal.* New York: Natural History Press.

Snyder, N. F., and J. W. Wiley. 1976. Sexual size dimorphism in hawks and owls of North America. *Ornith. Monogr.* 20:1–96.

Solheim, R. 1983. Bigyny and biandry in the Tengmalm's owl *Aegolius funereus. Ornis Scand.* 14:51–57.

———. 1984a. Breeding biology of the pygmy owl *Glaucidium passerinum* in two biogeographical zones in southeastern Norway. *Ann. Zool. Fenn.* 21:295–300.

———. 1984b. Caching behavior, prey choice, and surplus killing by pygmy owls *Glaucidium passerinum* during winter: A functional response of a generalist predator. *Ann. Zool. Fenn.* 21:301–308.

———. 1987a. Song activity and nest predation on a pair of hawk owls. Paper presented at Symposium on Biology and Conservation of Northern Forest Owls, 3–7 Feb., Winnipeg, Manitoba.

———. 1987b. Song activity and territorial defense behaviour of pygmy owls and Tengmalm's owls during winter and breeding. Paper presented at Symposium on Biology and Conservation of Northern Forest Owls, 3–7 Feb., Winnipeg, Manitoba.

Solis, D. M., and R. J. Gutiérrez. 1990. Summer habitat ecology of northern spotted owls in northwestern California. *Condor* 92:739–748.

Sonerud, G. A. 1985. Risk of nest predation in three species of hole nesting owls: Influence on choice of nesting habitat and incubation behaviour. *Ornis Scand.* 16:261–269.

———.1986. Effect of snow cover on seasonal changes in diet, habitat, and regional distribution of raptors that prey on small mammals in boreal zones of Fennoscandia. *Holarct. Ecol.* 9:33–47.

———. 1992. Search tactics of a pause-travel predator: Adaptive adjustments of perching times and move distances by hawk owls (*Surnia ulula*). *Behav. Ecol. Sociobiol.* 30:207–217.

———. 1997. Hawk owls in Fennoscandia: Population fluctuations, effects of modern forestry, and recommendations on improving foraging habitats. *J. Raptor Res.* 31:167–174.

Sonerud, G. A., J. O. Nybo, P. E. Fjeld, and C. Knoff. 1987a. A case of bigamy in the hawk owl *Surnia ulula:* Spacing of nests and allocation of male effort. *Ornis Fenn.* 64:144–148.

———. 1987b. Polygyny in the hawk owl: Allocation of male effort. Paper presented at Symposium on Biology and Conservation of Northern Forest Owls, 3–7 Feb., Winnipeg, Manitoba.

Soucy, L. J., Jr. 1980. Three long-distance recoveries of banded New Jersey barn owls. *N. Amer. Bird Bander* 5:97.

———. 1985. Bermuda recovery of a common barn-owl banded in New Jersey. *J. Field Ornith.* 56:274.

Southern, H. N. 1970. The natural control of a

population of tawny owls *Strix aluco. J. Zool.* (London) 162:197–285.

Sparks, E. J., J. R. Belthoff, and G. Ritchison. 1994. Habitat use by eastern screech-owls in Kentucky. *J. Field Ornith.* 65:83–95.

Sparks, E. J., G. Ritchison, and J. R. Belthoff. 1986. Home range and habitat utilization by eastern screech-owls in central Kentucky. Proc. of ann. meeting, Raptor Research Found., 20–23 Nov., Gainesville, Fla. (abstract).

Sparks, J., and T. Soper. 1970. *Owls: Their natural and unnatural history.* Newton Abbot, Eng.: David and Charles.

Spiers, J. M. 1961. Courtship of great horned owls. *Canadian Field-Nat.* 75:52.

Spreyer, M. 1987. A floristic analysis of great gray owl habitat in Aitkin County, Minnesota. *In* Nero et al., 1987, pp. 96–100.

Stacey, P. B., R. D. Arrigo, T. C. Edwards, and N. Joste. 1983. Northeastern extension of the breeding range of the elf owl in New Mexico. *Southwest. Nat.* 28:99–100.

Stahlecker, D. W., and J. W. Rawinski. 1990. First records of the boreal owl in New Mexico. *Condor* 92:517–519.

Steadman, D. W. 1981. Review of paleontological papers by C. J. O. Harrison and C. A. Walker. *Auk* 98:205–207.

Stepney, P. H. R. 1986. Status and distribution of the screech owl in Alberta, Saskatchewan, and Montana. Paper presented at 19th International Ornithological Congress, 22–29 June, Ottawa (abstract).

Stevens, D. A., and S. H. Sturts. 1997. Idaho bird distribution mapping by latilong. 2nd ed. Idaho Mus. of Nat. Hist., Pocatello.

Stevenson, H. M., and B. H. Anderson. 1994. *The birdlife of Florida.* Gainesville: Univ. Press of Florida.

Stewart, P. A. 1952. Dispersal, breeding behavior, and longevity of banded barn owls in North America. *Auk* 69:227–245.

———. 1969. Movements, population fluctuations, and mortality among great horned owls. *Wilson Bull.* 81:155–162.

Stewart, R. E. 1975. *Breeding birds of North Dakota.* Fargo: Tri-College Center for Environmental Studies.

Stewart, R. E., and C. S. Robbins. 1958. Birds of Maryland and the District of Columbia. U.S. Dept. of Interior, Fish and Wildl. Serv., North American Fauna no. 63.

Stiles, F. G., and A. F. Skutch. 1989. *A guide to the birds of Costa Rica.* Ithaca: Cornell Univ. Press.

Stillwell, J., and N. Stillwell. 1954. Notes on the call of a ferruginous pygmy owl. *Wilson Bull.* 66:152.

Stone, E., J. Smith, and P. Thornton. 1994. Sea-sonal variation and diet selection from pellet remains of short-eared owls (*Asio flammeus*) in Wyoming. *Great Basin Nat* 54:191–192.

Storer, R. W. 1972. The juvenal plumage and relationships of *Lophostrix cristata. Auk* 89:452–455.

Stotz, D. F., J. W. Fitzpatrick, T. A. Parker III, and D. K. Moskovits. 1995. *Neotropical birds: Ecology and conservation.* Chicago: Univ. of Chicago Press.

Strahlecker, D. W., and R. B. Duncan. 1996. The boreal owl at the southern terminus of the Rocky Mountains: Undocumented longtime resident or recent arrival? *Condor* 98:153–161.

Sumner, E. L., Jr. 1928. Notes on the development of young screech owls. *Condor* 30:333–338.

———. 1933. The growth of some young raptorial birds. *Univ. Calif. Pub. Zool.* 40:277–308.

Sutton, G. M., and D. F. Parmelee. 1956. Breeding of the snowy owl in southeastern Baffin Island. *Condor* 58:273–282.

Swengel, A. B. 1990. How to find saw-whet owls. *Bird Watcher's Digest* 12:68–75.

Swengel, S. R., and A. B. Swengel. 1987. Study of a northern saw-whet owl population in Sauk County, Wisconsin. *In* Nero et al., 1987, pp. 199–208.

———. 1992a. Diet of northern saw-whet owls in southern Wisconsin. *Condor* 94:707–711.

———. 1992b. Roosts of northern saw-whet owls in southern Wisconsin. *Condor* 94:699–706.

Swindle, K. A., W. J. Ripple, C. Meslow, and M. Schafer. 1999. Old-forest distribution around spotted owl nests in the central Cascades Mountains, Oregon. *J. Wildl. Manage.* 63:1212–1221.

Takats, D. L. 1998. Barred owl habitat use and distribution in the Foothills Model Forest. Master's thesis, Univ. of Alberta, Edmonton.

Taylor, A. L., Jr., and E. D. Forsman. 1976. Recent range extensions of the barred owl in western North America, including the first-records for Oregon. *Condor* 78:560–561.

Taylor, I. 1994. *Barn Owls: Predator-prey relationships and conservation.* Cambridge: Cambridge Univ. Press.

Taylor, P. S. 1973. Breeding behavior of the snowy owl. *Living Bird* 12:137–154.

Terrill, L. M. 1931. Nesting of the saw-whet owl in the Montreal District. *Auk* 48:169–174.

Thomas, J. W., ed. 1979. *Wildlife habitats in managed forests: The Blue Mountains of Oregon and Washington.* USDA, Forest Service, Agricultural Handbook no. 553.

Thomas, J. W., E. D. Forsman, J. B. Lint, E. C. Meslow, B. R. Noon, and J. Verner. 1990. A conservation strategy for the northern spotted owl: A report of the interagency scientific committee to address the conservation of the northern spotted owl. USDA and U.S. Dept. of Interior, Portland, Oreg.

Thome, D., C. J. Zabel, and L. V. Diller. 2000. Spotted owl turnover and reproduction in managed forests of north central California. *J. Field Ornith.* 71:140–156.

Thompson, C. D., and S. H. Anderson. 1988. Foraging behavior and food habits of burrowing owls in Wyoming. *Prairie Nat.* 20:23–28.

Thomsen, L. 1971. Behavior and ecology of burrowing owls on the Oakland Municipal Airport. *Condor* 73:177–192.

Thraillkill, J. A., R. G. Anthony, and E. C. Meslow. 1997. An update of demographic estimates for northern spotted owls (*Strix occidentalis caurina*) from Oregon's central coast ranges. *In* Duncan, Johnson, and Nicholls, 1997, pp. 432–448.

Toombs, T. P. 1997. Burrowing owl nest-site selection in relation to soil texture and prairie dog colony attributes. Master's thesis, Colo. State Univ., Fort Collins.

Torre, J. de la. 1990. *Owls: Their life and behavior.* New York: Crown Pub.

Tryon, C. A., Jr. 1943. The great gray owl as a predator on pocket gophers. *Wilson Bull.* 55:130–131.

Tulloch, R. H. 1968. Snowy owls breeding in Shetland, 1967. *Brit. Birds* 61:119–132.

Tyler, H. A. 1979. *Pueblo birds and myths.* Norman: Univ. of Oklahoma Press.

Tyler, H. A., and D. Phillips. 1978. *Owls by day and night.* Happy Camp, Calif.: Naturegraph Books.

U.S. Dept. of Interior. 1990. Endangered and threatened wildlife and plants: Determination of threatened status for the northern spotted owl. *Fed. Register* 55:26114–26194.

VanCamp, L. F., and C. J. Henny. 1975. The screech owl: Its life history and population ecology in northern Ohio. U.S. Dept. of Interior, Fish and Wildl. Serv., North American Fauna no. 71.

van der Weyden, W. J. 1975. Scops and screech owls: Vocal evidence for a basic subdivision in the genus *Otus* (Strigidae). *Ardea* 63:65–77.

Van Dijk, T. 1973. A comparative study of hearing in owls of the family Strigidae. *Neth. J. of Zool.* 23:131–167.

Van Rossem, A. J. 1927. Eye shine in birds with notes on the feeding habits of some goatsuckers. *Condor* 29:25–28.

van Woudenburg, A. M., and D. A. Christie. 1997. Flammulated owl (*Otus flammeolus*) population and habitat inventory at its northern range limit in the southern interior of British Columbia. *In* Duncan, Johnson, and Nicholls, 1997, pp. 466–475.

Verner, J., and A. S. Boss. 1980. *California wildlife and their habitats: Western Sierra Nevada.* USDA, Forest Service, Gen. Tech. Rep. PWS-37.

Verner, J., R. J. Gutiérrez, and G. I. Gould. 1992. The California spotted owl: General biology and ecological relations. In *The California spotted owl: A technical assessment of its current status,* coord. J. Verner et al., pp. 55–78. Gen. Tech. Rep. PSW-GTR133, U.S. Forest Serv., Albany, Calif.

Village, A. 1981. The diet and breeding of long-eared owls in relation to vole numbers. *Bird Study* 28:215–224.

————. 1985. The response of *Asio* owls to changes in vole numbers. Proc. Raptor Research Found. Symp., 9–10 Nov., Sacramento (abstract).

Voous, K. H. 1964. Wood owls of the genera *Strix* and *Ciccaba. Zool. Mededelinger* 39:471–478.

————. 1988. *Owls of the northern hemisphere.* Cambridge: MIT Press.

Voronetsky, V. I. 1987. Some features of long-eared owl ecology and behavior: Mechanisms maintaining territoriality. *In* Nero et al., 1987, pp. 229–230.

Wahlstedt, J. 1969. (Hunting, feeding, and vocalizations of the great gray owl *Strix nebulosa.*) *Vår Fågelvärld* 28:89–101. (In Swedish, English summary.)

Walk, J. W., T. L. Esker, and S. A. Simpson. 1999. Continuous nesting of barn owls in Illinois. *Wilson Bull.* 111:572–573.

Walker, L. W. 1974. *The book of owls.* New York: Alfred A. Knopf.

Wallace, G. J. 1948. The barn owl in Michigan. *Mich. State Coll. Agr. Exp. Sta. Bull.* 208.

Wallin, J., and M. Andersson. 1981. Adult nomadism in Tengmalm's owl *Aegolius funereus. Ornis Scand.* 12:125–126.

Walls, G. L. 1942. *The vertebrate eye and its adaptive radiation.* Bloomfield Hills, Mich.: Cranbrook Inst. of Science.

Walsh, P. J. 1990. Nest of northern pygmy-owl in southeast Alaska. *Northwest. Nat.* 71 (3): 97.

Walters, P. M. 1981. Notes on the body weight and molt of the elf owl (*Micrathene whitneyi*) in southeastern Arizona. *N. Amer. Bird Bander* 6:104–105.

Warnock, R. G., and P. C. James. 1997. Habitat fragmentation and burrowing owls (*Speotyto cunicularia*) in Saskatchewan. *In* Duncan, Johnson, and Nicholls, 1997, pp. 477–486.

Waters, W. T. 1983. Otoe Missouria oral narratives. Master's thesis, Univ. of Nebraska–Lincoln.

Watson, A. 1957. The behavior, breeding, and food ecology of the snowy owl, *Nyctea scandiaca. Ibis* 99:419–462.

Wauer, R. H., P. C. Palmer, and A. Windham.

1993. The ferruginous pygmy-owl in south Texas. *Amer. Birds* 47:1071–1076.

Webb, B. 1982a. Distribution and nesting requirements of montane forest owls in Colorado. Pts. 1–2. *Colo. Field Ornith. J.* 16:26–32, 58–64.

———. 1982b. Distribution and nesting requirements of montane forest owls in Colorado. Pt. 3. Flammulated owl (*Otus flammeolus*). *Colo. Field Ornith. J.* 16:76–81.

———. 1983. Distribution and nesting requirements of montane forest owls in Colorado. Pt. 4. Spotted owl (*Strix occidentalis*). *Colo. Field Ornith. J.* 17:2–8.

Wedgwood, J. A. 1976. Burrowing owls in south-central Saskatchewan. *Blue Jay* 34:26–44.

Weir, R. D., F. Cooke, M. H. Edwards, and R. B. Stewart. 1980. Fall migration of saw-whet owls at Prince Edward Point, Ontario. *Wilson Bull.* 92:475–488.

Weinstein, K. 1989. *The owl in art, myth, and legend.* New York: Crescent Books.

Wellicome, T. I. 2000. Effects of food on reproduction in burrowing owls during three stages of the breeding season. Ph.D. diss., Univ. Alberta, Edmonton.

Wellicome, T. I., and G. L. Holroyd, eds. 2001. Proceedings of the 2nd International Burrowing Owl Symposium, 29–30 Sept. 1998, Ogden, Utah. *J. Raptor Res.* 35:269–401.

Wells, D. R. 1986. Further parallels between the Asian bay owl *Phodilus badius* and *Tyto* species. *Bull. Brit. Ornith. Club* 106:12–15.

Wesemann, T. 1986. Factors influencing the distribution of the burrowing owl (*Athene cunicularia*) in Cape Coral, Florida. Proc. ann. meeting, Raptor Research Found., 20–23 Nov., Gainesville, Fla. (abstract).

Weske, J. S., and J. W. Terbaugh. 1981. *Otus marshalli*, a new species of screech-owl from Peru. *Auk* 98:1–7.

Wetmore, A. 1968. The birds of Panama. Pt. 2. Columbidae (pigeons) to Picidae (woodpeckers). *Smithsonian Misc. Coll.* 150:1–605.

Weyden, W. J. van der. 1974. Vocal affinities of the Puerto Rican and vermiculated screech-owls (*Otus nudipes* and *Otus guatemalae*). *Ibis* 116:369–372.

———. 1975. Scops and screech owls: Vocal evidence for a basic subdivision of the genus *Otus* (Strigidae). *Ardea* 63:65–77.

White, C. M. 1994. Population trends and current status of selected western raptors. In *A century of avifaunal change in western North America,* ed. J. R. Jehl Jr. and N. K. Johnson, pp. 161–172. Stud. Avian Biol. no. 15.

White, T. H. 1954. *The book of beasts.* New York: Putnam.

Wiklund. C. G., and J. Stigh. 1983. Nest defense and evolution of reversed sexual dimorphism in snowy owls. *Ornis Scand.* 14:58–62.

———. 1986. Breeding density of snowy owls *Nyctea scandiaca* in relation to food, nest sites, and weather. *Ornis Scand.* 17:268–274.

Willey, D. W., and C. van Riper III. 2000. First-year movements of juvenile Mexican spotted owls in the canyonlands of Utah. *J. Raptor Res.* 34:1–7.

Wilson, K. A. 1938. Owl studies at Ann Arbor, Michigan. *Auk* 55:187–197.

Wink, M., and P. Heidrich. 1999. Molecular evolution and systematics of the owls (Strigiformes). *In* König, Weick, and Becking, 1999, pp. 39–57.

Winter, J. 1979. The status and distribution of the great gray owl and the flammulated owl in California. *In* Schaeffer and Ehlers, 1979, pp. 60–85.

———. 1980. Status and distribution of the great gray owl in California. Resources Agency report, Calif. Dept. of Fish and Game.

———. 1986. Status, distribution, and ecology of great gray owls in California. Master's thesis, San Francisco State Univ.

———. 1987. Prey ecology of great gray owls in Yosemite National Park. Paper presented at Symposium on Biology and Conservation of Northern Forest Owls, 3–7 Feb., Winnipeg, Manitoba.

Wright, A. L., and G. D. Hayward. 1998. Barred owl range expansion in the central Idaho wilderness. *J. Raptor Res.* 32:77–81.

Wright, J. S., and P. C. Wright. 1997. Stygian owl in Texas. *Field Notes* 51:950–952.

Yaffee, S. L. 1994. *The wisdom of the spotted owl: Policy lessons for a new century.* Washington, D.C.: Island Press.

Yannielli, L. C. 1991. Preferred habitat of barred owls (*Strix varia*) in Litchfield County, Connecticut. *Conn. Warbler* 11:12–20.

Young, K. E., P. J. Zwank, R. Valdez, J. L. Dye, and L. A. Tarango. 1997. Diet of Mexican spotted owls in Chihuahua and Aguascalientes, Mexico. *J. Raptor Res.* 31:376–380.

Zabel, C. J., K. McKelvey, and P. Ward Jr. 1995. Influence of primary prey on home range size and habitat use patterns of spotted owls (*Strix occidentalis*). *Canadian J. Zool.* 73:433–439.

Zarn, M. 1974a. Burrowing owl (*Speotyto cunicularia hypugaea*). Habitat management series for unique or endangered species, U.S. Bur. Land Manage. Tech. Note 250, Denver.

———. 1974b. Spotted owl (*Strix occidentalis*). Habitat management series for unique or endangered. species, U.S. Bur. Land Manage. Tech. Note 242, Denver.

Index

This index includes primarily the species and subspecies of owls that are found in North America, including Mexico. The index follows the nomenclature used in the book; cross-referencing has been provided for the most commonly encountered alternative English vernacular names. Complete indexing for each owl species is found under its main English vernacular name. Primary species accounts are shown in **bold,** and maps and drawings are indicated in *italics.* Extinct owl taxa are identified by asterisks following their names. Other bird and mammal taxa are also indexed, at least to genus, but lower vertebrates and other organisms are not. The appendixes are not indexed.